AN INTRODUCTION TO POWDER METALLURGY

Dedicated to our friends:
Prof. Malcolm B. Waldron († 1988)
Prof. Dr. Gerhard Zapf

Contents

Preface

After the death of Professor Dr. M.B. Waldron the undersigned authors were honoured to write this introductory book on Powder Metallurgy which should have appeared much earlier than it did. It was our aim to provide a fairly comprehensive introduction with some emphasis of the fundamentals of the different processing steps, but, nevertheless, to pay attention also to the main powder metallurgy products including their applications. Since the boundaries between powder metallurgy and advanced ceramics are not definitely fixed and since they are sometimes competitive products, the description of processing includes, to some extent, ceramic materials. A short paragraph on advanced ceramics has been included.

In practice, powder metallurgy processes are often very complicated and sometimes yield unexpected results, which are difficult to understand and which may be hard to predict by theory. The main reasons are the numerous raw material and processing parameters. Not even single steps of the PM process have been susceptible to theoretical prediction in the past. Consequently, the majority of the PM processes and products have been developed empirically. Perhaps this has been changed a little bit recently. For further development it is essential to understand the processes in detail and to recognise the controlling parameters. The main future objective in practice is to increase the consistency of the raw materials as well as of the processing steps, in order to obtain further quality improvement and property scatter reduction of powder metallurgy products.

We hope this book will be useful for educating students with scientific or technological orientation which have some background in materials science and technology and a special interest in powder metallurgy. This book may also be useful to people in powder metallurgy research and development, as well as for production staff.

The authors enjoyed the help of several persons for discussing special paragraphs, namely Dr. H. Kleykamp and Dr. D. Vollath. Many industries provided properties or illustrations of their products, as indicated in the text. Dipl.-Ing. J. Schneider and Dipl.-Ing. D. Werner have prepared the majority of the line drawings, Mrs. A. Gottschalk has continuously prepared the manuscript. The authors gratefully acknowledge this support. We enjoyed excellent cooperation with the Series Editors, Dr. I. Jenkins and Dr. J.V. Wood, as well as with the Books Editor, Mr. P. Danckwerts.

F. Thümmler
R. Oberacker

Figure Acknowledgements

Figure numbers are given in sanserif type.

Powder Metallurgy International, Freiburg, Verlag Schmidt; **2**(3) 1970, p. 77–80 (2.7); **15**(3) 1983, p. 130 (7.46); **23**(5) 1991, p. 286, 318, 319, 321 (2.10; 8.2; 8.3; 8.4a); **13**(3) 1981, p. 121 (9.1); p. 123 (9.2)

Physical Chemistry of Metals, 1953, p. 349, Mc. Graw-Hill Book Comp. and: *J. of Iron and Steel Inst.* **168**, 1948, p. 261 (2.19)

J. Electrochem. Soc. **105**, 1958, p. 398–402; (2.20)

J. Metals **8**(6) 1956, p. 695–704 (2.21)

Metals Handbook, Powder Metallurgy, 9th Edition, Vol. 7, ASM International, Ohio, 1986, p. 161, (2.23d); p.298 (6.5)

The International Journal of Powder Metallurgy and Powder Technology **20**(2) 1984, p. 88 (2.25); p. 89 (2.26); p. 92 (2.27)

Advanced Materials and Processing (Eds.: H.E. Exner and V. Schumacher), Euromat '89, DGM Informationsgesellschaft-mbH, D 6370 Oberursel, Vol. 1, p. 632 (2.29)

Powder Technology **72**, 1992, p. 223 (2.30)

Journal of Material Science, Chapman and Hall, London, **20**(3), 1985, p. 1165 (2.33)

T. Allen: *Particle Size Measurement,* 4th Edn., Ed. B. Scarlett. Powder Technology Series, Chapman and Hall, London, New York 1990, p. 211 (3.7)

K. Dialer, U. Onken, K. Leschonski: *Grundzüge der Verfahrenstechnik und Reaktionstechnik,* Carl Hanser Verlag, München, Wien 1984, p. 49 (3.10)

Proceedings of the Institution of Mechanical Engineers **196**(18), 1982, p. 108 (6.4)

Hot Isostatic Pressing – Theories and Application (Proc. Int. Conf. on HIP, Luleå, Sweden, 1987), Ed. Tore Garvare, CENTEC Publishers, Luleå 1988, p. 29–40 (6.6, 6.27)

Powder Metallurgy **34**(2), 1991, p. 109–111 (6.43, 6.44)

E. Ernst: *Axiale Preßvorgänge in der Pulvermetallurgie,* VDI Forschrittsberichte, Reihe 2 No. 259, p. 99 (6.9)

Pulvermetallurgie In Wissenschaft und Praxis, Vol. 7, Ed. H. Kolaska, VDI Verlag Düsseldorf, 1991, p. 7–32 (6.13)

Pulvermetallurgie In Wissenschaft und Praxis, Vol. 1, Ed. H. Kolaska, H. Grewe, Verlag der Deutschen Keramsichen Gesellshaft, Düsseldorf, 1985, p. 191 (6.33)

J. appl. phys. **21**, 1950, p. 1205 ff (7.6); p. 632 ff (7.7)

Neue Modelle zur theoretischen Beschreibung von Sintervorgängen, H.E. Exner, inaugural dissertation, Universität Stuttgart, 1976, p. 76 (7.8a); p. 85 (7.8b)

W. Schatt: *Pulvermetallurgie, Sinter- und Verbundwerkstoffe,* 1986, Dr. Alfred Hüthig-Verlag, Heidelberg, p. 156 (**7.9**)

Acta met. **5**, 1957, p. 670, 671 (**7.10**)

W. Schatt: *Sintervorgänge, Grundlagen,* 1992, VDI-Verlag Düsseldorf, p. 91 (**7.12 a, b**); p. 118 (**7.19**); p. 137 (**7.41**); p. 136 (**11.19a**)

Powder Metallurgy: an Overview, (eds.: I. Jenkins, J.V. Wood) 1991, The Institute of Metals, p. 174, (**7.13**); p. 175 (**7.24, 7.18**)

J. Inst. Metals **83**, 1954–1955, p. 329 ff (**7.21**); **62**, 1938, p. 239 ff (**7.38b**)

Zhur. Tekhn. Fiziki **26**, 1956, p. 2076–2085 (**7.28**)

Planseeber. Pulvermet. **6**, 1958, p. 2 ff (**7.30**)

Stahl u. Eisen **79**, 1959, p. 1187 ff (**7.35**)

Z. Metallkunde **70**, 1979, H. 12, p. 794 (**7.39**)

Metal Powder Report, Jan. 1986, p. 40, (**11.18**), (**11.16**) March 1986, p. 184, (**7.40**)

General Electric Report 76 CRD 216, 1976 (**7.26**)

Transactions ASM **57**, 1964, pp. 443–463 (**7.33**)

Science of Sintering (eds.: D.P. Uskokovic, H. Palmour III, R.M. Spriggs), Plenum Press, New York, London. p. 342 (**7.43**); p. 363 (**7.44**)

R. Prümmer: *Explosivverdichtung pulvriger Substanzen,* Grundlagen, Verfahren, Ergebnisse, Springer-Verlag, Berlin, Heidelberg, London, New York, Paris, Tokyo, 1987, p. 34, 81 (**6.38; 6.39**)

J. of Material Science **25**, 1990, p. 20 (**6.42**)

J. of Iron and Steel Inst. **168**, 1948, p. 261 (**8.1**)

Handbuch der Fertigungstechnik (Hrsg. G. Spur) Carl Hanser, München, Wien 1980, Bd. 1, Kap 4: G. Zapf, Pulvermetallurgie, p. 832 (**9.3a**); p. 902 (**9.3b**)

DIN Deutsches Institut für Normung e.V. Beuth-Verlag GmbH, D-1000 Berlin 30, DIN-Taschenbuch 247, Pulvermetallurgie: DIN 30911, Teil 6 (**9.6; 9.7**); DIN 30912, Teil 6 (**9.4**); DIN 30912, Teil 3 (**9.5**); DIN 30910, Teil 1 (**11.2**); DIN 3995 (**3.29**); DIN 3923 (**3.27**)

Metallurgical Transactions A, Physical Metallurgy and Material Science, **23A** 1992, pp. 211–219 (**11.9**)

Pulvermetallurgie in Wisssenschaft und Praxis, Bd. 4, 1988, Schneidstoffe heute und morgen, Hartmetalle und beschichtete Hartmetalle, p. 147 (**11.10e**)

Friedr. Krupp GmbH, Krupp Widia, Drucksache Nr. 2.2–32.2 d 380 (**11.12**)

Technische Mitteilungen Krupp, Essen, 1988 (1), p. 17 (**11.15**)

Reviews on Powder Metallurgy and Physical Ceramics, **3** (3, 4) (1987), p. 233 (**11.20**)

New Materials by Mechanical Alloying Techniques, DGM Informationsgesellschăft mbH, D 6370 Oberursel, pp. 185–200 (**11.21b**)

Neue Werkstoffe (Herausg. A. Weber) 1989, VDI-Verlag Düsseldorf, p. 119, (**11.25**); p. 120 (**11.24**); p. 124 (**11.26**); p. 128 (**11.23**); p. 134 (**11.31**)

Zähigkeitsverbesserung von Sonderkeramiken, Radex-Rundschau, 1977, p. 110 (**11.29**)

1: Introduction

Metal powders are used in industry for a diversity of products, including catalysts, welding electrodes, paints, printing inks, explosives, etc, in all of which the powder particles retain their identities. Traditional powder metallurgy, on the other hand, in its simplest definition, is a process whereby a solid metal, alloy or ceramic in the form of a mass of dry particles, normally less than 150 microns in maximum diameter, is converted into an engineering component of predetermined shape and possessing properties which allow it to be used in most cases without further processing. The basic steps in the traditional process are those of powder production; compaction of the powder into a handleable preform; and 'sintering', which involves heating the preform to a temperature below the melting point of the major constituent when the powder particles lose their identities through inter-diffusion processes and the required properties are developed.

There can be no doubt that certain metals in the form of powders were produced by the carbonaceous reduction of local oxide ores, consolidated by sintering and forged to meet man's needs, by the early civilisations and long before means were discovered for melting them. The iron pins keying the marble blocks of the Parthenon in Athens are believed to have been made in this way whilst, in more massive form, the Delhi Pillar, dating from 300 A.D. is made of more than 6 tons of sponge iron. With the advent of melting practices for iron and copper, powder metallurgy, as an industrial process, remained dormant for many centuries, excepting probably in some of the more remote regions of the world. There was a brief revival around the beginning of the 19th century arising from the need for malleable platinum which could be fabricated into chemical laboratory ware. The problem was resolved by the production of a fine platinum powder which was pressed into a cake and hot forged. By the mid-19th century this process had been superceded by one of fusion. Further major industrial development in powder metallurgy had to await the advent of the electric lamp and the search for a filament material of high melting point, a low evaporation rate and adequate conductivity. Tungsten met these requirements but the sintered material was brittle. The difficulty was resolved in 1910 when it was discovered that tungsten bars, produced by sintering at 3000°C by direct electric resistance heating, when worked sufficiently hot, developed enough ductility to permit of continued working at progressively lower temperatures until a wire of the required filament diameter could be cold drawn. The process is still the standard method used all over the world for the production of incandescent lamp filaments.

The mid-1920s saw the emergence of two powder metallurgy products which set the patterns for the future industrial development of the technology. The first of these was a hard, wear-resisting product known as cemented carbide, which is produced by compacting and sintering a mixture of powders of tungsten carbide with up to 15% of cobalt. At a temperature of about 1400°C, a molten phase is formed between the cobalt and some of the tungsten carbide and this promotes rapid and virtually complete densification to give a final structure of carbide particles in a tungsten carbide/cobalt alloy matrix. Developed originally for wire-drawing dies, cemented carbides have found extensive application in metal cutting, rock drilling and hot working dies. Many new compositions have been developed including additions of other refractory metal carbides such as tantalum and titanium.

The foregoing principle of sintering in the presence of a liquid metal phase was later applied to the 'heavy alloys' which contain about 90% of tungsten, with a balance of either nickel/iron or nickel/copper. The alloys were developed originally for radium therapy screening but later also found extensive use for mass balancing in aircraft and in gyroscopes. Molten metal is also used in the process of 'infiltration' in which a porous sintered metal is brought into contact with and is wetted by a molten metal of lower melting point, which then becomes rapidly absorbed to fill the pores. Tungsten/silver and tungsten/copper heavy duty electrical contacts are produced in this way.

The second significant development was the porous bronze self-lubricating bearing, which was invented in the late 1920s, and which is made from a mixture of 90% copper, 10% tin powders, usually with a small graphite addition. The powder mixture is pressed and sintered to form a tin-bronze alloy with little dimensional change, and containing up to 35% of fine, interconnected, porosity. The final close dimensional tolerances required in the bearing are achieved by pressing the sintered compact in a sizing die after which it is impregnated with oil. When in use, the bronze reservoir supplies oil to the bearing interface. Other porous products today include filters for separation processes made from lightly compacted and sintered powders of, for example, bronze or stainless steel.

The first sintered iron structural parts were produced towards the end of the 1930s when it was recognised that the process for producing the porous bearing was a basis for making net-shape engineering components. The development was of considerable significance in that, hitherto, the powder metallurgy process had only been used to make products which could not be made by any other technique. Now the process was to challenge conventional fusion methods. The early products, which still retained 10% or more porosity, were required to meet a modest engineering specification and the main market attraction was the ability of the process to produce components to dimensional specifications at a relatively low cost. These features remained for many years as the primary selling points as the product base was gradually extended, but during the past twenty years the growth in this sector of the industry has been substantial. There are many reasons for this. Stimulated by a better understanding of the under-pinning science, the manufacturing technology has progressed steadily with improved control of processing parameters and automation of the pressing and sintering operations. The level of mechanical properties attainable has increased considerably as a result of improved densification of the sintered component and by the use of alloying, yielding products which can be heat treated, and with properties, homogeneity and reproducibility which challenge those achieved in fusion products. Not least has been the contribution from the design engineer who, with growing confidence in the sintered product, has responded to an increasingly effective dialogue with the powder metallurgy technologist.

The major tonnage growth of the industry has been in iron and alloy steel structural parts, while the major economic value is in the field of cemented carbides. The materials base has been extended to include aluminium and metals of higher intrinsic value such as copper, nickel, titanium and other refractory metals. Soft magnetic parts for pole pieces, relay cores, etc, and high energy magnets, including the more recent rare-earth types, are important products of the industry. The versatility of the process has long been demonstrated in the manufacture of metal/non-metal products of which important engineering examples are metal/diamond abrasive wheels; metal/graphite brushes for electric motors; friction materials for vehicle brake linings comprising a dispersion of compounds, usually silicates, in a metal matrix; and light duty electrical contacts of which the primary example is silver/cadmium oxide. A class of products known as 'Cermets' include nuclear materials such as the oxide or carbide of uranium dispersed in a metallic matrix, and molybdenum/zirconium oxide which is used for hot working dies. Metal powders are also processed by more traditional fabrication processes which include rolling of a lubricated powder or a metallic paste into strip; extrusion of powder, vacuum-sealed in metal cans, into shapes; and mechanical working of sintered preforms.

The powder metallurgy process is neither energy nor labour intensive, it conserves material, it is ecologically clean, and it produces components of high quality and with homogeneous and reproducible properties. In more recent years, these attributes have attracted studies into the application of the technique to the development of materials for high technology uses. Hitherto, the scientific input into the process and its products had been confined largely to the specialist areas of the refractory metals and nuclear materials. The technology of the traditional sintered engineering components sector, on the whole, failed to stimulate an adequate scientific challenge. The situation has changed markedly, however, during the past 15 years or so, arising from the challenging demands for new and improved processes and materials of high integrity for advanced engineering applications, including those of aero-space. During this time there has been substantial growth world wide in scientific research into all aspects of the production, characterisation, handling and consolidation of powders of a wide range of engineering alloys. The traditional powder metallurgy industry has not been ignored in such scientific activity and continues to benefit from the overall accumulation of scientific knowledge of the process and of related process and product developments.

Important developments arising from the upsurge in scientific effort have been sintered and forged superalloys for gas turbines and dispersion-strengthened alloys in which a very fine dispersion of an oxide in a metal matrix such as nickel or lead confers improved microstructural stability and enhanced resistance to creep. More recently, the process designated 'mechanical alloying' has been developed in which metal powders are subjected to high energy milling. Non-metallic additives such as stable oxides or carbides become coated with the softer metal and are re-distributed as a very fine dispersion through repeated fracture and re-welding of the composite powder particles. Developed originally in alloys for high temperature service, the process has been extended to light alloys for aero-space applications and other engineering materials.

Novel microstructures are also produced by the consolidation of very fine powders produced by the gas 'atomisation' of molten metal. During quenching by the atomising gas, the metal droplets solidify so rapidly that they can exhibit non-equilibrium behaviour such as high solubility extensions of certain alloying elements, together with microcrystallinity or even an amorphous structure. Several aluminium alloys processed in this way have reached a commercial stage whilst investigations continue on magnesium, titanium and copper alloys.

An important extension of rapid solidification technology is that of spray-forming, in which the atomised metal stream is directed onto a substrate on which it solidifies and is allowed to build-up to the required form. The process is applicable to a wide range of engineering materials including steels and light alloys.

The foregoing techniques are used in the production of metal matrix composites in which small diameter fibres, whiskers and/or particulates, such as silicon carbide or aluminium oxide, are used to reinforce the metallic base. The reinforcement combined with the refined microstructure confers enhanced strength and stiffness together with improved elevated temperature properties. Aside from conventional powder metallurgy techniques and mechanical alloying, the manufacturing processes include spray-forming in which the non-metallic fibre or particulate is injected into the molten metal as it is sprayed onto a substrate. Aluminium alloy matrix composites are already in production, whilst those based on magnesium, titanium and certain high temperature alloys are under development.

The high performance materials also receiving much attention is a new class of ceramics produced from powders based on oxides, nitrides, borides or carbides. Manufactured from powders of high purity and carefully processed to avoid both contamination and the introduction of structural defects, certain products are already challenging the traditional markets of wear-resisting materials in heat exchangers, pump parts in corrosive environments and components and special parts in petrol and diesel engines. Many ceramics are now well-established in the electrical and electronic fields where their applications include dielectrics, piezo electrics, electrical and thermal insulators, magnetics, ion conductors and sensors. The processing technology which has developed is likely to be of particular value in the development of the new family of high temperature ceramic superconductors.

This volume, which is one in a series on Powder Metallurgy, is concerned with the basic scientific and technological concepts of this important and rapidly developing sector of the materials industry. Volume 2 is a companion volume which illustrates through a number of Case Studies, the technological applications of the basic principles and procedures followed in the industry. Volume 3 provides state-of-the-art reviews by international experts of the important advances which are being made in the technology. The student, particularly, is advised to study at least this volume and *Selected Case Studies in Powder Metallurgy* in order to acquire a balanced view of both the science and the practise of the subject.

FURTHER READING

1. W.D. Jones, *Fundamental Principles of Powder Metallurgy*, Edward Arnold (Publishers) Ltd, London, 1960.
2. F. Eisenkolb and F. Thümmler (Eds), *Fortschritte der Pulvermetallurgie*, Akademie Verlag, Berlin, 1963.
3. R.M. German, *Powder Metallurgy Science*, Metal Powder Industries Federation, Princeton, New Jersey, 1984.
4. H.H. Hausner and M. Kumar Mal, *Handbook of Powder Metallurgy*, 2nd edn, Chemical Publishing Company Inc, New York, 1982.
5. *Powder Metallurgy: State of the Art*, Vol 2, W.J. Huppmann, W.A. Kaysser and G. Petzow (eds), Verlag Schmid GMBH, Freiburg, 1986.
6. F.V. Lenel, *Powder Metallurgy: Principles and Applications*, Metal Powder Industries Federation, Princeton, New Jersey, 1980.
7. *Powder Metallurgy: the Process and Its Products*, British Powder Metallurgy Federation, Wolverhampton.
8. W. Schatt (Ed.), *Pulvermetallurgie, Sinter- und Verbundwerkstoffe*, 2nd edn, Deutscher Verlag fur Grundstoffindustrie, Leipzig, 1985.

2: Powder Production Techniques

2.1 METAL POWDERS

The production of metal powders is one of the important bases for the entire field of powder metallurgy and the total shipment of metal powders is controlled by their use in manufacturing sintered products. The most important metal powders are: iron and steel, copper and copper base, aluminium, nickel, molybdenum, tungsten, tungsten carbide and tin. The global consumption of metal powders in 1992 is estimated of about 750 000 tons. Iron and steel are absolutely dominating and summarize to about 570 000 tons worldwide and 246 000 tons in USA in 1992. These are 76 and 78% respectively, of the entire metal powder tonnage. Powders are used, of course, also for other than powder metallurgy applications, namely for welding electrodes, submerged arc welding (iron), painting, pyrotechnical applications and others. These non-powder-metallurgy applications are very dependent on the specific metal and range in several countries for iron powder at about 15% of the entire shipment. The total metal powder shipment in Western Europe for powder metallurgy applications was in 1992 more than 100 000 tons and for iron and steel 89 500 tons. The figure for Japan in iron and steel powder (1992) is about 98 000 tons. Thus, the powder production represents an important section of metal industry.

The following treatment contains as well mechanical as chemical (including electrochemical) production processes. Besides metallic powders also carbide, nitride, boride and special ceramic powders as well as whishers and short fibres are considered.

2.1.1 Mechanical Processes

The size reduction of metal powders by mechanical means is performed in the solid state as well as of melts. The latter is called atomisation and is of special significance for mass production of metal powders.

2.1.1.1 Grinding and Milling

Principal Aspects

Grinding and milling are the oldest processes in powder metallurgy and ceramics for the production of particulate materials. They are extensively applied in the field of ceramics and in the cement industry for size reduction of brittle materials, often accompanied by intensive mixing effects. Their use in powder metallurgy is more limited, since the disintegration of metallic materials, which mostly exhibit considerable plasticity, is less effective. Nevertheless, materials such as intermetallic compounds and ferroalloys can be effectively comminuted by mechanical means. Additionally, there are important mechanical disintegration processes in powder metallurgy, especially those involving high energy milling procedures, during which severe embrittlement of the metal occurs. This aspect can be turned to advantage in the process known as mechanical alloying (see below).

The general phenomena during size reduction in the solid state are based on fracture mechanics: the nucleation of cracks, followed by crack propagation and fracture, by which new surfaces are formed. A further decrease of mean particle size can take place only when these processes occur. The kinetic energy within the milling aggregate is partially transformed into mechanical stresses in the material to be disintegrated. The forces acting in these processes cause mainly compression and shear stresses, applied as impact or slow-acting stresses by the milling balls during vessel rotation or vibration, or by the rotating arms in an attritor. In plastic materials, gross dislocation formation and movement, i.e. plastic deformation, is induced before cracking and fracturing occur. In ceramic materials, plastic deformation does not take place in coarse particles, but becomes significant for mean particle sizes in the micron range. The main process phenomenon for all grinding and milling processes, however, is crack propagation and fracture, which is accompanied by local, not well defined, temperature rises. These are often referred to as thermal 'spikes' and can result in anomalous structures.

The limit of the minimum obtainable particle size depends on the conditions of the mechanical process as well as on the material itself. This so-called *grinding or milling equilibrium*, has been investigated for many metallic and non-metallic powders. It ranges between roughly 0.1 and 1 μm and can be explained by several independent processes as follows:

- In a solid material there is no constant binding force within the volume but only a binding spectrum, dependent on the local types of defects and defect concentration. These can be macro, micro or on an atomic scale. Only the weaker part of the binding spectrum can be destroyed by mechanical means, and a further reduction in size of small particles becomes increasingly difficult.
- The smaller the particle size the more difficult is the application of the necessary shear stress on each particle in order to achieve further particle

fracture. With small particles the elastic or elasto-plastic response to mechanical stresses becomes dominating.

- Small particles exhibit higher surface activity than larger ones and therefore, have a higher probability of being re-welded. This was previously believed to be the main reason for the 'milling equilibrium' observed (see Fig. 2.5).

The efficiency of mechanical size-reduction processes is generally very low. Only about 0.1% of the spent energy in the conventional ball milling process is found in the generated new surfaces of the fine particles. The energy E necessary for the physical process of size reduction is given by

$$E = \gamma \,.\, \Delta S \qquad (2.1)$$

where γ = the specific surface energy and ΔS = the increase of specific surface. The efficiency may be somewhat higher in high energy milling processes (see below), but is still less than 1%. The percentage figure becomes larger when, instead of the energy defined in equation (2.1), the energy necessary for crack propagation is considered — being about one order of magnitude larger. The sources of 'lost energy' are many: they include the elastic and plastic deformation of the particles, the kinetic energy of the particles in motion, impact and friction energy in the form of heat outside the powder, etc. The main cause of lost energy, however, is the generation of heat.

In order to *optimize milling efficiency* special conditions have to be established.

- in conventional steel ball mills, defining the size and number of milling balls which will take up 40–50% of the vessel volume. The size of the balls generally ranges between 12 and 16 mm, about 10–20 times larger than the initial size of the particles to be milled.
- using vibration milling instead of milling by vessel rotation. The impact intensity of the milling balls is much higher in vibration milling than in the 'falling' process in rotational ball milling. The noise level, however, of a vibration mill is very high. Such mills were used for a long time in the fabrication of cemented carbides, but have now mostly been replaced by attritors. For laboratory purposes discs are used successfully instead of balls in vibration milling.
- other types of fine-milling equipment, such as planetary ball milling, centrifugal milling and attritor milling are widely used.
- wet milling instead of dry milling is often advantageous because liquids tend to break up agglomerates as well as reducing re-welding of powder particles. As milling media, hydrocarbons (hexane, heptane), ethanol and other organic liquids, which help to avoid oxidation of the milled product, are often used. Special effects have been achieved by using surface active additives (e.g. oleinic acid or other substances with a long carbon chain and a surface active group), which reduce the interfacial energy between particles and the milling media (Rehbinder-Effect) and, more importantly, increase the crack propagation during milling by reducing the cohesion at crack tips.

During the milling of hard and abrasive powders the *level of impurities* sometimes rises considerably, because of severe wear loss from walls and milling media. Iron contamination can be removed by subsequent acid leaching, when the milled powder itself is acid-resistant. Rubber liners and ceramic balls (Al_2O_3) which cause less contamination are sometimes advisable. The best but most expensive solution is to use liners and balls of similar composition to that of the material being milled (e.g., the milling of special ceramic powders, like Si_3N_4).

Equipment

There is a wide variety of grinding and milling equipment, but the following are selected for particular mention:

Ball and Vibration Milling

Figure 2.1 shows the principles of conventional steel ball mills, which are available from a small laboratory size (less than 1 litre) to very large sizes used mainly for the ceramic and cement industry. At too low rotation speed the milling balls are mainly rolling over each other, but at higher speed some of the balls are in free fall at each moment (Fig. 2.1), which is the optimum condition for the milling process. At too high speed, however, the centrifugal forces hinder the balls in free fall. Ball mills are sometimes used in powder metallurgy for mixing purposes only.

In vibration mills the design of the vessel is similar to that in rotation ball mills, but the vessel is mounted on special steel springs. The amplitude as well as

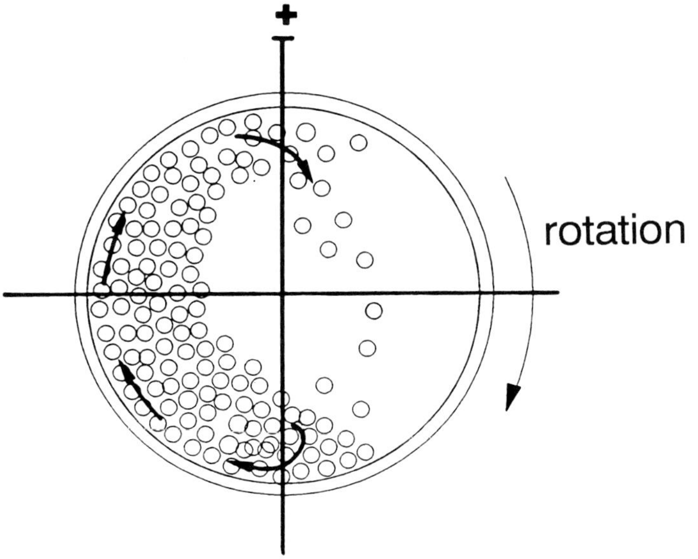

Figure 2.1 Milling balls in action in a conventional ball mill vessel

the frequency of vibration can be adjusted according to the characteristics of the vibrating mechanism and the springs. Vibration mills are often used with only short milling times in research and development for size reduction of inorganic and organic products.

Figure 2.2 shows a planetary mill. The difference from the conventional ball mill is the superposition of the single vessel rotation by rotation of the table supporting the fixed vessels, which accelerates the movement of the milling balls. Centrifugal accelerations up to 20 g (acceleration due to gravity) are possible.

Attritor milling

An attritor is a ball milling system in which the balls, together with the material to be milled are set in motion by a shaft with stirring arms, rotating at 100–300, small units up to 2000 rpm. The containing cylindrical vessel is usually water cooled because of the considerable heat generated by the process. The process may be undertaken dry, or wet with water or organic liquids. The balls are generally smaller (0.5–2 mm) than in conventional ball mills. For production scales the process can be run continuously by feeding the vessel from the bottom while unloading the milled product from the top. A closed cycle can be provided for the milling liquid, including distillation apparatus for recycling. For processing of materials with special oxidation sensitivity (e.g., powders for Co-rare-earth-magnets) the attritor can be equipped with an inert gas supply, which is generally advisable in dry milling. A definite ratio of milling balls to milling material is essential for optimum efficiency. Figures 2.3 and 2.4 show two types of attritors. Discs with holes or profiled discs are also used instead of arms as rotating units in so-called agitation ball mills.

The milling intensity of an attritor is generally much higher than that of conventional ball mills, the difference being at least one order of magnitude in

Figure 2.2 Planetary ball mill with 4 vessels, 500 ml each (Fritsch, Idar-Oberstein)

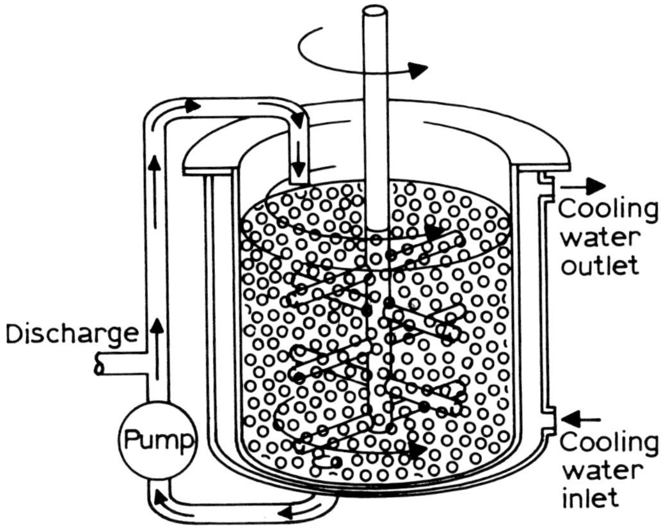

Figure 2.3 Batch attritor for wet grinding (Union Process, Inc., Akron, Ohio)

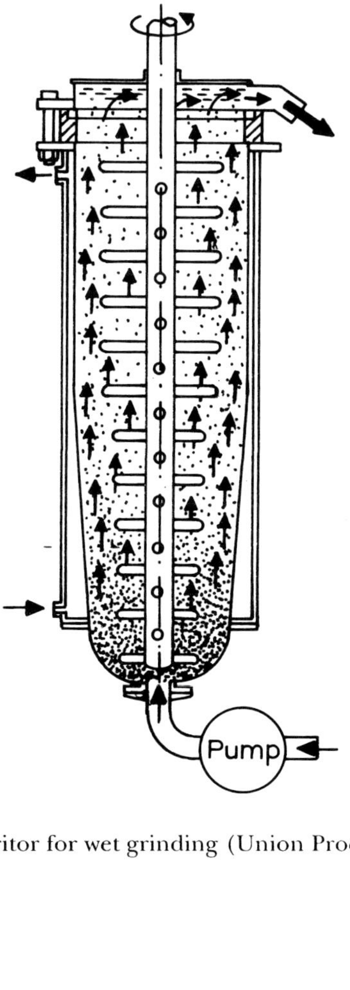

Figure 2.4 Continuous attritor for wet grinding (Union Process, Inc., Akron, Ohio)

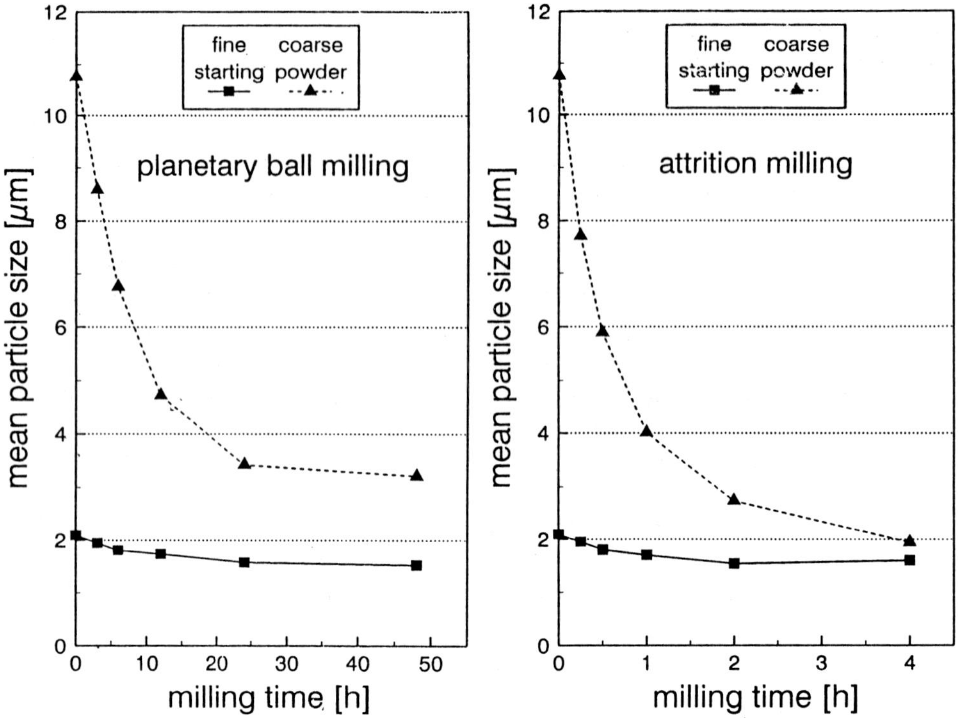

(a) (b)

Figure 2.5 Time to reach similar milling effect of TiB_2 powder in (a) Planetary ball mill (b) Attritor (Jüngling)

time for achieving similar results (see Fig. 2.5). In addition to compression stresses high rate shear stresses are involved leading to lower energy losses compared with other processes.

Dry and wet attritors are widely employed for inorganic pigment and paint production, for limestone, and oxides, and in the hard-metal industry, where they have replaced vibration mills. In these fields of application very fine particulates are required. In the powder mixtures for cemented carbide manufacture (e.g. WC–Co), a certain degree of 'coating' of the brittle carbide by the ductile cobalt particles during milling–mixing occurs which is desirable for further processing. Several metals and alloys may become amorphous during extended milling in an attritor. Investigations, mainly during the last decade, have resulted in the development of tailor-made amorphous alloys (glassy metals), by this method.

Roller Milling

For brittle materials a new size reduction technology may become important, the high-compression roller mill, Fig. 2.6, which operates with profiled rollers in the pressure range 50–500 MPa. According to the working conditions products with narrow particle size ranges, between 200 and about 5 μm are achievable. The service life is claimed to be 10 to 20 times that of ball mills.

The Cold Stream Process

In this process coarse particles (less than 2 mm) are transported in a high-pressure, high-velocity gas stream and shot on to a target of a highly wear-resistant material. A typical pressure of the system is 70 bar with a gas velocity up to Mach 1 or even higher. Due to the drop in pressure of the gas stream on passing through a venturi-nozzle before entering the collision chamber at lower than atmospheric pressure, a rapid cooling of the particles occurs in the nearly adiabatic system. This leads to considerable particle embrittlement and allows easy particle fracturing during collision with the target. In Fig. 2.7a, the particle fines are extracted from the continuously repeated process and collected in special chambers, while the coarser fractions stay in the cycle. Figure 2.7b shows the blast chamber in operation. By using inert gas (Ar) the process provides fine powders (a few μm) with very low oxygen content. The surface oxides of the raw material can be spalled off and separated during the first few cycles. The cost of the process is high partly due to high energy consumption. This results in economic powder production only for materials with a high value, such as superalloys, cemented carbides, etc.

The Hametag (Eddy-Mill) Process and Jet Milling

In this process a ductile material, mainly iron in the form of fine scrap (wires, sheets, etc.) is comminuted by rotating propellers in a vessel under a protective gas. The high speed (up to 85 ms^{-1}), generates gas streams in which the particles undergo the main part of their size reduction as self-disintegration. Up to 1950 much of the iron powder used for machine parts production was made

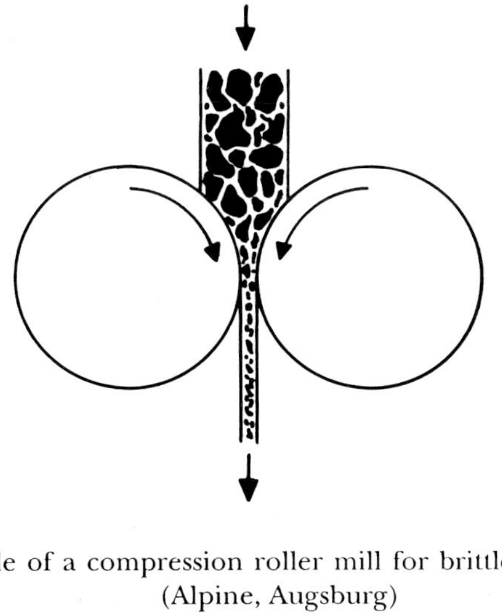

Figure 2.6 Principle of a compression roller mill for brittle material size reduction (Alpine, Augsburg)

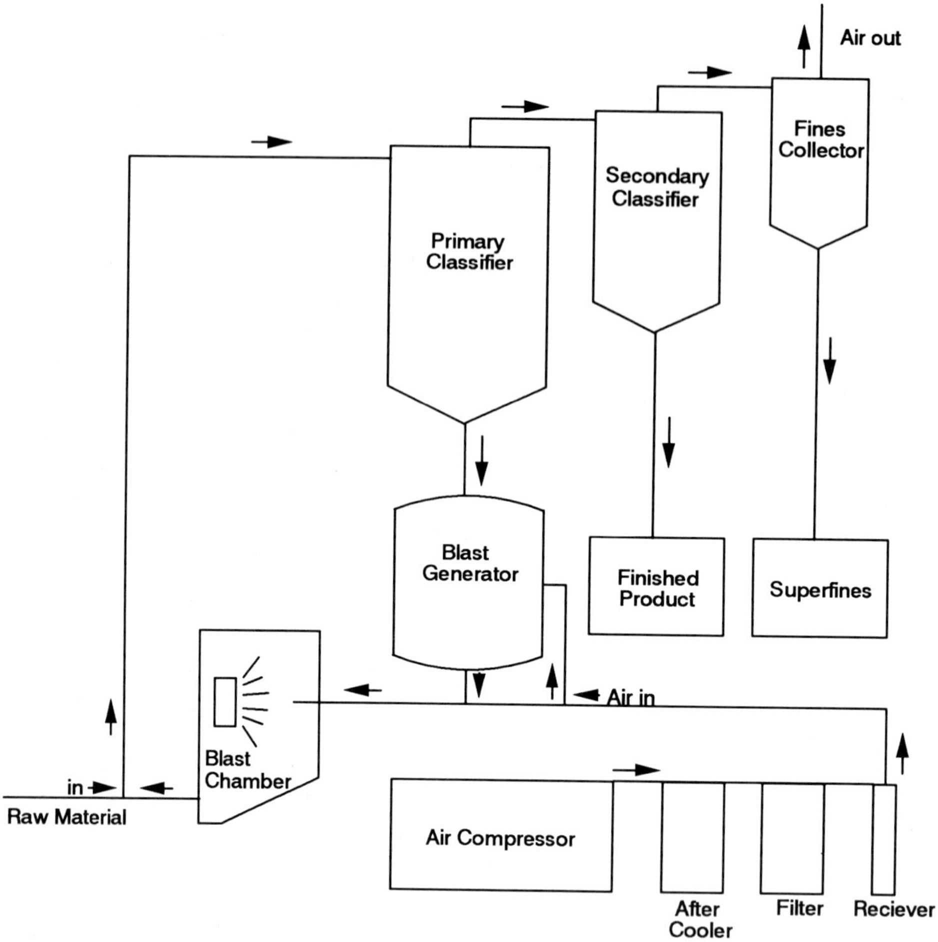

(a)

(b)

Figure 2.7 The Cold Stream process (Walraedt) (a) flow sheet (b) blast chamber in action by this method. The plate-like particles (0.05 to 0.5 mm) with a relatively smooth surface have only a moderate compressibility and the process has a low efficiency with high energy consumption. This type of mill is no longer in service. New types of jet-mills of different design are widely used, but mainly outside the field of powder metallurgy.

Mechanical Alloying

Mechanical alloying is a high-energy ball milling process for producing composites with a controlled, even distribution of a second phase in a metallic matrix. It was first published in 1970 and introduced in the context of developing dispersion-strengthened alloys (see section 11.9.3), in which strengthening by precipitations and dispersed oxides is combined. The process enables the development of special microstructures essential for achieving good high-temperature mechanical properties in multiphase powder metallurgy materials. In particular, the distribution of a non-metallic phase in a metallic, ductile matrix can be homogenized to a degree, which can be achieved otherwise only by chemical means. Also chemical reactions and formation of solid solutions can be obtained, in such cases the term 'reaction milling' is used.

The process consists of long period milling of mixtures, in which the main component (later the matrix) is ductile. As a result of the permanent high energy ball–powder interaction (Fig. 2.8) the ductile phase undergoes a continuous cycle of plastic deformation, fracture and re-welding processes, by which the fine dispersoids are implanted step-by-step into the interior of the ductile phase. This can be understood as a three stage process. In the first stage strong particle deformation ('miniforging') and welding of particles are dominant, while the dispersoid is forced to cover the steadily increased surface of the ductile phase, which embrittles continuously. During the second stage the large lamellar-like particles formed are fractured, forming again new surfaces, which may pick up more of the dispersoid particles, re-weld and form finer lamellae (Fig. 2.9). An equilibrium between deformation and welding on the one hand and fracturing on the other is reached (more or less) during the third stage, while homogenisation proceeds, and the mechanical

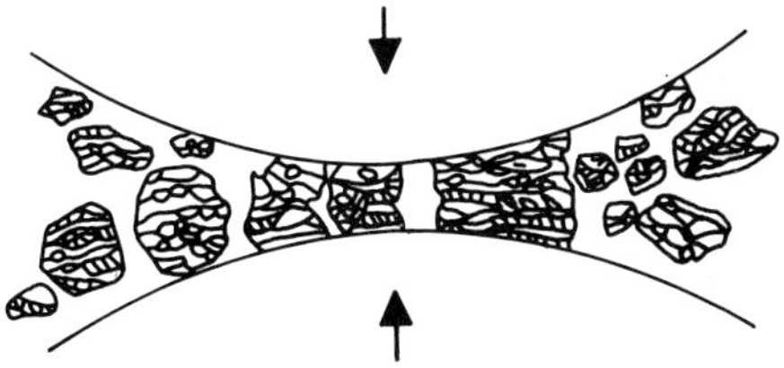

Figure 2.8 Ball powder interaction during mechanical alloying, including 'miniforging' and fragmentation

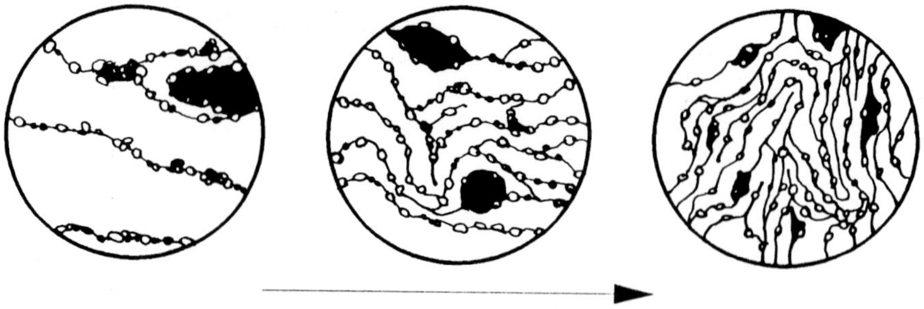

Figure 2.9 Implanting of disperse particles into a ductile matrix and formation of lamellar structures during mechanical alloying (Benjamin)

alloying process terminates. A considerable initial coarsening of the ductile phase can take place, when the starting particle size of the metal powder is very fine (see Fig. 10), while cristallite size decreases.

Very effective mechanical alloying has been achieved by attritor milling, but other, more economic, milling processes can also be successful. Longer times, however, have to be allowed with low energy ball mills. Mechanical alloying is applicable to nearly all combinations of brittle phases (oxides, carbides, nitrides, carbon, intermetallics) and ductile metallic powders, which demonstrates the possibilities for extensive composite materials development.

The process has also been used for the amorphisation of metal or alloy powders, which is of some interest for the manufacture of metallic glasses. The sintering process, however, has to be carefully adjusted to avoid recrystallisation and the temperature of the powders must always be below the glass transition temperature (Tg). Mechanical alloying yields amorphous materials over broader composition ranges than other processes such as splat cooling.

Processing of Metal Chips

Metal chips produced by cutting or milling during mass production are often treated as scrap, because the handling charges exceed their value as raw material. Nevertheless, attempts have been made by mechanical processing to re-cycle ferrous and non-ferrous chips, from which powders can be produced. This may be economic with expensive non-ferrous metals and alloys such as copper, brass or bronze, when the chips can be delivered with low impurity levels, in sufficient quantities and with uniform quality.

The process of powder manufacturing from chips involves first cleaning from impurities. The following fragmentation, which is essential in order to get press-and-sinter-grade powders, is done by high energy (e.g. vibration) ball milling. The cold deformed powders have to be annealed in order to improve their

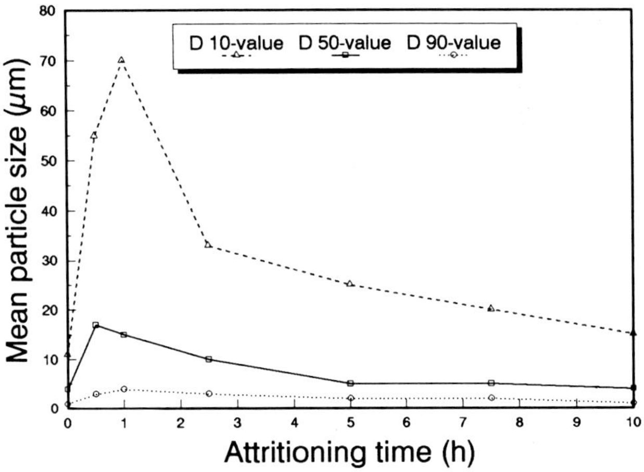

(a)

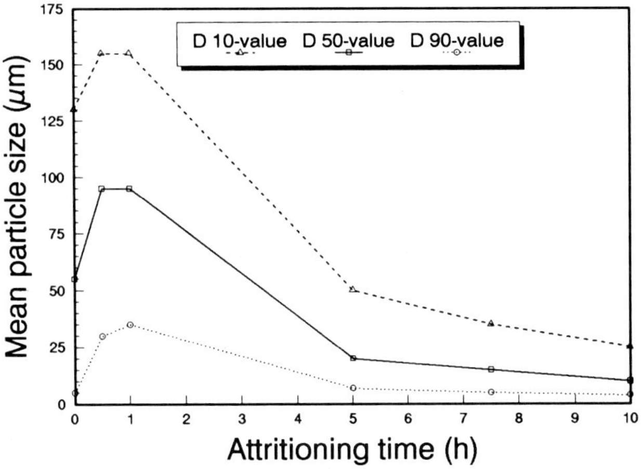

(b)

Figure 2.10 Time-dependent mean particle size in Fe-10 vol-% NbC mixture during mechanical alloying in the attritor (a) Carbonyl powder (b) Water atomised powder (Gutsfeld)

compactibility. The process may become more significant in future because of the need to conserve raw materials.

2.1.1.2 Atomisation

Principal Aspects

Melt atomization is the most important production method for metal powders. The process generally consists of three stages:

- melting
- atomisation (disintegration of the melt into droplets)
- solidification and cooling.

In most cases, additional processing becomes necessary before the powders attain their desired properties, e.g., reduction of surface oxides, degassing, size classification, etc.

For each of the process stages different methods can be used, resulting in a large number of variants. Table 2.1 provides an overview of the present situation. The main classification criterion is the way in which the energy for disintegration is introduced into the melt. This can be by capillary forces (*melt drop process*), mechanical impact (*impact disintegration*), electrostatic forces (*electrodynamic atomisation*), liquid or gas streams or jets (*liquid or gas atomisation*), centrifugal forces (*centrifugal atomisation*), gas supersaturation of the melt (*vacuum atomisation*), or ultrasonics (*ultrasonic atomisation*).

In the melting stage the most important criterion is whether melting and melt distribution require a crucible system or not. Crucibles are one of the main sources for contamination of atomised powders. The second criterion is the heating source. Essentially all melting techniques known in metallurgy can be used, e.g., induction, arc, plasma, and electron-beam melting, but some of these may also contribute to contamination, as for example in arc melting.

During the soldification and cooling stage, the cooling rate is the controlling parameter. This depends, of course, on the dimensions of the liquid droplets or solid powder particles, and also on the type of heat transfer from the particles to the surrounding medium. The undercooling prior to nucleation and cooling rate are the controlling factors for determining the microstructure of the powder particles, as well as for the dimensions of the atomisation unit, which has to provide a path, on which the droplets can solidify without touching the unit walls or structure. The various melt disintegration methods are explained in more detail below.

Methods Based on Capillary Drop Formation

In the *melt drop process*, the liquid metal flows vertically through the outlet capillary of a tundish, which controls the melt stream diameter, d_S. The melt stream breaks up into droplets of a diameter d_D, which depends mainly on d_S, and only slightly on the melt viscosity ν_S and the surface tension γ_S of the liquid metal. Within the usual range of properties of liquid metals, the correlation between d_S and d_D can be approximated by equation (2.2).

$$d_D \approx 1.88\ d_S \qquad (2.2)$$

As the resulting particle size is relatively large and the production capacity quite small, the melt drop process has only been used in laboratory scale production.

A smaller particle size can be achieved by *impact disintegration* of the resulting

Table 2.1 Variables in the processing steps of atomization

Melting		Melt disintegration		Solidification/Cooling
crucibles/liners	heating source	melt volume	energy input	heat transfer
● with	● induction	○ melt pool	● liquid/gas stream	● radiation
● without	● arc	● melt stream	● rotational energy	● convection
	● plasma	● melt film	● gas supersaturation	○ contact cooling
	● electron beam	○ melt droplet	○ ultrasonic vibrations	
			○ mechanical impact	
			○ hydrostatic pressure	
			○ electrostatic forces	

● commercially used; ○ experimental state

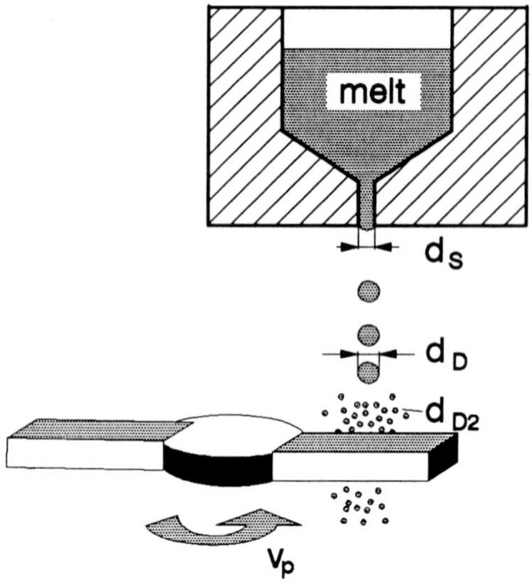

Figure 2.11 Capillary drop formation and subsequent impact disintegration

droplets (Fig. 2.11). The relation between the diameter of the secondary droplets d_{D_2} and d_D is given by:

$$d_{D_2} = d_D \,(v_c/v_p) \tag{2.3}$$

v_p being the impact velocity and v_c the minimum impact velocity at which droplets of the size d_D can be broken down. The application potential of this process is for the production of rapidly quenched powders. With a cooled impact surface a cooling rate of more than 10^7 Ks^{-1} necessary in the production of amorphous powders is said to be possible.

The *electrodynamic atomisation* process has some similarities to the methods mentioned above, because it is also based on droplet formation out of a capillary orifice (Fig. 2.12). A DC voltage of 3–20 kV is applied between this orifice and a perforated electrode plate in front of it. Melt droplets, which have a strong positive charge, are emitted from the orifice. With a capillary diameter of 76 μm, a particle size in the range of 0.1–10 μm has been achieved. This is a very fine powder compared to that from other atomisation processes. The production capacity of existing units is only in the order of grams per hour and the process has not yet progressed beyond laboratory scale.

Atomisation by Liquids

Liquid (generally water) and gas atomisation are the most important process variants for industrial powder production. Water atomisation is mainly used for the production of iron base powders. Figure 2.13 shows a scheme for a water atomisation unit. The starting material is melted and metallurgically treated in

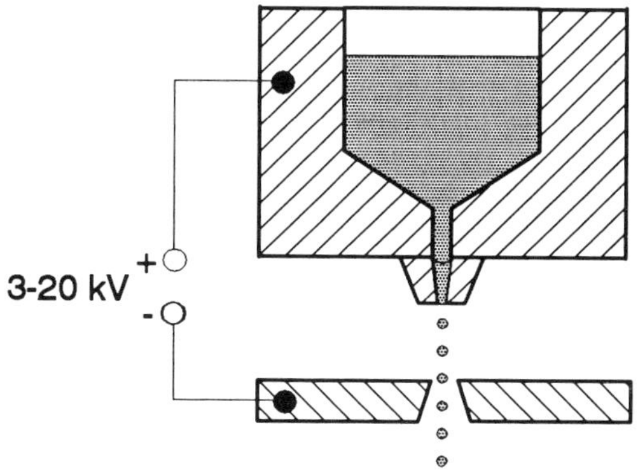

Figure 2.12 Electrodynamic melt atomisation

a separate furnace and then fed into a tundish. The tundish provides a uniformly flowing vertical melt stream, which is disintegrated into droplets in the focal area of an arrangement of several water jets. Configurations of the water jets may differ in the number of jets, the angle α between the jets and the metal stream, and the focusing of the jets into a point or line. The impact from the high pressure stream of water leads to disintegration of the flowing metal. A high energy input to the water stream is needed. The usual pressure range is 6–21 MPa, resulting in a velocity of 70–250 ms^{-1} of the water jets. Throughput is typically 10–100 kg min^{-1} of metal for 0.1–0.4 m^3 min^{-1} water consumed. The overall process efficiency is much better than in the mechanical disintegration of solid materials, but still remains $\leq 1\%$. Atomisation occurs under an unsteady state of turbulent flow conditions, which makes a theoretical treatment of the process very complicated. From empirically derived relations it can be concluded, however, that the most important parameters which control the average powder particle size, d_D are the water pressure P_W, and velocity v_W, the angle α, the melt stream diameter d_S, the melt viscosity η_S, melt density ρ_S, melt surface tension γ_S, and the ratio of molten metal to water flow rate (q_S/q_W). The particle size decreases with increasing component of the jet velocity perpendicular to the melt stream

$$d_D = \text{const}/(v_W \sin\alpha) \tag{2.4}$$

From the empirically derived equation (2.5) it can be seen that small melt stream diameters, high flow rate ratios (q_S/q_W), high density, and low viscosity and surface tension of the melt also favour the production of fine particles.

$$d_D = d_S(d_S\rho_S v_S/\eta_S)^{-0.57}\,(d_S\rho_S v_S/\gamma_S)^{-0.22}\,(q_S/q_W)^{-0.043} \tag{2.5}$$

The overall range of the mean particle size attained by water atomisation is about 30–1000 μm. Quite different particle shapes can be produced, depending on the process parameters. With decreasing superheating of the melt, increasing jet velocity, and decreasing flow rate ratio the particle shape changes

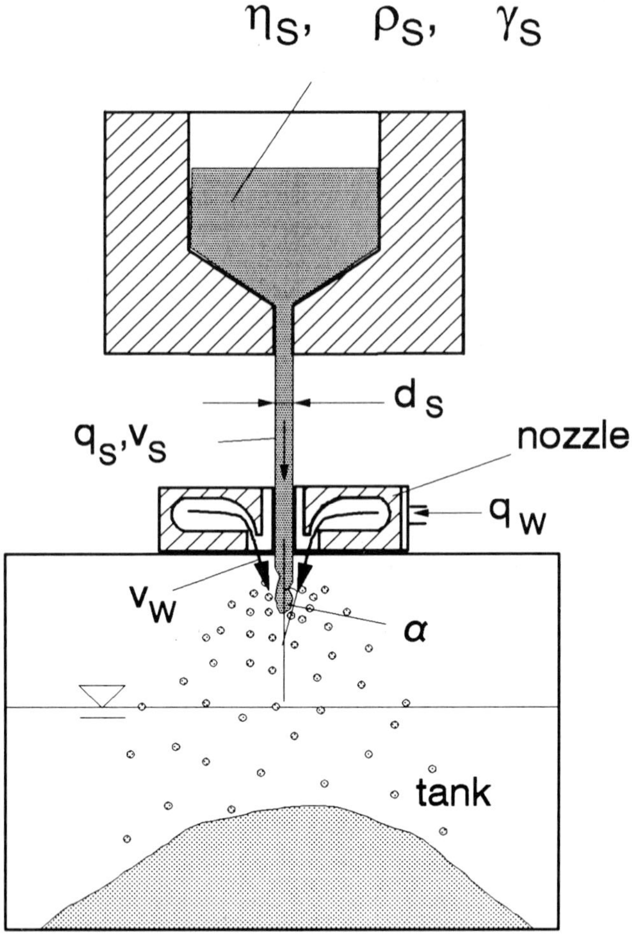

Figure 2.13 Water atomisation unit (schematically, symbols see text)

from nearly spherical to irregular. Irregular particle shapes provide a good green strength of the cold compacted powders and are therefore preferred in the production of iron and steel powders for structural part applications.

Due to its low capital and operating costs, water atomisation is, in terms of output, the most important atomisation method. The limiting factor is the reaction of the atomised metal with water. Water atomisation is therefore restricted to metals and alloys of low oxygen affinity, which pick up only tolerable amounts of oxygen or which can be reduced easily in a subsequent step. Iron and low alloy steel powders can usually be reduced in hydrogen-containing atmospheres. In some special cases, even iron base powders containing substantial amounts of highly reactive elements are produced by water atomisation, e.g., high speed tool steels. In such processes, special measures are necessary, such as adjustment of the carbon content of the melt and vacuum reduction annealing of the powder.

The use of high purity liquid steel with low concentrations of interstitial elements ensures the production of powders with highest compressability ('supercompressable powders'), which are used e.g. for low-alloy Mn–Mo–Ni–steels. Their residual oxygen content is, depending on the alloying element, only 0.08–0.15 wt-%.

More recently, synthetic oils have been used as the atomisation liquid, thereby reducing the oxidation problem. In oil atomisation, however, carbon pick-up occurs in the powders. Decarburization of powders containing reactive elements is, however, less of a problem than is the need for reduction in the case of water atomisation. Industrially oil atomisation is used for the production of low alloy Mn–Cr steel powders.

Gas atomisation

Gas atomisation is the second most important atomisation process. Air, nitrogen, argon or helium are used, depending on the requirements determined by the metal to be atomised. Atomisation is undertaken either in vertical or horizontal units. In Fig. 2.14, a vertical unit for inert gas atomisation is schematically represented. The process principles are similar to those of water atomisation

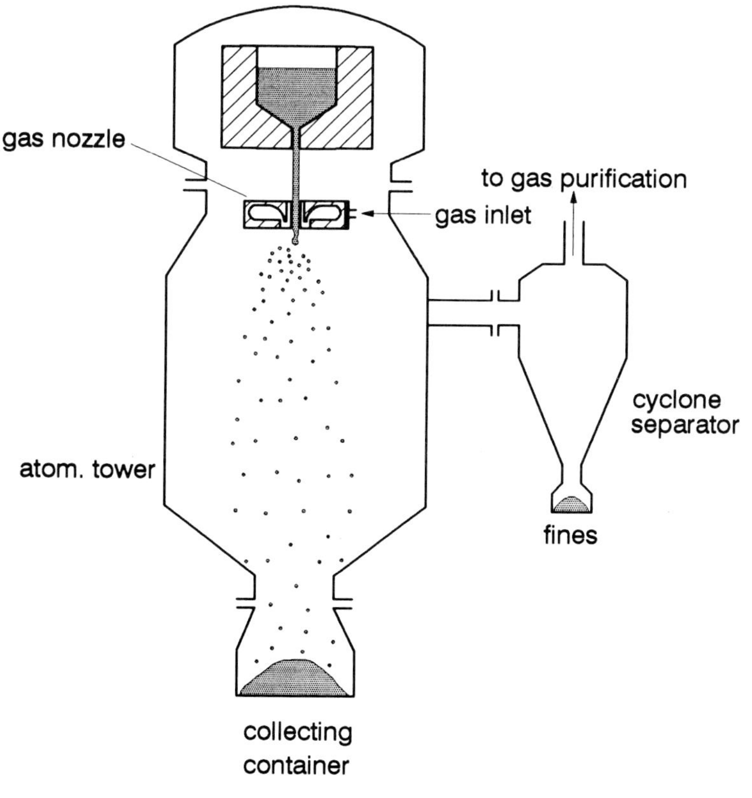

Figure 2.14 Vertical gas atomisation (schematically)

including the problem of mathematical treatment. An empirically derived relation between the process variables and the mean particle size is given in equation (2.6).

$$d_D = C\, d_S \sqrt{\frac{\nu_S \gamma_S}{v_G\, \nu_G\, \rho_S\, d_S}\left(1 + \frac{q_S}{q_G}\right)} \tag{2.6}$$

where ν represents the kinematic viscosity, and G is the index for the atomisation gas. Again, a high velocity of the atomising fluid, small melt stream diameters, high density, and low viscosity and surface tension of the melt favour the production of fine particles. The factor C, is dependent on the geometrical design of the nozzle which forms the gas stream or jets. The nozzle design seems to be an important key to success or failure of a gas atomisation system.

The overall process efficiency is similar to that of water atomisation (i.e. ≤ 1%), but the process costs are substantially higher. The throughput of single melt stream installations is up to 50 kg/min. Gas pressures up to 12 MPa are applied, resulting in supersonic gas velocities of up to Mach 2 and a gas flow rate of up to 40 $m^3\ min^{-1}$. The gas leaves the atomisation chamber through a cyclone, where the fine powder particles are separated before the gas is recycled into the process. Because of the relatively low cooling rates, the atomisation chambers are quite large. A complete vertical atomisation installation can be up to 20 m in height.

The mean particle size of gas atomised powders is in the range of 20–300 μm. The particle shape is spherical or close to spherical. Irregular particle shapes can only be produced in systems where reactions between the gas and the liquid metal cause the formation of solid surface layers. This is the case, for example, in the air atomisation of aluminium.

Melt atomisation by air is used in the production of the so-called 'Roheisen-Zunder' (RZ) iron powder. This process starts from a cast iron melt. The surface oxides formed during atomisation are reduced by the inherent carbon of the cast iron particles during a simple subsequent annealing treatment

$$3Fe_3C + Fe_3O_4 \rightarrow 12Fe + 2CO + CO_2 \tag{2.7}$$

A powder low in oxygen and carbon is obtained only when their concentrations within the powder are balanced according to equation (2.7). Since this is normally not the case, a final hydrogen reduction is advisable. As a result of increasing quality demands, the RZ process has now been almost entirely replaced by water atomisation. Air atomisation is still used in the production of aluminium and aluminium alloys, copper and copper alloys, precious metals, tin and lead powders.

Inert gas atomisation is applicable for all metallic alloys which can be melted. The main application is for high alloy products such as stainless steel, tool steels, iron, nickel- or cobalt-base superalloy powders, as well as aluminium alloy powders. Limiting factors are the availability of suitable crucible and auxiliary melting process materials. Powders from refractory metals with high melting

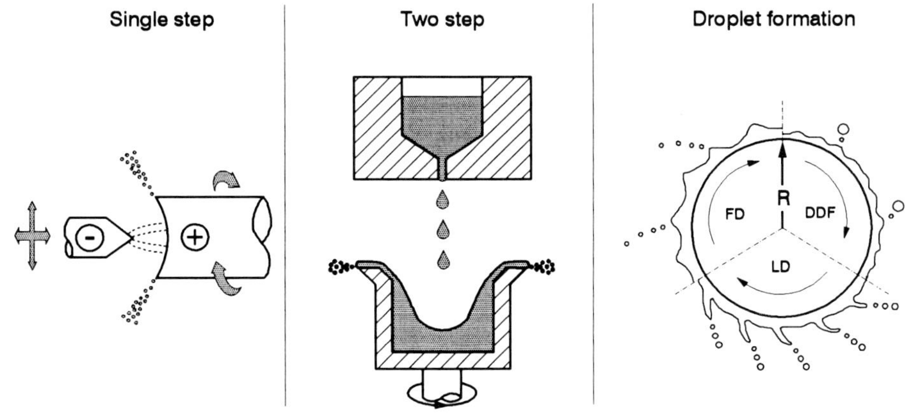

Figure 2.15 Centrifugal atomisation, principles of processing and droplet formation

temperature and highly reactive materials, such as titanium alloys, are usually produced by other methods.

Centrifugal Atomisation

The centrifugal atomisation processes use centrifugal force for the melt disintegration and involve one or two process steps (Fig. 2.15). In the single step method, a liquid film is produced on the surface of a rapidly rotating consumable alloy bar by an arc (rotating electrode process), plasma, or electron beam source. The electrode or alloy bar, which is rotating at up to 15,000 rpm has to meet stringent requirements as to stability, uniformity and surface quality. The melt cannot be highly superheated, as the molten material then rapidly runs off from the electrode. The two-step process overcomes these problems by the separation of the melting and melt-disintegration stages. The latter is performed by a rapidly revolving cooled wheel, disc or cup. The problem with the two-step process is to transfer high rotational velocities to the melt.

At low melting rates the melt disintegration occurs by the direct drop formation mechanism DDF (Fig. 2.15). A primary particle is emitted from liquid protuberances at the edge of the rotating disc or cup. This droplet is for a short time linked to the edge by a fine liquid thread, which is also then emitted as a smaller secondary particle. The particle size distribution, therefore, becomes bimodal in the DDF mode. At higher melting rates, the protuberances stretch into ligaments, which break down into strings of many particles by a ligament disintegration mechanism (LD). The proportion of secondary particles increases steadily with the melting rate. At even higher melting rates, a film is formed at the edge of the disc, directly emitting the droplets by a film disintegration mechanism (FD).

A mathematical treatment of centrifugal disintegration gives equation (2.8)

$$d_D = \frac{1}{2\pi n} \sqrt{\frac{6\gamma_S}{\rho_S R}} \qquad (2.8)$$

The liquid droplet diameter decreases with increasing rotational velocity, increasing melt density and decreasing melt velocity.

The particle shape is spherical. One advantage of centrifugal atomisation is the narrow range of particle size distribution. The mean particle size, however, is relatively coarse, in the range from 150–250 µm for single step processes. Powders with a mean size of 80–90 µm can only be prepared in two-step installations, where a crucible rotation speed up to 25,000 rpm is feasible.

The main advantage of single step centrifugal atomisation is that the melt has no contact with ceramic crucibles or liners. The process is therefore suitable for the production of refractory powders like tungsten, or highly reactive powders, e.g., titanium alloys. The production capacity is low (1–6 kg min^{-1}). Two-step centrifugal atomisation is used for the production of aluminium, titanium alloys, superalloys and refractory metal powders.

Vacuum Atomisation

The vacuum atomisation or melt explosion technique utilises the pressure dependence of gas solubility in metal melts. After vacuum induction melting, the melt is saturated at an enhanced pressure (1–3 MPa) with a gas, usually hydrogen. A ceramic nozzle system is then immersed into the melt to connect it to the evacuated atomisation chamber. The gas is spontaneously released and breaks the melt stream down into droplets. The resulting powder morphology is similar to inert gas atomised powders in both particle shape and size distribution.

Ultrasonic Atomisation

The main process variants for ultrasonic atomisation are melt disintegration by ultrasonic capillary waves or standing ultrasonic waves in a gas (Fig. 2.16). The

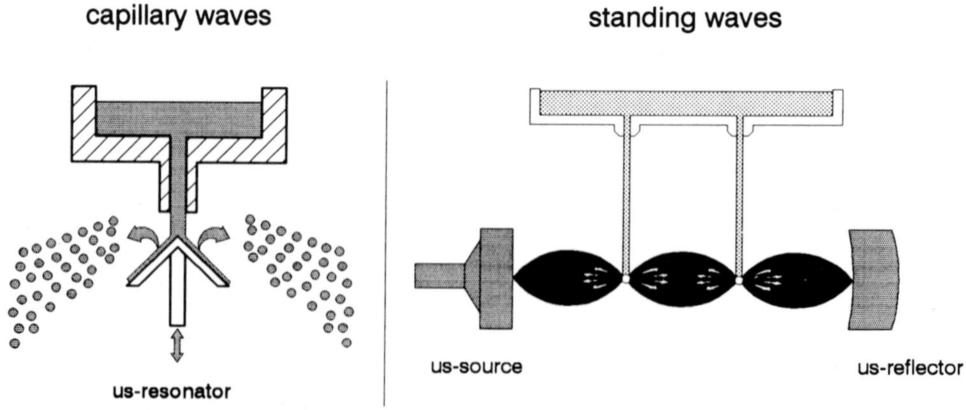

Figure 2.16 Atomisation by ultrasonic standing and capillary waves

former process utilises stationary capillary waves at the surface of a liquid metal film for its disintegration into droplets. As with centrifugal atomisation, single or two step processing is possible. In single step processing, the liquid film is produced by plasma heating a sample placed on the top of a resonating electrode. In two step processing, the separately melted metal is poured over a solid state resonator.

Standing ultrasonic waves can be generated in a gas volume between an ultrasonic source and a reflector (Fig. 2.16). A metal stream poured through the pressure points of this wave field is disintegrated into droplets by ultrasonic pressure.

The particle shape produced by the above methods is spherical. One advantage seems to be the relatively low in-flight speed of the ejected droplets, enabling small dimensions of the atomisation unit. At present, these methods are not used for large scale production.

There are other methods in which ultrasonic vibrations are used to modulate the high speed gas flow from gas atomisation nozzles. These methods are known as *ultrasonic gas atomisation.* Whether the superimposed ultrasonic vibrations really have a particle refining effect in gas atomisation has, however, still to be confirmed.

Solidification and Cooling

Depending on the different mechanisms of heat transfer (see Table 2.1), the processes described above lead to different cooling rates. The cooling rate itself determines the powder particle shape, its microstructure and, in combination with the in-flight speed, the dimensions of the atomisation chamber.

Figure 2.17 gives an order of magnitude overview of the cooling rates achieved by the different methods. Cooling occurs by heat transfer via radiation and heat flow through the medium which is in contact with the melt. The melt volume, together with the heat flow, determines the cooling rate during solidification. Small melt volumes favour rapid cooling. Atomisation under vacuum or reduced pressure conditions results in relatively low cooling rates. For the gas and liquid atomisation methods, cooling is enhanced by convection. The cooling rate is given as

$$\frac{dT}{dt} = \frac{6\,h\,(T_S - T_G)}{\rho_G\, C_G\, d_D} \qquad (2.9)$$

where C and ρ are, respectively, the heat capacity and density of the atomisation gas or fluid, respectively. h is the convective cooling coefficient, which closely depends on the relative velocity of the particles in the atomisation fluid. The highest cooling rates are reached by splat quenching, due to the low temperature and the high heat transfer and heat flow coefficients of the solid substrate, which is brought into direct contact with a thin liquid layer or particle.

The microstructure of the powder particles is determined by nucleation and growth factors. After nucleation has occurred, the relation of the growth rate of

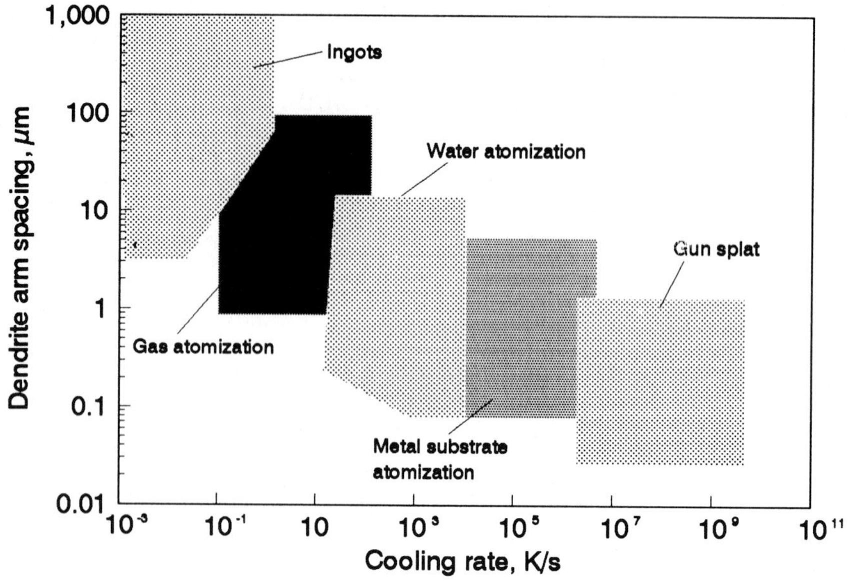

Figure 2.17 Range of cooling rates and secondary dendrite arm spacing for various atomisation techniques (steel, copper, aluminium)

the solid from the liquid and the diffusion rate in the liquid determines the microstructure. A large undercooling increases the growth/diffusion ratio and favours uniform crystal structures. Low growth/diffusion ratios result in dendritic microstructures as a consequence of segregation. For many common alloys the transition from dendritic to cellular microstructure occurs at a cooling rate of the order of 10^5K/s. The secondary dendrite arm spacing is a direct measure of the degree of segregation. It decreases with the cooling rate, as already shown in Fig. 2.17. With increasing cooling rate, the particles tend to form microcrystalline structures, combined with other non-equilibrium states such as high defect concentrations or metastable phases, which promote sintering effectively. Even noncrystalline (amorphous) structures can be obtained. Some processes which can achieve such rates are the *droplet impact disintegration* method, planar flow casting or *melt extraction* which uses a fine edged spinning wheel. These utilise chill block cooling and belong to the so called *rapid solidification* methods. They are widely used for steels, copper, magnesium and aluminium alloys. Rapid solidification, in its different forms, is steadily growing in importance.

The dimensions of the atomisation chamber are determined by the solidification time and the in-flight velocity of the liquid droplets. Therefore, large units are required for vacuum processes (e.g. centrifugal atomisation and vacuum atomisation) and for gas atomisation with high speed gas streams. This problem can be reduced by deflector plates, which divert and retard the droplet flow.

Figure 2.18 Atomised iron powders resulting from water atomisation (a), (b), gas atomisation (c), rotational atomisation (d)

Particle adhesion to the plates can be inhibited by coating or by regenerating liquid films. Small chamber dimensions are suitable for liquid atomisation, due to the high cooling rates, and in ultrasonic atomisation, where the droplet velocity remains moderate.

Figure 2.18 shows several powders produced by the most common atomisation techniques: water, gas and centrifugal atomisation. Spherical particles result from the solidifying droplets in gas and centrifugal atomisation as a result of surface tension. Water atomised powders can be produced with very irregular particle shapes for high green strength requirements, but particles close to spherical can also be produced if good flowability and tap density are required.

Trends for Future Development

The development of new atomisation technologies is directed towards the improvement of process economy and product quality. From the latter point of view, the major requirements are:

- ultra-high purity
- ultra-rapid cooling rate
- ultra-fine particle size.

Future development is targeted on at least one and often more than one of these requirements. Impurities can be reduced by avoiding contact between the melt and electrodes, crucibles or linings. This favours electron beam skull melting. Ultra-rapid cooling rates can best be achieved in combination with ultra-fine particle sizes. *Explosive evaporation* methods are the most suitable of the atomisation techniques, for the production of ultrafine powders. They are based on the introduction of an extremely large amount of energy in a very short time, causing melting and explosive separation of the material into droplets. Energy input is realised by electric arc discharge, laser or electron beam heating, or capacitor discharging by direct electrical current through thin wires (*wire explosion technique*). (Other methods of producing ultrafine powders are listed in section 2.4.)

2.1.2 Chemical Processes

2.1.2.1 Reduction

Principle aspects

The main chemical processes in powder metallurgy involve the reduction of metal compounds such as oxides, carbonates, nitrates or halogenides with gases (generally hydrogen) or solids (carbon or highly reactive metals). In most cases the metal compounds to be reduced are in the solid state. However, hydrometallurgical processes have been developed also, especially for the reduction of nickel and cobalt solutions by pressurized hydrogen. When the reaction takes place in the solid state or by solid–gas reaction, no subsequent purification is possible. A sufficiently pure raw material has, therefore, to be provided as a feedstock. For kinetic reasons, the compound to be reduced has to have a suitably small particle size, which should not be exceeded otherwise the controlling diffusion processes may lead to very long reaction times which are not acceptable in practice.

The process is controlled by its free energy of reaction which has to be negative. The more stable the compound, the stronger the reduction media have to be. Some basic thermodynamic data are given in Fig. 2.19. The free energies of oxidation for some of the reducing agents are included. Since alkali, and especially alkaline earth metals, form very stable oxides, they are very strong reducing agents. Figure 2.19 also shows the extreme reducing capability of Ca, which is needed when H_2, C, CO or Na cannot be used.

Hydrogen reduction

The reduction of oxides or other compounds with hydrogen is generally undertaken well below the melting temperature of the metal. Examples of technical importance are refractory metals such as tungsten and molybdenum, ferrous metals and copper, which form compounds with only moderate stability. Very pure and fine powders can be obtained. The process is often undertaken in tube furnaces, in which the oxide powder is moved in flat crucibles in the opposite direction to the hydrogen stream.

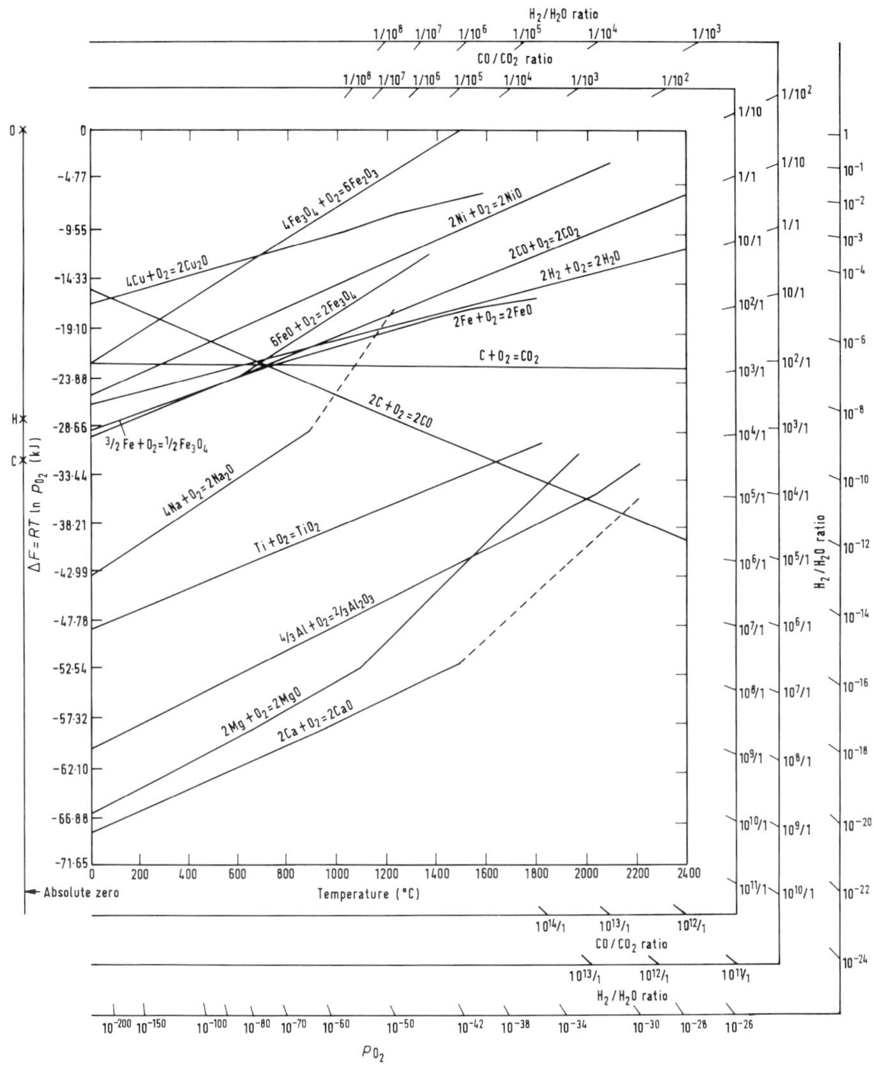

Figure 2.19 The standard free energy of formation of many metal oxides as a function of temperature including H_2O, CO and CO_2 (Darken and Gurry) (for utilising the edge scales see remark at Fig. 8.1)

The overall reaction equations such as

$$MeO + H_2 \leftrightarrows Me + H_2O \tag{2.10}$$

are simple. In reality, several different reaction steps can be involved, depending on the lower oxide compounds that are stable and the degree of Me-O solid solution. This is shown below for the reduction of WO_3. Furthermore, the properties of the resulting powder and, consequently, the reproducibility of the

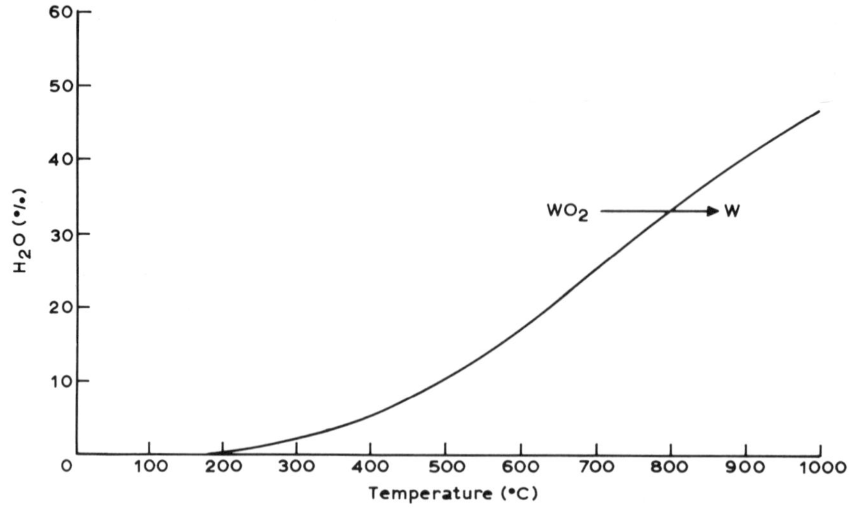

Figure 2.20 Temperature dependent H_2O vapour concentration during WO_3-reduction in the WO_2-W-equilibrium (Griffith)

product, are strongly dependent upon the reducing conditions. Some general rules can be stated as follows:

- a higher reduction temperature and a longer time provide: larger particle size, lower specific surface, lower residual oxygen content and (possibly) formation of a 'sintered cake'.
- a lower reduction temperature and a shorter time provide: smaller particle size, higher specific surface, higher residual oxygen content and (possibly) pyrophoric powders (see section 5.1).
- a high flow rate of hydrogen with a low dew-point results in a high reduction rate, low residual oxygen content and no, or very little, re-oxidation during the cooling period.
- a low flow rate of hydrogen with a high dew-point would produce the opposite results and would therefore be generally impracticable. The danger of re-oxidation of reduced powders at low temperatures is caused by the temperature dependence of the equilibrium according to equation (2.10) moving to the left hand side at low temperatures. This is shown for the $WO_2 \leftrightarrows W$ reduction in Fig. 2.20.

Chemical equilibria are generally defined by the equilibrium constant K in accordance with the law of mass action. *K* is the ratio of the concentrations of starting materials to those of reactants, given in this case as

$$K_{(T)} = p_{H_2O}/p_{H_2} \qquad (2.11)$$

in the form of the ratio of the partial pressures of the gaseous reactants, because the activities of the solid phases are constant (unity).

Table 2.2 W-O-phases formed during reduction of WO_3 under H_2

WO_3	triclinic	D = 7.27	yellow
$W_{20}O_{58}(WO_{2.9})$	monoclinic	D = 7.15	dark blue
$W_{18}O_{49}(WO_{2.72})$	monoclinic	D = 7.96	blue-violet
WO_2	monoclinic	D = 10.82	brown
β W	cubic	D = 19.0	grey
α W	cubic	D = 19.4	grey-metallic

By way of example the production of *tungsten powder* generally takes place by the hydrogen reduction of tungsten trioxide, according to the general equation

$$WO_3 + 3\ H_2 \rightarrow W + 3\ H_2O \qquad (2.12)$$

starting at 650°C, and increasing to 900–1000°C at the end of the process. The starting material for WO_3 is high purity ammonium paratungstate, 5 $(NH_4)_2O$, 12 WO_3, nH_2O. By calcination in air, a yellow oxide results, having the highest possible state of oxidation, namely WO_3. When the calcination occurs in hydrogen, a blue oxide with the formula between $WO_{2.90}$ and $WO_{2.97}$ results, depending on the calcination conditions. Within this range, the oxide may consist of two phases, WO_3 (alpha phase) and $WO_{2.9}$ (W_{20} O_{58}, beta phase) and the colour may vary between yellow and blue according to the content of $WO_{2.9}$.

Taking WO_3 as the starting material, the reduction process normally follows the steps shown in Table 2.2.

This entire process is combined with a vapour phase transport of tungsten, which is responsible for the morphology of the intermediate oxides (e.g. the needle-shape of $W_{18}O_{49}$) and the characteristics of the final alpha W powder. Under special reduction conditions, $WO_{2.72}$ is not observed as an intermediate phase. The beta W is considered as a metastable metallic phase, its appearance being dependent upon the impurities, doping elements and moisture content of the atmosphere. The final alpha W powder (Fig. 2.23c) is obtained with grain size between 1 and 20 µm, depending on process variables.

For the technical production of tungsten filaments for electric light bulbs, a specially doped tungsten powder with a particle size range of 3–5 µm is needed. Doping takes place, by the addition of aqueous solutions of potassium silicate, potassium chloride and aluminium trichloride each in concentrations between 0.2 and 0.6% to WO_3. A part of the doping elements enters the tungsten crystals during reduction *in statu nascendi*, especially during the formation of beta or the transformation of beta to alpha tungsten whilst the remainder is leached out from the final powder by hydrofluoric acid HF. Although the residual doping elements evaporate to a great extent during sintering and subsequent wire fabrication, they provide an optimised microstructure for the filaments (see section 11.5.1.).

The production of *molybdenum powder* is similar to that of tungsten. One difference is that it is possible to purify MoO_3 by sublimation, which is usual in practice. Impure MoO_3 concentrate containing silica and iron oxide impurities obtained by the roasting of sulphuric ores, is transformed by the sublimation of impure MoO_3 which melts at between 1000–1100°C in an air stream, into a high purity product. The reduction to Mo powder occurs in a two step process with MoO_{2+x} as an intermediate product.

Other metal powders produced by hydrogen reduction of chemical compounds are, for example, Cu, Co, Ni and Re. The latter is a high melting point metal used as an alloying element in high strength high temperature Mo or W based alloys. Rhenium powder may be obtained by the hydrogen reduction of ReO_2 or ammonium perrhenate, NH_4ReO_4 at 600° or 400°C, respectively. ReO_2 is volatile and can be obtained as an intermediate sublimation product during the re-cycling of the Re scrap.

Hydrochemical reduction

Several metals can be produced in powder form directly from aqueous (or organic) solutions by reduction with (preferably) gaseous hydrogen. This process has been developed by Sherrit Gordon Mines Ltd, Canada, especially for the production of *nickel and cobalt powder* from sulphidic ores containing (preferably) Cu, Ni, Co, Fe and S. A convenient technique is the separation of a Cu rich and a Ni + Co rich fraction by flotation of the fine ore with a subsequent leaching procedure in autoclaves with NH_4OH under air pressures of 7–9 bar. Ni, Co (and the residual Cu) form readily soluble ammines (e.g., Ni $[NH_3]_6$ SO_4: nickel hexammine sulphate), Fe forms insoluble $Fe(OH)_3$ and the sulphur is transformed to sulphates. Most of the other constituents of the ore are insoluble in ammonia and form a solid phase together with $Fe(OH)_3$. However, a few other metals, like Sn and Cd also enter the solution.

The clean solution is then treated with gaseous hydrogen under 28–35 bar pressure at 180–220°C in an autoclave when the following reaction occurs (ignoring the complex ions):

$$Me^{++} + H_2 \leftrightarrows Me + 2H^+ \tag{2.13}$$

Low $[H^+]$ and high p_{H2} shift the equilibrium to the right hand side and facilitate metal precipitation.

The electrochemical conditions for the process are summarized in Fig. 2.21. Gaseous hydrogen is able to precipitate metals from their ionic solution, when the H_2/H^+ potential is more negative than Me/Me^{n+}, i.e. in the field below the straight lines. There is a small dependence of the H_2/H potential on the H_2 partial pressure (as seen by the small distance between lines a and b), but a large dependence on pH value. Thus, most of the metals can be precipitated from alkaline solutions and only a few such as Cu and Ag from stronger acidic liquids. The precipitation potential of the complex ammine ion of Co and Ni is not very different from that of the simple hydrated ions shown in the figure. Furthermore, there is no large dependence of the potential on the ion concentration, (as shown on the left and right hand sides, respectively, of this figure), which provides the possibility of nearly full deposition of Co and Ni

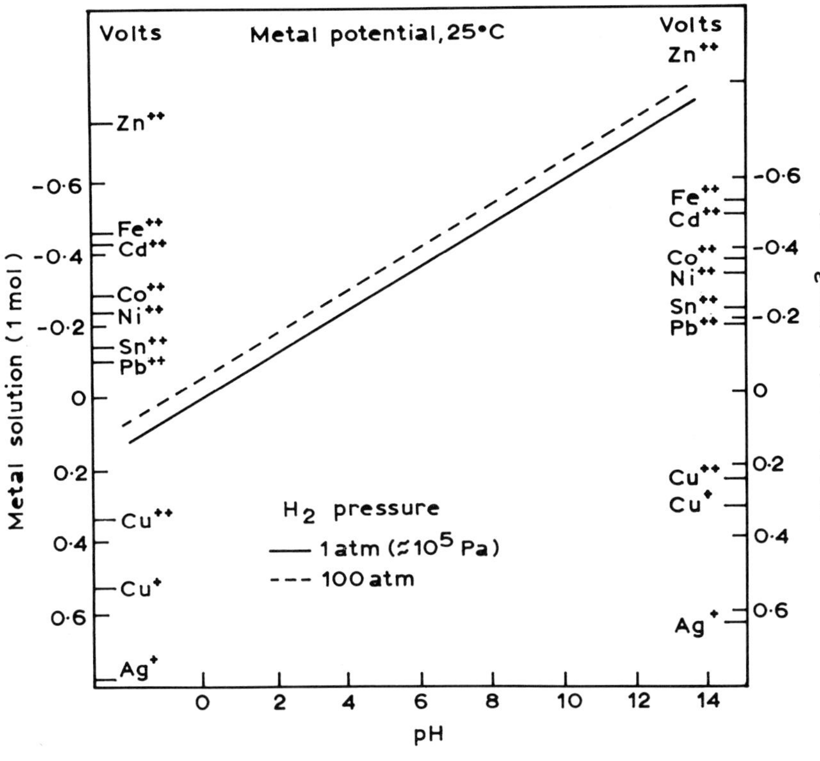

Figure 2.21 Metal and hydrogen potentials at different pH and different metal-ion concentrations (Schaufelberger)

from alkaline solutions. This also allows a selective precipitation of metals with a similar reduction potential ($\varepsilon_{Ni} = -0.23$ V, $\varepsilon_{Co} = -0.27$ V from molar solutions), at least in a certain concentration range when the H_2 partial pressure and pH are kept constant.

The process in practice is more complicated than Fig. 2.21. indicates, mainly because of the use of precipitation catalysts and by the necessary re-cycling procedures. The purity of the powders is nearly 99.8% metal content; Ni contains some Co and vice versa. Depending upon the stage of the process, coarser or finer particle size distributions can be obtained.

A special application of this process is the preparation of *composite powders*, when a 'seed powder' is dispersed in the liquid, which has to be kept in continuous motion in order to achieve a homogeneous dispersion. Each powder particle acts as a nucleus and is coated by the deposit. Examples are Ni-coated oxide or graphite and Co-coated diamond particles. The uses of composite powders (see section 4.3.) include avoiding segregation of the components; facilitating the even distribution of a dispersant in a matrix (to prevent particle-particle contacts); improving the wettability in melts etc. Composite powders can also be produced by other methods such as chemical vapour deposition

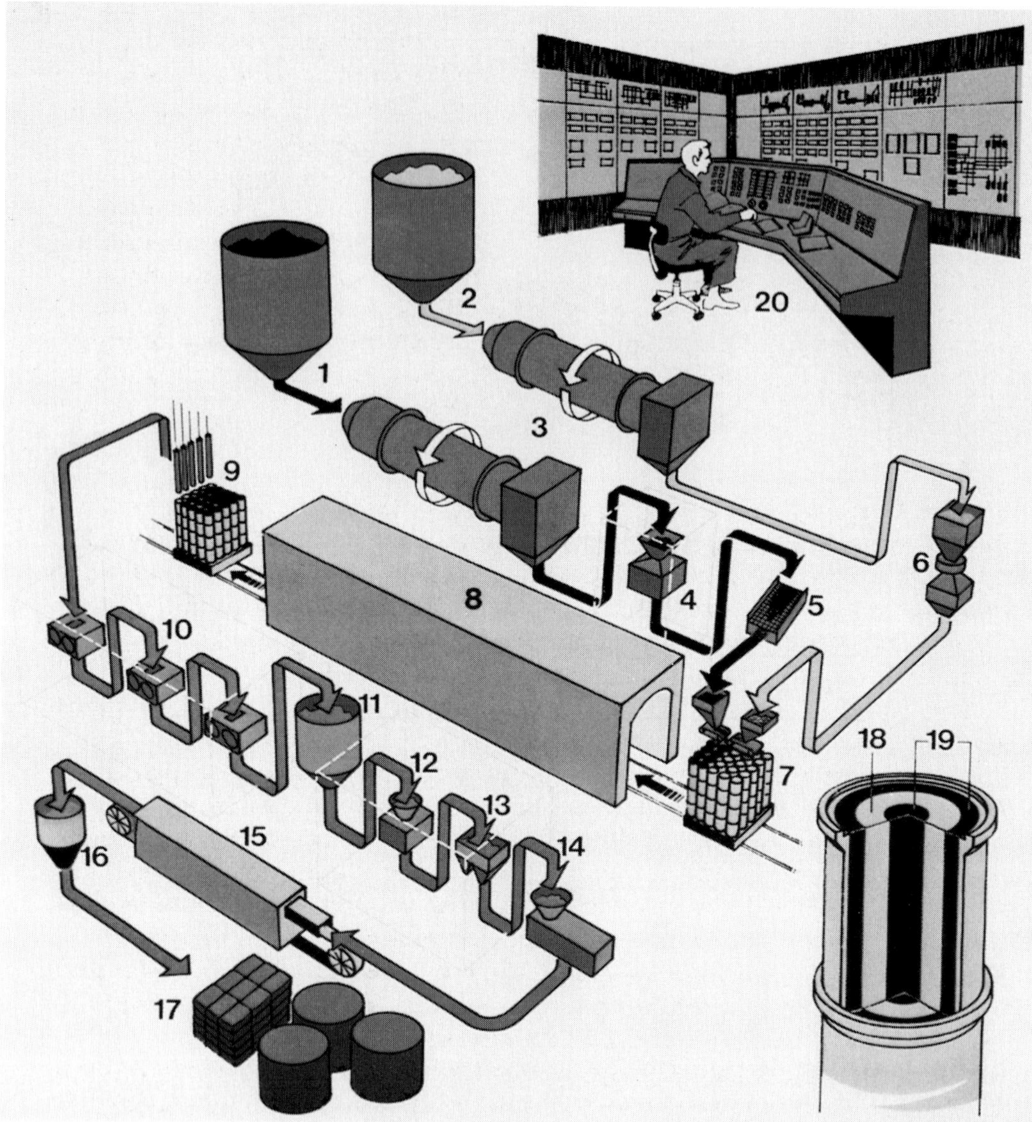

Figure 2.22 Sponge iron and iron powder production in Höganäs
1. Reduction mix of coke breeze and limestone; 2. Iron ore; 3. Drying; 4. Crushing; 5. Screening; 6. Magnetic separation; 7. Charging in ceramic tubes; 8. Reduction in tunnelkilns, approximately 1200°C; 9. Discharging; 10. Coarse crushing; 11. Storage in silos; 12. Crushing; 13. Magnetic separation; 14. Grinding and screening; 15. Annealing in belt furnace, approximately 800–900°C; 16. Equalizing; 17. Automatic packing of pressing powder, welding powder and cutting powder; 18. Iron ore; 19. Reduction mix; 20. Control room

(CVD), physical vapour deposition (PVD), sol/gel processes, etc. Recently, WC–Co composite powders have been produced in form of nanosized (20–40 nm) particles (see section 2.4) by a spray drying process for manufacturing of special hard metals.

Precious metal powders such as Ag, Au, Pd and Pt are also obtainable from aqueous media. While most of these powders are prepared by chemical precipitation from their respective aqueous salt solutions, a solvent extraction based process for refining the gold and platinum group of metals has been developed, which produces purified aqueous or organic solutions of Au, Pd and Pt. From these solutions, the three metals can be reduced separately. Pt and Pd can be precipitated by hydrogen reduction of $PtCl_4$– (or $PdCl_4$–) HCl–NaCl solutions at normal temperatures. Au can be reduced by hydrogen only from organic solvents.

Fine precious metal powders, especially Ag and Au, are used in the electronic industries for the manufacture of multilayer ceramic capacitors, for conductor pastes in thick-film applications, conductive tracks on glass and ceramics. Large amounts of Ag powders are used for conductive tracks in heatable car windows.

Carbon reduction

One of the most important powder production processes is the direct reduction of magnetite ores (Fe_3O_4) by carbon. High grade magnetite ores are found in large quantities in northern Sweden (Kiruna, Gällivare) and the process was developed in Sweden as early as 1910. It is now established industrially in Sweden, the USA and Japan. The final product, sponge iron powder, is one of the starting materials for the mass production of sintered iron and steel parts, and more than 50% of the iron powder used is produced by this process which is shown in Fig. 2.22.

The ore is ground to particle sizes of $\leq$ 0.5 mm and purified by magnetic separation, up to the iron content of about 71.5%. (the theoretical value is 72.5%.) After drying the ore is poured into cylindrical ceramic capsules, together with mixtures of fine coke and lime in concentric layers. The lime is used to extract the sulphur impurities. A number of capsules is arranged on a transportation unit which is moved through a reduction tunnel furnace, heated by natural gas.

The furnace is divided into a heating zone, a reduction zone with a maximum temperature of about 1250°C and a cooling zone, with an overall length of about 270 m. The reduction process, starting at 650–700°C, lasts two to three days, works continuously and is fully automated. The reaction product is an iron sponge formed by slight reaction sintering, which is then crushed, classified and subjected to a final treatment to adjust carbon and oxygen contents and to reduce residual stresses.

Most of the powder is milled to particle size distributions suitable for powder metallurgy parts production, whilst some (particle size > 150 μm) is sold for the manufacture of welding electrodes. It is very important to reduce the carbon as well as the oxygen content (and especially the non-metallic oxide inclusions) to levels acceptable for sintered parts.

The reduction process occurs, after an early solid state reaction, mainly via the gas phase, although no reducing gas is introduced: The overall reaction is

$$Fe_3O_4 + 4\,CO \leftrightarrows 3Fe + 4\,CO_2 \qquad (2.14)$$

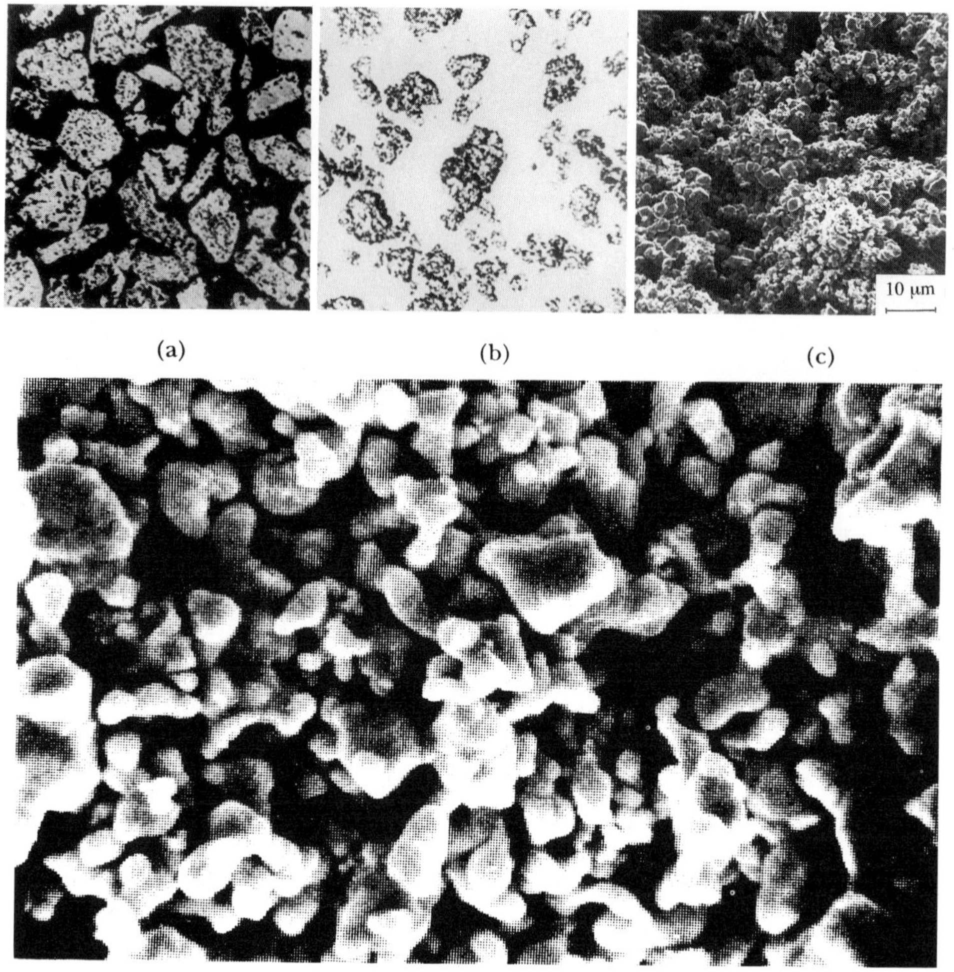

(a) (b) (c)

(d)

Figure 2.23 Höganäs iron powder (a) microstructure (b) loose powder (c) Hydrogen reduced W powder (Plansee, Reùtte) (d) Na reduced Ta powder (Lambert and Droegkamp)

while the carbon is in equilibrium according to the Boudouard equation

$$C + CO_2 \leftrightarrows 2CO \tag{2.15}$$

which is strongly dependent on temperature (see section 8.1). Side reactions, such as the formation of Fe_3C, are not important under the processing conditions. The sponge iron powder has some internal porosity, is of irregular shape (Fig. 2.23) and has a good compressibility, although the pressed density is influenced by its microporosity. This is a residue of the reduction process and is in contrast to the near pore-free microstructure of water atomised particles (see section 2.1.1.2).

In Table 2.3 a variety of commercial iron powders based on this process in addition to atomised powder is listed and some of their properties are shown. Most sintered iron and steel parts are manufactured from those powders, which can be partly obtained also with reduced density.

Attempts have also been made to *reduce milling 'scale'* for the production of iron powder. Scale is a high temperature oxidation product of iron and steel and has an overall composition close to Fe_3O_4. Reduction processing is simple in principle but technically and commercially feasible only when the scale is available in large quantities, i.e. from big milling plants, producing high purity and uniform quality over long periods. This cannot be achieved, when different (unalloyed and alloyed) steels are milled at the same plant, and which contain Mn, Si, Cr, Al, which form oxides in the scale and which cannot be reduced completely by carbon — the resulting iron powder containing uncontrollable amounts of non-metallic inclusions.

Reduction by metals

Metallo-thermic reduction is a standard procedure in metallurgy for the production of some refractory and highly reactive metals, e.g., Ti, Zr, U, Th, etc. The preferred reducing metals are Na, Ca and Mg. The Kroll process for Ti and Zr is well known, based on the reaction (for Ti)

$$TiCl_4 + 2Mg \rightarrow Ti + 2\ Mg\ Cl_2 \qquad (2.16)$$

At a suitable temperature, the metal melts down *in statu nascendi*, forming a sponge or sometimes an ingot. The sponge is the raw material for producing ingots by melting in a vacuum furnace, heated by electric arc or electron beam. Metals such as Ta with a high melting point are generally obtained in powder form by this method, since the free energy of reaction is not sufficient to heat the batch above the melting point of the metal. In principle, all metallo-thermic reactions can be adjusted (by inert additives, batch size and geometry, etc.) in such a way that metal powders can be obtained. However, powders are not mainly of interest for Ti and Zr, but they are for Be, Ta, Nb and Th.

The technically most important metallo-thermic process in powder metallurgy is the production of *Ta by reduction of K_2TaF_5 with sodium*, based on the exothermic reaction

$$\begin{gathered} K_2TaF_7 + 5Na = 2\ KF + 5NaF + Ta \\ (\Delta H = -2985\ KJ\ kg^{-1}\ \text{reactants}). \end{gathered} \qquad (2.17)$$

The reaction occurs in an argon atmosphere or in vacuum at 800–900°C and involves both liquid as well as gaseous sodium because of the high sodium vapour pressure in this temperature range (boiling temperature 881°C). The reaction product is processed by leaching of the fluorides and intensive washing of the Ta powder, which is obtained with an average particle size of 1–2 μm and a purity of about 99.6%.

Table 2.3 Properties of some commercial iron and iron alloy powders

Grade	Composition (wt %)	Approximate particle size range (μm; %)	Apparent density (g/cm³)	H_2 loss (%)	C (%)	Compressibility at 600 MPa 1.0% Zinc stearate[(1)] 0.8% Zinc stearate[(2)]
WP 200[(1)]	Iron	− 63 ; 25–45 + 63 ; 20–40 + 100 ; 25–45 + 160; < 15	2.8–3.1	0.20	0.02	7.02–7.07
WP 200 HD[(1)]	Iron			0.15	0.01	7.12–7.17
WP 200 SD[(1)]	Iron			0.15	0.01	7.17–7.20
Ultrapac A[(1,3)]	Cu 1.50; Ni 1.75 Mo 0.50; Mn < 0.2 + iron	− 63 ; 20–40 + 63 ; 25–45 + 100 ; 20–40 + 160 ; < 10	3.00 ± 0.1	< 0.20	< 0.02	> 7.02
MSP 2[(1,3)]	Ni 1.7–2.0; Mo 0.4–0.6 Mn 0.20 + iron	− 63 ; 20–50 + 63 ; 20–40 + 100 ; 15–40 + 160 ; < 15	2.8–3.0	0.20	0.02	> 6.80
SC 100.26[(2)]	Iron	20–180	2.65	0.12	< 0.01	7.10
ASC 100.20[(2)]	Iron	20–180	2.95	0.10	< 0.01	7.18
ABC 100.30[(2)]	Iron	30–200	3.0	0.05	< 0.01	7.25
PASC–45[(2)]	P 0.45 + iron	20–180	3.15	0.10	0.02	7.14
Distaloy AE[(2,3)]	Ni 4.0; Cu 1.5 + iron	20–180	3.0	0.10	< 0.01	7.15
Astaloy Mo[(2,4)]	Mo 0.5 + iron	20–180	3.25	0.10	< 0.01	7.11

[(1)] Mannesmann Demag; [(2)] Höganäs AB; [(3)] Diffusion stabilized; [(4)] Prealloyed powder

This process is closely associated with the production of high purity Ta powder with a very fine particle size, large specific surface and high purity for the manufacture of capacitors. The process competes with that of fused salt electrolysis, which is used mainly in the production of Ta-powders for semi-finished sheets as corrosion resistant material for the chemical industry. In 1989, nearly 90% of all Ta powder was produced by sodium reduction.

Several highly reactive metals (Ta, Nb, Zr, Hf) can also be obtained by *calcio-thermic reduction* of their oxides according to (for Zr)

$$ZrO_2 + 2Ca \rightarrow Zr + 2\,CaO \tag{2.18}$$

or from halogenides

$$ZrCl_4 + 2Ca \rightarrow Zr + 2\,CaCl_2 \tag{2.19}$$

The physical characteristics of the resulting powders as well as the oxygen content can be adjusted to some extent by addition of $CaCl_2$ to the reaction mixture as a 'diluting' medium. The use of calcium hydride (CaH_2) instead of elementary Ca as a reductant may also help to moderate the high exothermicity of the Ta_2O_5 and Nb_2O_5 reductions (because the dissosiation of CaH_2 is endothermic), and facilitate the production of finer powders by minimizing *in situ* sintering effects. The purity of calcio-thermic powders, however, seems to be generally not as high as that obtained by the Na reduction process, the oxygen content being between 0.5 and > 1%.

Micrographs of powders produced by reduction processes (sponge iron, W and Ta) are shown in Fig. 2.23.

2.1.2.2 Carbonyl Processes

Several metals react with carbon monoxide (CO) under certain conditions to form metal carbonyls ($Me(CO)_x$). Fine metallic powders can be obtained by their thermal decomposition. This is especially important with iron carbonyl ($Fe(CO)_5$, boiling point = 102.7°C) and nickel carbonyl ($Ni(CO)_4$, boiling point = 43°C), from which carbonyl iron and carbonyl nickel powders are obtained. Because of the low boiling point of the carbonyls (which are both toxic and flammable) they can be purified easily by distillation. The resulting metal powders are therefore very pure (see Table 2.4), except for some carbon and oxygen impurities.

For carbonyl iron production, spongy iron, having a large surface area with high surface activity, is required to obtain an adequate reaction rate. The reaction and decomposition occur according to the following equation

$$Fe + 5\,CO \;\underset{\substack{\text{Decomposition} \\ (1\text{ bar, }250^{\circ}\text{C})}}{\overset{\substack{\text{Reaction} \\ (\sim 200\text{ bar, }200^{\circ}\text{C})}}{\rightleftarrows}}\; Fe\,(CO)_5 \tag{2.20}$$

Table 2.4 Chemical composition of carbonyl iron and nickel powders

Impurities (%)	Carbonyl Iron (BASF)	Carbonyl Nickel (INCO, different types)
C	~ 1, H_2 reduced ≤ 0.01	0.06–0.02
Si	< 0.001	other elements (total) < 0.001
Mn	< 0.002	
P	< 0.002	
S	< 0.001	0.0002
N	variable	0.0002–0.003
O		0.05–0.07
Ni	< 0.002	balance
Fe	balance	0.005
Co		0.0003

The decomposition is undertaken in a cylindrical, vertical, tube-type reactor under atmospheric pressure, in a process invented by BASF (Ludwigshafen) in 1924. The CO is re-cycled continuously. Side reactions that occur include:

$$Fe + CO \longrightarrow FeO + C \tag{2.21a}$$

$$3\,Fe + 2\,CO \longrightarrow Fe_3C + CO_2 \tag{2.21b}$$

$$Fe + CO_2 \longrightarrow FeO + CO \tag{2.21c}$$

$$3\,Fe + C \longrightarrow Fe_3C \tag{2.21d}$$

These are responsible for the carbon and oxygen impurities of the raw powders. These reactions can be suppressed by introducing ammonia (NH_3) during decomposition, which decreases the carbon and oxygen, but increases the nitrogen content ($Fe_{1-x}\,N_{x-}$ phases) of the powders. A hydrogen reduction of the raw powder likewise decreases the carbon, oxygen and nitrogen content. The particle diameter range is normally 2–10 μm. The particle microstructure is onion shell like and spherical (Fig 2.24a,b).

Carbonyl iron powders are produced in different grades, mainly for electronic industries and powder metallurgical applications. They are used, for example, as a component in Fe–Ni and other soft magnetic alloys. Nowadays the metal injection moulding (MIM) technique (see section 6.1.1.3.) is an important and growing field of application.

High purity nickel powder is also made by the carbonyl process, mainly by INCO, Clydach, South Wales, UK., according to the following reaction:

$$Ni + 4\,CO \leftrightarrows Ni\,(CO)_4 \tag{2.22}$$

The reaction takes place at atmospheric pressure and proceeds more easily than with iron. The decomposition of the $Ni(CO)_4$ occurs under similar condi-

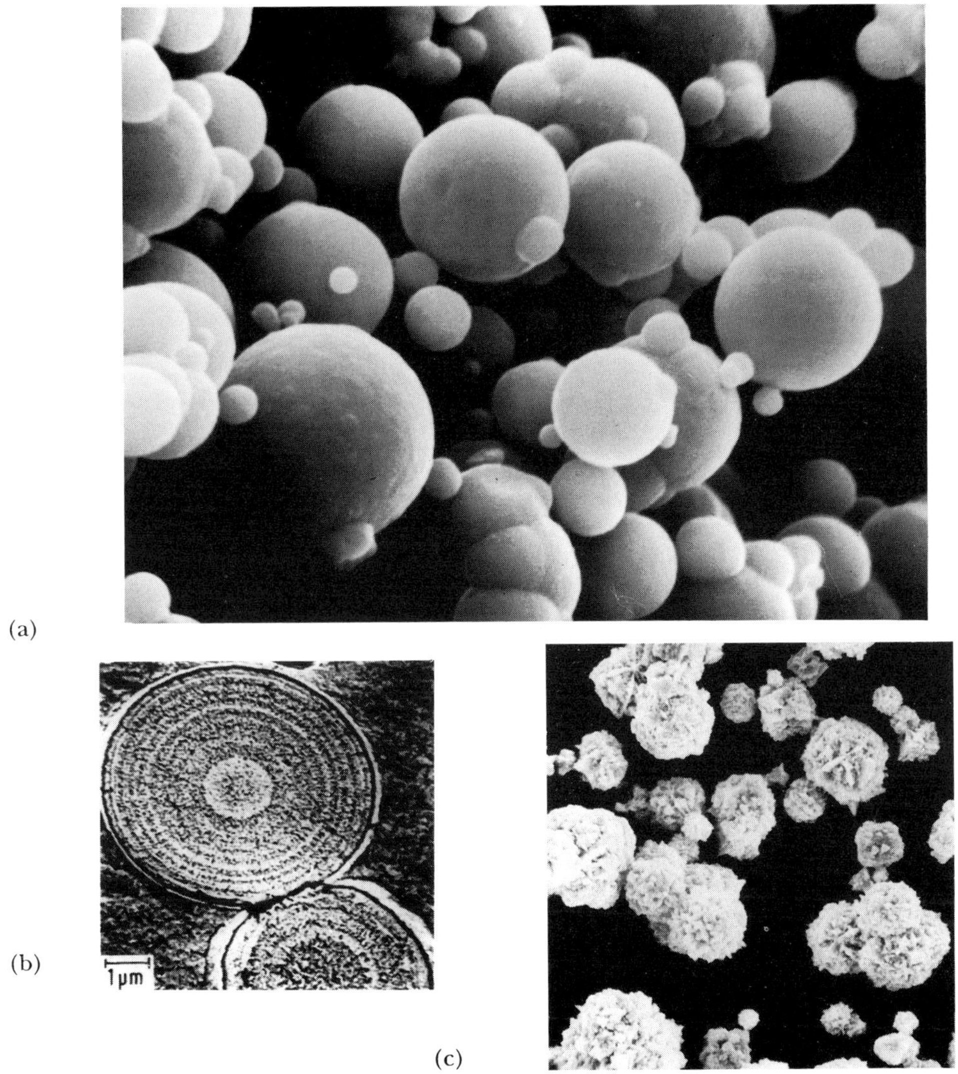

Figure 2.24 Carbonyl powders; (a) Iron 1–10 μm (BASF, Ludwigshafen); (b) Iron, onion shell microstructure; (c) Nickel 13 μm mean particle size (INKO, Clydach)

tions to that of $Fe(CO)_5$. The process produces fine powder or pellets, the powder being generally not spherical, as shown in Fig. 2.24c. Carbonyl nickel is used for many applications in powder metallurgy.

2.1.2.3 Hydride–Dehydride Processes

The hydride route enables very fine and reactive powders to be produced from metals which are capable of forming solid hydrides. The hydrides are obtained by the reaction of metal with molecular hydrogen at elevated temperatures under a suitable partial pressure. Since they are brittle they can be powdered easily and dissociated in vacuum at similar or somewhat higher temperatures than those at which they are formed.

For powder metallurgy, the hydrides of Ti, Zr, Hf, Ta, U, Th, Y and perhaps Mg are of interest. As an example, the hydride formation and decomposition is given below for thorium, which forms two hydrides:

$$\mathrm{Th} + \mathrm{H_2} \xrightarrow{600\text{–}650°\mathrm{C,\ H_2}} \mathrm{ThH_2}\ \text{(no disintegration)} \qquad (2.23a)$$

$$4\mathrm{ThH_2} + 3\tfrac{1}{2}\,\mathrm{H_2} \xrightarrow{250\text{–}300°\mathrm{C,\ H_2}} \mathrm{Th_4H_{15}}\ (\textit{in situ}\ \text{powderisation}) \qquad (2.23b)$$

$$\mathrm{Th_4H_{15}} \xrightarrow{\sim 500°\mathrm{C,\ H_2}} 4\mathrm{ThH_2} + 3\tfrac{1}{2}\,\mathrm{H_2} \qquad (2.23c)$$

$$\mathrm{ThH_2} \xrightarrow{\sim 700°\mathrm{C,\ Vac}} \mathrm{Th}\ \text{(powder)} \qquad (2.23d)$$

Recently a process known as hydrogen decrepitation has allowed fine scale milling of Fe-Nd-B hard magnetic alloys.

Metal powders produced from hydrides are highly active and often pyrophoric, depending on dissociation conditions. Special precautions, such as inert gas absorption on the powder surfaces, are often necessary in order to avoid instantaneous ignition on exposure to air. Compacts of zirconium powder produced from the hydride can be sintered nearly to full density. Tantalum hydride powder is of some importance in obtaining very pure tantalum carbide powder by carburisation, using carbon black. The hydride can be economically obtained from Ta scrap or waste materials.

The hydride–dehydride process does not involve any purification of the metal phase. On the contrary, an increase of oxygen and nitrogen can be expected in the final metal powder, depending on the quality of hydrogen and the vacuum system used.

2.1.2.4 *Electrochemical Processes*

Principal Aspects

Powder production by electrolysis is a reduction process in which the metal ions are neutralised by the cathodic current. The cathodic potential necessary for deposition depends on the position of the metal in the electromotive series and on special inhibiting electrode processes, which can produce a voltage surge necessary for deposition. Whenever possible the process is carried out in aqueous solutions (e.g. for Cu, Fe, Ni, Co, Zn, etc). Highly reactive metals (e.g. Be, Ta, Nb, Th, etc.) which form highly stable oxides, have to be deposited from molten salt electrolytes. While aqueous solutions are employed at slightly elevated temperatures (up to 60°C) the molten salt process needs higher temperatures, depending on the melting point of suitable (binary or ternary) salt eutectics which should contain a reasonable concentration of the metal to be deposited. The powder can be obtained directly or indirectly, the latter by milling of the often hydrogen-embrittled deposits.

The main process parameters to be considered are: metal ion concentration, electrolyte conductivity, temperature, voltage, current density and bath circulation, and, in special cases, the addition of colloids in order to inhibit the growth of nuclei. In aqueous solutions, low temperature and high current density facilitate the precipitation of powder rather than larger deposits.

Electrolysis of aqueous solutions

Electrolysis of aqueous solution is the most important process in the production of copper powder. The electrolysis takes place in plastic lined steel tanks, with a volume of ~ 5 m^3. A group of these tanks form a parallel working unit with a circulating electrolyte system. The electrolyte operates at a temperature of about 50°C and contains 5–35 gl^{-1} Cu^{++} and 120–250 gl^{-1} sulphuric acid, whilst the current is 7500–10 000 A. The cathodic current density ranges up to 4000 Am^{-2}, the anodic density being only about one tenth of this value. Thus, the total surface area of the negative electrodes is about one tenth that of the

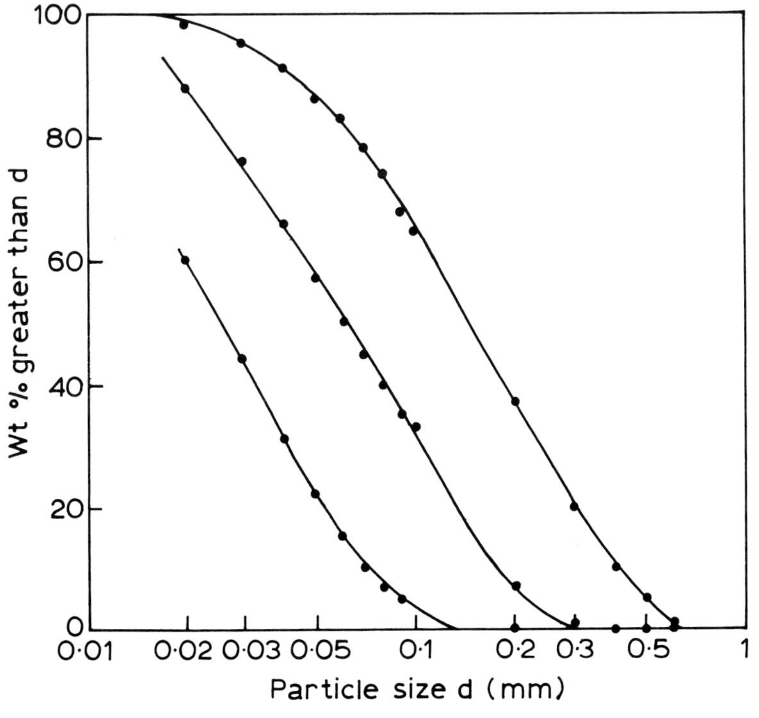

Figure 2.25 Particle size distribution of three electrolytic copper powders (Peissker)

(a) (b)

Figure 2.26 Electrolytic copper powder (Norddeutsche Affinerie, Hamburg) (a): fine fraction (–63 μm dendritic) (b): coarse fraction (–200 μm)

positive electrodes. Anodes as well as cathodes are made from dense, pure electrolytic copper. Most of the powder deposit falls to the tank bottom due to the movement of the electrolyte and by tapping the cathodes. When a certain quantity of powder has filled the bottom of the tank, the electrolyte is removed and the powder is washed to clean it of acid. Final drying is undertaken in a continuous flow of air at about 100°C.

Three examples of particle size distribution of primary deposits, depending on deposition conditions, are shown in Fig. 2.25. The finer powders are obtained at higher current densities. Figure 2.26 shows micrographs of such powders. Fine powders have a more pronounced dendritic shape than coarser ones, which is the reason for their different apparent densities. When powders with a high specific surface (i.e. a small particle size) are deposited, hydrogen begins to form, since the cell voltage increases. There is a decline in current efficiency (Fig. 2.27) along with an increase of specific energy consumption.

Clean Cu powder has a strong tendency to oxidize, especially when exposed in air of high humidity. This is a result of a high concentration of lattice and surface defects and is a problem for prolonged storage and during handling, but can be reduced by annealing at a moderate temperature. Some stabilizing treatment is often applied to provide a very thin protective coating, which is not harmful for the subsequent pressing and sintering process. Typical data for some commercial powders are presented in Table 2.5.

The finer powders with the highest specific surface are not free flowing and are therefore not suitable for die filling in production processes. For this purpose, however, they can be mixed with other powders. They have very low impurity levels, often below 5 ppm for Al, Mg, Zn, Sn, As, Sb, Bi, Fe, Co and Ni. For Pb, Ca and S values between 10 and 100 ppm have been measured, depending on the specific surface. The high purity derives from the *in situ* refining process, which the starting Cu anode undergoes during powder production.

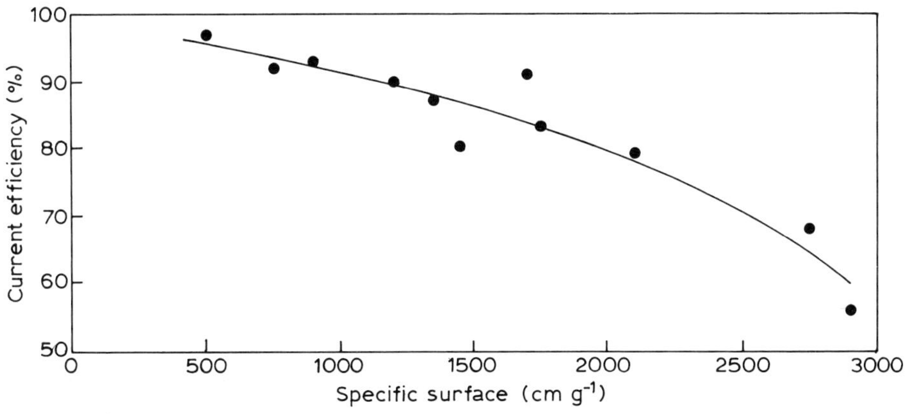

Figure 2.27 Decrease of current efficiency with decreasing powder fineness (= increasing specific surface) during electrolytic copper production (Peissker)

Table 2.5 Typical data for commercial copper powder (Norddeutsche Affinerie Hamburg)

Code	FFL	FL	FS	SSM	M	GF
Sieve analysis: + 0.315 mm						./.
+ 0.200 mm				./.	./.	1
+ 0.080 mm	./.	./.	./.	26	40	94
+ 0.063 mm	1	1	2			
+ 0.040 mm	4	9	13	33	33	4
– 0.040 mm	95	90	85	41	27	1
Apparent density (g cm^{-3})	1.00	1.25	1.90	2.40	2.40	2.40
Flow rate (sec 50g^{-1})	∞	∞	∞	33	19	27
Specific surface (cm^2 g^{-1})	1800	1400	850	500	450	300
Oxygen content (wt %)	0.09	0.08	0.07	0.04	0.03	0.02

The dendritic particle shape offers excellent compressibility resulting in high strength of the green compacted powder. This is of interest for the manufacture of thin-walled bronze bearings, copper-containing carbon brushes, friction materials and diamond tools.

Electrolytic iron powder with its high purity and good compressibility was industrially important until the 1970s. Its significance was reduced, however, by the development of highly compressible sponge iron powder and by the use of carbonyl iron as an even purer raw material. The electrolytic powder is produced from $FeSO_4$ or $FeCl_2$ solutions and is obtained mostly as a solid, brittle deposit which has to be milled to the desired particle size distributions (Fig. 2.28). The cathodic voltage and current density have to be adjusted to obtain simultaneous hydrogen production. Atomic hydrogen easily enters the α-iron lattic, leading to substantial embrittlement. The final powder is annealed to

Figure 2.28 Electrolytic iron products (Glidden Metals)

reduce internal stresses as well as hydrogen content and, because of it's high price, has only special applications.

Fused Salt Electrolysis

For economic and technical reasons fused salt electrolysis is used only when deposition from aqueous solutions is impossible. The electrolyte consists of salt mixtures, mostly chlorides or fluorides, which have near-eutectic compositions with low melting points. As a result of the high specific electrical conductivity of alkali and alkaline earth metal chlorides they are often a main constituent of the melts. The voltage has to be adjusted below the (more negative) deposition potential of the alkali-metal which is easily achieved. The metallic deposit has to be carefully protected from moisture and oxygen, either in closed cells or in an argon atmosphere. In order to clean-up the melts from traces of moisture (which are present even at several hundred degrees Celsius) a pre-electrolysis at low voltage (1.1 to 1.5 V) is advisable, and provides in addition a pre-deposition of noble metal traces.

Of some industrial importance is the deposition of *beryllium powder* (or flakes) from $BeCl_2$–KCl–NaCl melts at 320–380°C with a current density of 700 Am^{-2} in airtight Ni cells with Ni cathodes and graphite anodes. The process is run semi-continuously by post-charging the main constituent. After isolation from the melt, the flakes are carefully washed with water and dilute acid, resulting in complete purification from salt residues. The purity of the powder is about 99.5%, and can be increased to 99.8% or even higher by repeated electrolysis.

For *tantalum powder* production, which is the most important fused salt electrolysis product, the electrolyte consists of, for example 50–70% KCl, 20–35% KF, 5–10% K_2TaF_7 and 4–5% Ta_2O_5. The process works at 700–900°C and 3000

Am^{-2} in cast iron vessels, which serve as cathodes. The anodic material is graphite, located in the centre of the vessel. The Ta compounds are re-charged during the process until the Ta deposit has penetrated most of the electrolyte volume. After cooling, the mixture of metal and salt is crushed and washed with acid. The dendritic powder has an average particle size of about 20 μm with a purity of 99.8%.

Niobium powder is produced by a very similar process, using K_2NbF_7 based melts.

2.2 TRANSITION METAL CARBIDE, NITRIDE, BORIDE AND SILICIDE POWDERS

Transition metal carbides of the 4a, 5a and 6a group of the Periodic Table, e.g. WC, TiC, TaC, NbC are the basic materials for hard metals (cemented carbides), which form a large group of powder metallurgy materials (see section 11.6.). Together with the nitrides, borides and silicides, they represent the group of hard metallic compounds. The non-metallic hard materials (SiC, Si_3N_4, B_4C, BN, AlN, diamond, cubic BN) which belong to the field of ceramics, are partly treated in section 2.3. Hard compounds were, in earlier years, a classical field for chemists. The first book, on hard compounds was written by the French chemist H. Moissan and appeared around 1900. O. Ruff and D. Hönigschmidt are the authors of a book on carbides and silicides which appeared in 1914.

2.2.1 Carbide Powders

Transition metal carbide powders can generally be prepared *from elemental powders* by high temperature reaction, especially from metals which can be obtained easily with adequate purity and low oxygen content such as W and Mo. As an example, the reaction

$$W + C \rightarrow WC \tag{2.24}$$

is processed in graphite tube furnaces in hydrogen at about 1700°C, using soot for carburisation. The above equation disregards the partial formation of hydrocarbons, which may take part in the carburisation process. This is the main route for producing WC powders for cemented carbides, although direct reduction of WO_3 with carbon is also possible.

Another possible route is the *reaction of metals with hydrocarbons*, e.g.

$$W + CH_4 \leftrightarrows WC + H_2 \tag{2.25}$$

where an intermediate formation of free carbon may occur. The critical parameter is the adjustment of the hydrocarbon partial pressure, so that no significant amount of free carbon remains after the end of process. For this reason it is not a routine industrial process.

Yet another process is the *direct carburisation of oxides,* which is used mainly to produce TiC

$$TiO_2 + 3\,C \leftrightarrows TiC + 2\,CO \tag{2.26}$$

undertaken at temperatures as high as 1800–1900°C. When hydrogen is employed as a protective gas, side reactions such as

$$2\,H_2 + C \leftrightarrows CH_4 \text{ and} \tag{2.27}$$

$$TiO_2 + CH_4 \leftrightarrows TiC + 2\,H_2O \tag{2.28}$$

take place, which influence the total reaction considerably. ZrC, VC, NbC and TaC are also produced by carburisation of their oxides.

Other processes of less significance are the arc melting of metals using graphite electrodes (the original method of Moissan) and the electrolysis of molten salts (sodium borate, carbonate, fluoride with WO_3 or MoO_3), resulting in cathodic precipitates of carbides (W_2C, WC, etc.). During the arc melting process temperatures of 2500–4000°C are achieved, which are well above the stability range of most carbides. In this process mixtures of carbides together with larger amounts of free carbon are obtained, which cannot be used as raw materials for powder metallurgy. However, wear resistant coatings on metallic surfaces can be produced by this method. Of some interest is the reaction of metals with carbon in iron melts, which can be used for many carbides whose free energy of formation is more negative than that of Fe_3C. This so-called *Menstruum process* is practised by several producers of cemented carbides.

In addition to the single compounds, solid solutions of two or three carbides, nitrides, etc. are also of great interest, especially for use in cemented carbides for cutting or milling purposes. *Mixed carbides* can be obtained by the following processes:

- Carburisation of carefully mixed oxide or metal powders with soot or graphite powder under conditions similar to single carbide preparation. The use of oxide and metal powder can be also employed.
- Annealing of carefully mixed, very fine carbide powders at temperatures for solid solution formation (1600–2000°C). The homogenisation rate is a function of the relative chemical diffusion coefficients, which are related to the melting temperatures of the compounds. Sometimes repeated grinding, mixing and annealing is necessary. The homogenisation can be facilitated by the addition of small amounts (0.5–1%) of Co, Ni or Fe powders.

Typical analyses of some carbide powders, important for hard metal production, are given in Table 2.6. The total carbon and free carbon, and the oxygen content are of particular interest, as well as the particle size.

Table 2.6 Properties of some commercial carbide powders (H.C. Starck, Goslar) for hard metal production

	WC	TiC	TaC	NbC
Mean particle size (FSSS† in μm)	0.7–10	1–4	0.5–5	0.5–5
Typical purity (%)	> 99,8	Ti 79,7*)	> 99%⊗	> 99%⊗
Typical impurities (except free C and O) (ppm)	< 250	< 2400	< 900	< 900
Oxygen (ppm)	500–2500⊕	1500–5000⊕	1000–3000⊕	1000–3000⊕
Carbon, total (%)	6.14	19.3–19.6	6.1–6.3	11.0–11.3
Carbon, combined (%)	6.11	19.0–19.3	> 6.0	> 10.9
Carbon, free (%)	0.03	0.01–0.05	< 0.15	< 0.15

*) Because of the substoichiometry of TiC (TiC_{1-x}) generally observed, the Ti-content instead of TiC is given.
⊗ TaC/NbC mixtures are obtainable in all compositions.
⊕ Dependent on particle size.
† Fischer sub-sieve size

2.2.2 Nitride, Boride and Silicide Powders

Many *nitrides* in powdered form can be prepared by reaction of the metal (or metal hydride) powders with pure N_2 or NH_3, (e.g., TiN, ZrN, VN, NbN, TaN, WN, etc). The reaction temperatures are about 1200°C or higher, depending on powder particle size. Very pure nitrides can be obtained by this method, provided the starting materials are pure. A more economic route is the nitridation of oxides in the presence of carbon, again with N_2 or NH_3. The overall reaction for TiN is:

$$2\,TiO_2 + 4\,C + N_2 \rightarrow 2\,TiN + 4\,CO \qquad (2.29)$$

The reaction temperatures are in the range of 1250–1400°C; the purity of nitrides obtained in this way is not as high as from metal powders. Also, metal chlorides or oxichlorides can be reacted to from nitrides.

Metal *boride powders* have been obtained from the elements by melting or by reaction sintering below the melting temperature. Borides are formed by reaction of metal (Me) halogenides (X) with boron halogenides according to the formula:

$$MeX_4 + 2BX_3 + 5H_2 \rightarrow MeB_2 + 10\,HX \qquad (2.30)$$

These methods are convenient for laboratory scale production and very pure borides can be obtained. The industrial methods, however, are the carbothermal or boro-thermal reduction of mixtures of metal oxides, B_2O_3 and C or metal oxides and B, respectively. The reaction equations for TiB_2 are:

$$TiO_2 + B_2O_3 + 5\,C \rightarrow TiB_2 + 5\,CO\ (1700\text{–}2000°C) \qquad (2.31)$$

$$TiO_2 + 4\ B \rightarrow TiB_2 + 2\ BO\ (1500\text{–}1700°C) \qquad (2.32)$$

Because of the high price of boron the latter process is less economic than the former. The carbothermal reduction, however, yields products with some 0.1% free carbon and oxygen as the main impurities. The boron-carbide method

$$2\ TiO_2 + B_4C + 3\ C \rightarrow 2\ TiB_2 + 4\ CO\ (1700\text{–}2200°C) \qquad (2.33)$$

is another type of carbothermal reaction of some significance. The TiB_2 obtained from this reaction also contains free carbon and some oxygen.

Many of the methods mentioned for the preparation of carbides and borides can also be used for *silicides.*

2.3 CERAMIC POWDERS

2.3.1 Principal Aspects

Since the boundaries between powder metallurgy and ceramics are no longer strictly defined, some information on ceramic powders is useful in the context of powder metallurgy. This applies mainly to advanced ceramics. While for the traditional silicate ceramics in general use are natural raw materials (clay, kaolin, etc.), the processing of advanced ceramics requires pure and well-defined powders normally made synthetically. Also the manufacture of ceramic parts is to some extent similar to powder metallurgy processing. It is convenient to subdivide this field into oxide and non-oxide ceramics.

The synthesis of ceramic powders employs a wide variety of methods, yielding powders of very different particle size distribution and specific surface, chemical and phase composition, purity, compactibility, sinterability, etc. Sinter grade powders for manufacturing dense, fine-grained parts must be produced in the lower micron or even sub-micron range with specific surfaces between 5 and 15 m^2g^{-1} or higher. Applications of coarser powders (Al_2O_3, SiC) include abrasives, refractories (Al_2O_3, ZrO_2, Mullite), etc.

The principle methods for production are as follows:

- *Solid state reactions.* Hydroxides, carbonates, sulphates, oxalates and other compounds can be thermally decomposed to oxides by well known chemical reactions. Depending on temperature–time, loose or agglomerated powders or a sintered cake result, which have to be milled to the desired particle size. In principle the majority of the oxide powders can be produced by this method. The precursor compounds are obtained by either liquid or gas processes. Carbides are made by reaction of metals or oxides with carbon. One of the most important industrial processes is the Acheson method for the production of α SiC from SiO_2 (see section 2.3.4), in which it is possible that part of the reaction occurs via the gas phase.

- *Solid–gas reactions.* Oxides, carbides and nitrides may be synthesised by the reaction of metals with oxygen, hydrocarbons, nitrogen or ammonia respectively. While this is less important for oxides, it is a technique used, for example, for the synthesis of β SiC and Si_3N_4 (see section 2.3.4), the latter sometimes under increased N_2 pressure.
- *Gas-phase reaction.* Vapour phase decomposition or hydrolysis in a flame is a common technique for the production of TiO_2 and SiO_2 from halogenides $TiCl_4$ and $SiCl_4$, according to the reaction (for SiO_2):

$$SiCl_4 + 2\ H_2O \rightarrow SiO_2 + 4\ HCl \tag{2.34}$$

  High purity powders with extremely high specific surface areas (100 m^2g^{-1} and higher) with only very little agglomeration of the powder particles, are obtained in this continuous process.
- *Melting processes.* Oxides such as Al_2O_3 and ZrO_2 are manufactured on a large scale by arc-melting with subsequent milling to make powders of different particle size ranges. The process is mainly used to produce coarse grade powders for the manufacture of refractories.
- *Reactions from solution* are treated in more detail in section 2.3.2.

2.3.2 Powders from Solutions

In several techniques an aqueous or non-aqueous solution of salts is the precursor for preparing powders. After solvent removal, the solid residue is thermally converted to ceramic or in some cases metallic powders. The solutions are normally homogeneous on an atomic scale, but the degree of homogeneity of the residue, however, depends on the solvent removal technique. Extremely fine powders, powder mixtures and powders with uniformly distributed additives (e.g., to improve sintering), are obtainable.

This field may be subdivided as follows:

- *Precipitation and filtration* of single compounds or mixtures. This is a 'classical' method for the preparation of fine ceramics or, after hydrogen reduction, metallic powders. The latter step is described in section 2.1.2.1.
- *Hydrothermal reactions.* This is a special process for manufacturing pure and doped ZrO_2 based powders. The precipitation or hydrolysis reactions are operated either in $ZrOCl_2$ solutions at an elevated temperature (200–250°C) and a high pressure (2–6 MPa), or precipitates obtained under normal conditions are treated hydrothermally. These powders can have a very fine and narrow range of particle size distribution, are well crystallized and are suitable for further processing at a sintering temperature of about 1400°C.
- *Solution combustion.* Alcohol based or organo-metallic solutions are burned in air or oxygen to produce oxide particulates.

- *Solvent vaporisation or dehydration*, either by direct evaporation (mainly for single component solutions), *spray drying* (preventing segregation in multicomponent solutions by dispersing the liquid into very small droplets), *spray roasting* (where the spray is dried and decomposed at higher temperatures in a single step), or *freeze drying* (in which the solution is sprayed as small droplets into a bath of immiscible cold liquid or into a gas chamber where a rapid freezing of the droplets occurs).

The Sol-gel Process

In a separate sub-group are gelation–dehydration techniques in which liquid droplets are transformed into a gelatinous state from which the powder is obtained. These sol–gel techniques now play an increasing role in the preparation of advanced ceramics. The techniques consist in the formation of a three dimensional network of inorganic matter (gel) from colloidal or molecular solutions of the precursor (sol). This reaction takes place in aqueous or non-aqueous media, e.g. by diminishing the water content, by changing the P_H-value or the surface charge (zeta potential) of the sol, or by other means which can lead to gelation of the liquid precursor. A wide variety of methods may be used, producing small uniform spheres or powders. The process was developed in the early 1960s for the preparation of micro-spheres of oxide nuclear fuels, consisting of highly reactive agglomerates of very fine particles, which can be sintered at low temperatures to very high densities.

The main uses of sol-gel technology are shown in Fig. 2.29. The decisive step in powder formation is a controlled aggregation of the colloid particles. For transformation into powders, the gel particles have to be dried and calcined. A disadvantage of the process is that some microporosity may be formed in the range of seveal nm within the powder particles. Advantages are the high sintering-activity of the powder and, in multicomponent systems, the complete homogenisation of elements found in mixed sols. Finally these are experience production routes and many are still only partly understood.

2.3.3 Oxide Ceramic Powders (Al_2O_3 and ZrO_2)

Alumina (Al_2O_3) is by far the most important oxide ceramic and its powder is produced on a very large scale and to a wide variety of specifications. Almost 35 million tonnes per year are produced throughout the world, of which 93% is used as 'smeltering grade material' for the production of aluminium. The remainder is mostly calcined and milled for use in abrasive and polishing applications and in the production of refractories and ceramics. Except for powders of the highest purity (99·99%) all qualities are produced from Bauxite via the Bayer process which is a wet alkaline route, separating the Al_2O_3 from Fe_2O_3 and other oxides by forming sodium aluminate. This is transformed by hydrolysis into $Al(OH)_3$ and subsequently calcined to Al_2O_3. The highest purity

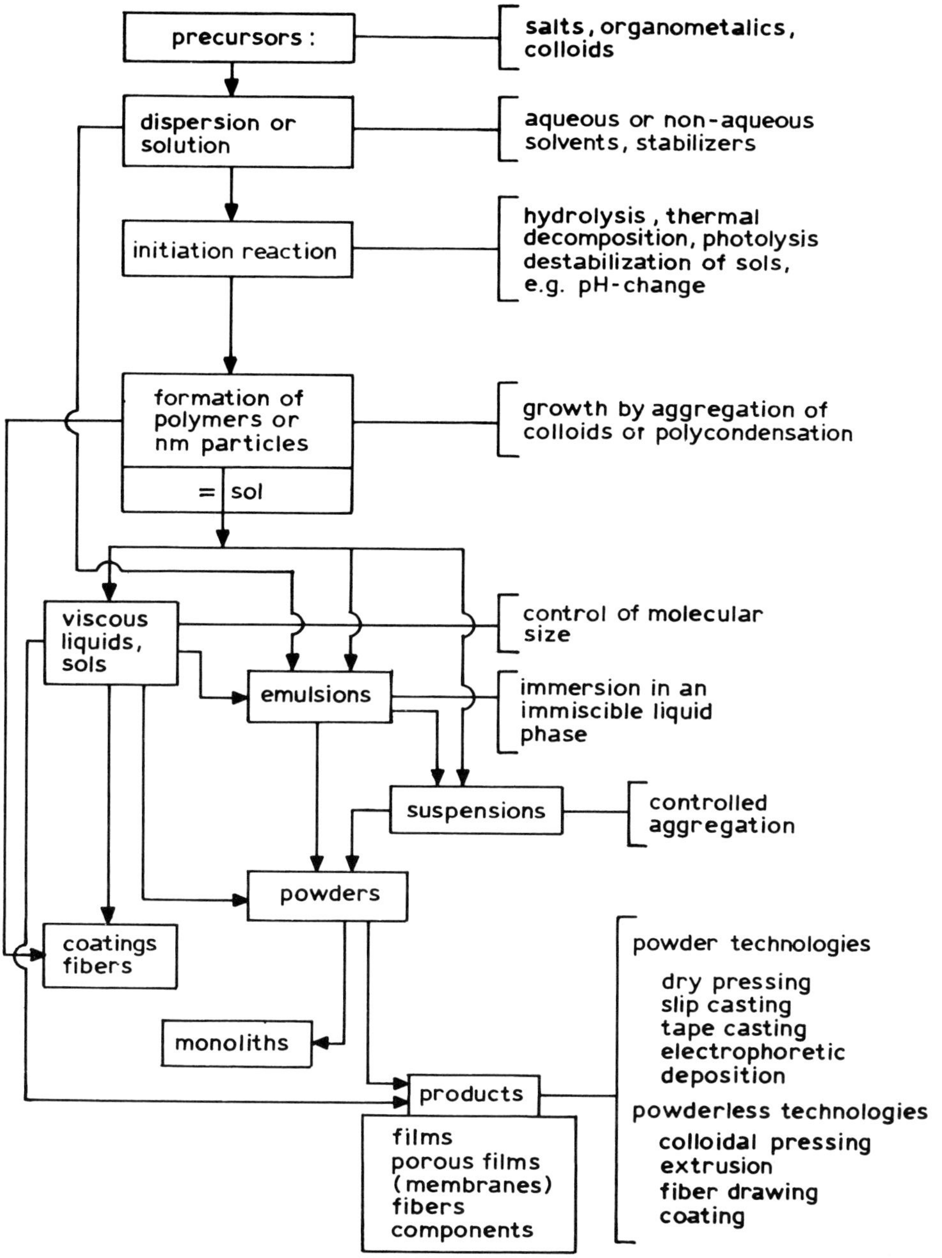

Figure 2.29 The Variability of Sol-Gel Technologies (H. Schmidt)

aluminas are predominantly obtained by the decomposition of high purity aluminium-based salts (sulphates, chlorides, nitrides). Calcined aluminas, consisting mainly of α Al_2O_3, are categorised into groups with different Na_2O content and total impurities. Examples are given in Table 2.7.

Table 2.7 Composition of Al_2O_3 powders with different impurity levels

Al_2O_3-content (%)	99.0[1]	99.6[2]	99.8[2]	> 99.95[1]	99.99[1]
Surface area (m^2/g)	6	0.3–0.7	0.7	7–10	6
Crystal size (μm)	1	6	4.5	< 0.5	< 0.5
Na_2O (wt-%)	0.3	0.25	< 0.1[3]	0.0075[3]	0.0005[3]
SiO_2 (wt-%)	0.02	0.03	0.03	0.02	0.0060
Fe_2O_3 (wt-%)	0.03	0.03	0.03	0.01	0.0020
α Al_2O_3 (%)	> 80	> 95	> 95	Mean particle size 0.5 μm Sintering after compaction at 345 MPa to density of 3.91 g cm^{-3} at 1540 °C, 1h	

1: Alcoa, AR, USA; 2: Martinswerk, Germany; 3: low Na_2O content

The ceramics market encompasses a wide variety of applications including spark plugs, machine parts, electronic parts, bio-ceramics, cutting tools, high temperature parts and, not least, special glasses and chinaware. Electronic applications require low Na_2O, high purity and easily sinterable powder. Sodium vapour lamp tubes are manufactured from 99·99% Al_2O_3, the translucency of which would not be possible without this high-grade material.

Zirconia (ZrO_2) is another highly refractory material with a melting point of about 2950 K. The tetragonal-monoclinic transformation at 1370 K occurs with a considerable volume change, which makes it impossible to sinter pure ZrO_2 to dense, crack-free bodies. Ceramic grade ZrO_2 powder has therefore to be manufactured with some additive content, (CaO, MgO, Y_2O_3, CeO) which ensures a fully or partly stabilized material.

The raw materials for manufacturing ZrO_2-powder are Baddeleyite (crude ZrO_2) and Zircon sands ($ZrSiO_4$). In one of the possible processes $ZrSiO_4$ is transformed by melting with NaOH or Na_2CO_3 to sodium zirconate by either:

$$ZrSiO_4 + 4\,NaOH \xrightarrow{870\ K} Na_2ZrO_3 + Na_2SiO_3 + 2\,H_2O \qquad (2.35)$$

or

$$ZrSiO_4 + 2\,Na_2CO_3 \xrightarrow{1270\ K} Na_2ZrO_3 + Na_2SiO_3 + 2\,CO_2 \qquad (2.36)$$

After separation from Na_2SiO_3 by water rinsing, the Na_2ZrO_3 is transformed with HCl into $ZrOCl_2$. This is soluble in water, can be mixed with the above mentioned additives in the form of chlorides, and then is hydrolized to $Zr(OH)_4$ and calcined to ZrO_2. Submicron powders with an extremely homogeneous distribution of the doping elements can be obtained using milling and spray drying steps. From these powders (see Fig. 2.31) the highest strength oxide ceramics so far obtained are manufactured (see section 11.10).

2.3.4 Non-Oxide Ceramic Powders

The most important non-oxide ceramic is *silicon carbide* (SiC), existing as hexagonal α SiC and cubic β SiC. α SiC is the stable form and appears in many polytypes with different stacking sequences. SiC was manufactured industrially as early as 1891 by the Acheson process

$$SiO_2 + 3\ C \rightarrow SiC + 2CO \tag{2.37}$$

yielding mainly α SiC. The process which passes through several stages with gaseous intermediate products is carried out on a large scale producing about 800 000 tonnes per year worldwide. The starting material contains pure sand, petroleum coke, a percentage of sawdust to decrease packing density and 1–3% sodium chloride for purification. This is heated with graphite electrodes, with temperature reaching 2000–2300°C during the 30 h cycle which is necessary for good crystallisation of the SiC. The raw material has a purity of about 99%, and undergoes extensive wet milling to produce a large variety of grain size fractions, from the coarsest, in the range of half a millimetre, up to submicron powders. Much of the material is used for abrasives, although another major use is as clay-bonded or porous recrystallised high temperature or wear resistant materials. The finest powders in the micron and sub-micron range (surface area 12–25 m^2g^{-1}) are used for manufacturing dense sintered parts (see section 11.10), in conjunction with sintering additives, such as carbon or boron to obtain near full densities.

β SiC is formed mainly by gas-phase processes, e.g. by reaction of silicon halogenides ($SiCl_4$) with hydrocarbons or by pyrophysis of trimethylsilane ($(CH_3)_3$ SiH), trimethylchlorsilane ($(CH_3)_3$ SiCl) or tetramethylsilane ($(CH_3)_4$ Si). The reaction temperatures are between 800 and 1500°C; either Ar or H_2 is used as a carrier gas. Particle size and morphology depend considerably on the reaction temperature and the composition of the gas phase: at 1000°C an nanosized powder (see section 2.4) of 10–30 nm size results. Powders obtained below 1100°C are amorphous.

Silicon nitride (Si_3N_4) exists in two polymorphous forms, α and β, which are both hexagonal. The c-axis in α Si_3N_4 is nearly double the spacing of β Si_3N_4. α Si_3N_4 transforms into β Si_3N_4 at high temperatures, but the reverse β–α transformation has not yet been observed.

Powders of Si_3N_4 are generally obtained by reaction of Si or Si compounds with N_2 or NH_3, processes according to the following exothermic reactions:

$$3\ Si + 2\ N_2 \rightarrow Si_3N_4 \text{ or} \tag{2.37a}$$

$$3\ Si + 4\ NH_3 \rightarrow Si_3N_4 + 6H_2 \tag{2.37b}$$

run between 1000 and 1400°C, leading to both α and β Si_3N_4. The reaction rate depends on temperature as well as on Si particle size. A cheaper process starts from SiO_2:

$$3\ SiO_2 + 6\ C + 2\ N_2 \rightarrow Si_3N_4 + 6\ CO \text{ or} \tag{2.38a}$$

$$3\ SiO_2 + 6\ C + 4\ NH_3 \rightarrow Si_3N_4 + 6\ CO + 6\ H_2 \tag{2.38b}$$

leading to α Si_3N_4 powders with some residual oxygen and carbon content. Both products have to be milled and eventually purified to obtain sinter grade powders. Very fine and pure powders are obtainable by the Diimid process starting with $SiCl_4$:

$$SiCl_4 + 6\ NH_3 \rightarrow Si\ (NH)_2 + 4\ NH_4Cl \tag{2.39a}$$

$$3\ Si\ (NH)_2 \rightarrow Si_3N_4 + 2\ NH_3 \tag{2.39b}$$

which is a process used commercially.

A direct reaction of $SiCl_4$ to Si_3N_4 is also possible:

$$3\ SiCl_4 + 4\ NH_3 \rightarrow Si_3N_4 + 12\ HCl \tag{2.40}$$

Sintering to high density can be achieved by the use of special additives (MgO, Al_2O_3, Y_2O_3, etc.).

Boron carbide (B_4C) is usually synthesized by the reduction of boric oxide with carbon at above 2500°C (ESK electro-furnace process)

$$2\ B_2O_3 + 7\ C \rightarrow B_4C + 6\ CO \tag{2.41}$$

followed by subsequent milling and purification. The synthesis is also possible from the elements or by a magnesio-thermic reduction of boric oxide in the presence of carbon

$$2\ B_2O_3 + 6\ Mg + C \rightarrow B_4C + 6\ MgO \tag{2.42}$$

Boron carbide shows a rhombohedral lattice structure with a wide homogeneity range between 8.6 and 18.8 at-% carbon which corresponds to a formula range between $B_{10\cdot6}C$ and $B_{4\cdot3}C$. Boron carbide with 18·5 at-% carbon melts congruently at 2450°C. The hardness at room temperature is about 3700 HV. B_4C can be densified by hot pressing or hot isostatic pressing (200 MPa; > 2000°C). It is also possible to sinter with carbon precursors to about 95% of the theoretical density. The powder is used for grinding and finishing purposes, but dense B_4C has yet to be developed as a structural ceramic.

2.4 ULTRAFINE POWDERS (NANOCRYSTALS)

Although the terms in the finest particle size range are not clearly defined it may be distinguished as follows: submicron particles are per definition below 1 μm, preferably down to 0.5 μm, ultrafine particles between 0.5 μm and 0.1 μm (100 nm) and nanosized (nanocrystalline) powders less than 100 nm, many of them range between 50 and 20 nm. Such powders have become of increasing interest in the last ten to fifteen years for both scientific and technical reasons. Their processing, and sintering behaviour (see section 7.2.4.), is unique due to their high surface to volume ratio. Several methods have been used in the past in order to obtain ultrafine powders, including:

- Gas phase reactions in a DC, RF or microwave plasma. For example, Si_3N_4 powder with a spherical particle size between 30 and 60 nm has been obtained from $SiCl_4$–NH_3 mixtures, and a SiC-powder (mean particle size 260 nm) from $SiCl_4$-H_2-CH_4 mixtures or from CH_3 $SiCl_3$ in an argon plasma. By selecting proper conditions it is possible to obtain powders consisting of crystallized particles. Metallic powders have also been obtained by plasma reaction.
- Gas phase reaction powered by a laser. SiC-powders with different amounts of free silicon (particle size <20 nm) have been obtained by CO_2 laser pyrolysis using SiH_4 or SiH_2Cl_2 and C_2H_4 as gaseous precursors. These powders are nearly spherical and have been found to be amorphous in the as-received state.
- Hydrogen reduction of metal chloride vapours has yielded powders of Ag, Cu, Ni, W, etc. with particle size ranges from 20–1000 nm. The ratio of the reaction temperature to the melting point of the metal seems to be an important factor in determining the mean particle size.
- Evaporation and condensation techniques have produced ultrafine metallic powders, such as those of Fe and FeNi. The metal is melted by induction heating and vaporizes in a low pressure inert gas atmosphere where the nanoparticles are formed by homogeneous nucleation. For explosive evaporation techniques see section 2.1.1.2, p. 30.
- Processing of solutions (see section 2.32) may also yield ultrafine or nanosized particles under special conditions.

Such powders are often strongly agglomerated and form linear arrays when they are ferromagnetic and are chained together according to their magnetic dipoles.

The powders react with oxygen and may be pyrophoric. However, under controlled oxygen exposure, they can be passivated. Amorphous powders have relatively low recrystallisation temperatures, e.g. SiC at 1200°K. Figure 2.30 shows a nanosized SiC powder, Fig. 2.31 a commercial, ultrafine ZrO_2.

Ultrafine powders are used to some extent for high density sintered products, but more often in the form of bonded powders for products such as catalysts (Ni, Pd) conductive pastes (Ag, Ni, Cu) magnetic recording media (Fe-Co-Ni-alloy powders) and filters with micro-porosity (Ni). The application of nanosized WC and WC–Co composite powders for fine grained hard metals (see section 11.6.4.) is now established to some extent. Their use for dispersion hardening or ultrafine composites is still to be fully exploited.

2.5 WHISKERS AND SHORT FIBRES

Whiskers are short metallic or non-metallic, usually monocrystalline fibres, which are deposited generally by a vapour-phase process. Whiskers of metallic elements

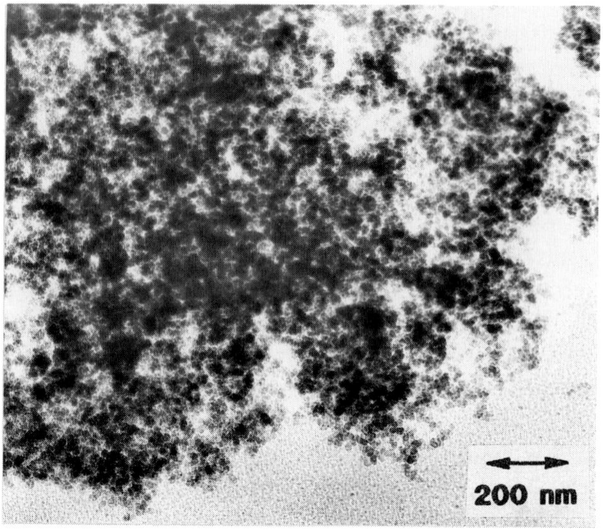

Figure 2.30 Nanosized SiC-powder (Stöver)

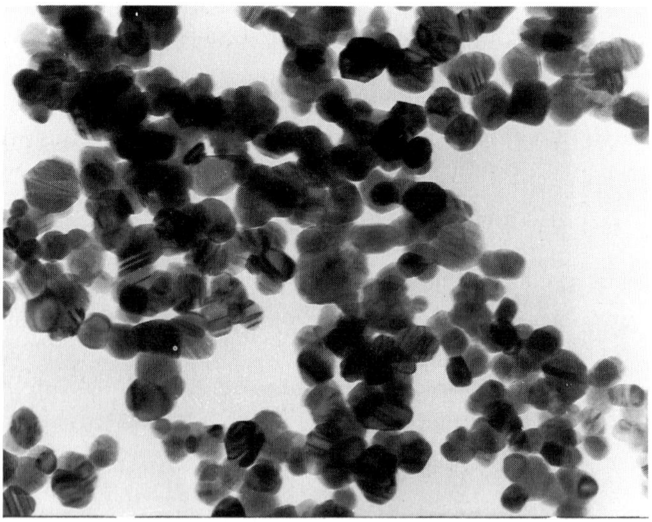

Figure 2.31 Ultrafine ZrO_2 powder with particle size less than 0.5 μm (Tosoh, Tokyo)

such as Ag and Cu and also of several compounds have been known for a very long time as naturally grown minerals. Scientific and technical studies of whiskers began in the 1940s. The condition for whisker formation (instead of 'normal' crystals or crystallites) is a strongly anisotropic unidirectional growth, which often appears under undisturbed growth condition within a certain concentration and temperature range. Whiskers can be easily obtained from most metals and several ceramic materials (such as Al_2O_3 and SiC). Some are manufactured industrially in greater than experimental quantities although there is considerable concern about the safety of handling whiskers which has held back keen exploitation.

For most whisker manufacturing processes volatile compounds are needed from which the deposition occurs on solid or liquid substrates. Metallic whiskers can be obtained by the reduction of a metal halogenide vapour with hydrogen in a carrier gas at normal reduction conditions, e.g.:

$$2\,FeCl_3 + 3H_2 \rightarrow 2Fe + 3HCl \tag{2.43}$$

At the present time non-metallic whiskers, such as Al_2O_3, SiC and Si_3N_4 are more important in the development of high performance materials than metallic ones. Cubic β-SiC whiskers are of particular interest. They are commercially produced by the so-called vapour-liquid-solid (VLS) process: a hydrocarbon compound (e.g. CH_4) reacts at 1400°C with a volatile Si-compound (e.g. $SiCl_4$, SiO) on a liquid iron alloy, which acts catalytically. Another inexpensive, process is the pyrolytic treatment of rice husks, which contain SiO_2. This reacts with the organic material to form SiC-whiskers.

Whiskers can be obtained in a wide range of diameters (from 0.1 to several μm) and lengths (10 to several hundred μm; in some cases up to several millimetres), depending on the material and the processing conditions. The length/diameter ratio ('aspect' ratio) ranges from 20 to several hundred and is an important figure. The cross section of whiskers can be hexagonal, rectangular or even irregular, depending upon crystal structure and growth conditions. Hollow whiskers, stacking faults, kinks and impurities are frequent and determine the quality of the material and its efficiency for reinforcement purposes. SiC whiskers are produced as several types: ARCO and Tateho materials less than 1 μm in diameter and up to 25 μm in length, VLS whiskers with 4–5 μm diameter and length in the mm range. Whiskers with defects are shown in Fig. 2.32, SiC whiskers in Fig. 2.33.

The most outstanding property of whiskers is their extremely high strength, which is strongly dependent on diameter and length: the smaller the diameter, the greater the strength. The reason for this is that a smaller section will have fewer or no glide dislocations. This is important for the strength of metallic whiskers, while in ceramic whiskers minimizing of the flaw size, including surface flaws, is the decisive feature. Strength values of metallic and ceramic whiskers in the range of 10 GPa and elastic moduli of 400–700 GPa for ceramics can be obtained. These values approach to some extent the theoretical strength σ_{max} according to the Orowan equation:

$$\sigma_{max} = \frac{2\,\gamma E}{a} \tag{2.44}$$

(γ = surface energy, E = elastic modulus, a = atomic distance.)

A major advantage of whiskers for reinforcement, including short chopped fibres in the length range of mm, is that they can be incorporated into normal powder metallurgy processes, i.e. mixing, compacting and sintering. However,

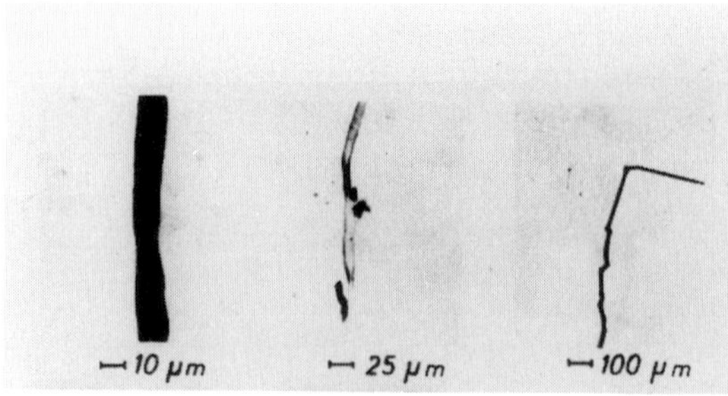

Figure 2.32 Growth anomalies of Fe whisker

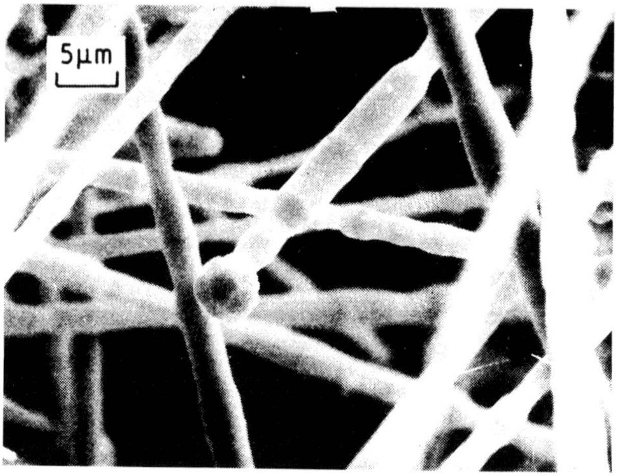

Figure 2.33 SiC whisker (Milenski *et al.*)

the danger of cluster formation has to be carefully considered and liquid processing is often advisable. Longer or continuous fibres [which are outside the scope of this book] are even more important reinforcement materials than whiskers, but have to be processed in completely different ways, similar to textile techniques. Whiskers in the diameter range below 3 μm, however, involve a considerable health hazard (see section 5.1.) and rigorous protection has to be provided especially during handling and processing. One of the important applications of whisker reinforcement is in Al_2O_3 cutting tools (see section 11.10.3).

A completely different type of fibrous material in the cm length range with 1–2 mm diameter, is obtained by a melt extraction process with rotating profiled rollers. The metallic melt wets the V-shaped edges of the rotating roller and is pulled out in irregular fibrous form, solidifying during extraction. Diameter, length and shape of the fibres are adjustable. Several metals and alloys have been produced in that way including heat resistant steel fibres which are utilized for reinforcing refractories. Also such techniques can be emplo produce fine directional particulate or platelets which can be extende conventional powders to give anisotropic properties.

FURTHER READING

W.M. Small (ed.), *Physical Chemistry of Powder Metals*, Production and Processing, MPIF, TMS, 1989.

K. Dialer, U. Onken and K. Leschonski (eds.), *Grundzüge der Verfahrenstechnik und Reaktionstechnik*, Chapter 5: K. Schönert: Zerkleinern, München, Wien, Carl-Hanser-Verlag, 1986.

M.J. Rhodes (ed.), *Principles of Powder Technology*, Chichester, Wiley, 1990.

S. Bradbury (ed.), *Powder Metallurgy Equipment Manual*, MPIF, Princeton, NY, 1986.

J.M. Capus and R.M. German (eds.), *Advances in Powder Metallurgy and Particulate Materials*, Princeton NJ, MPIF, 1992. Vol 7: Novel Powder Processing; Vol. 9: Particulate Materials and Processing.

E. Arzt and L. Schultz (eds.), *New Materials by Mechanical Alloying Techniques*, Oberursel, DGM Informationsgesellschaft, 1989.

P.H. Shingu (ed.), 'Mechanical Alloying', *Materials Science Forum*, Vol 88–90, Zürich, Trans. Tech. Publ., 1992.

D.R. Maurice and T.H. Courtney, 'The Physics of Mechanical Alloying', *Metallurg. Trans. A.*, **21**, 289–303, 1990.

W.A. Kaysser and K. Rzesnitzek, Principles of Atomisation in *Science of Sintering: New Directions for Materials Processing and Microstructural Control*, (ed. D.P. Uskokovic *et al.*), 157–175, New York, Plenum Press, 1989.

S. Jönsson, Present Status and Future Trends of PM Technology, in *New Trends in Materials Technology*, (ed. Y.U.A. Ossipyan and A. Hauff), pp 165–254, Leybold AG, Mashinostroenie Publishers, 1989.

A. Lawley, Modern Powder Metallurgy Science and Technology, *J. Met.*, 15–25, August 1986.

A. Lawley, *Atomisation the Production of Metal Powders*, Monograph in PM Series No. 1 MPIF, Princeton, NJ, 1992.

E. Pink and L. Bartha, *The Metallurgy of Doped, Non-Sag, Tungsten*, Chapter VIII, E. Lassner, W.D. Schubert, R. Haubner, B. Lux: Mechanism of Tungsten Reduction, 119–140, Elsevier Applied Science, 1989.

W. Köck and P. Paschen, Tantalum Processing, Properties and Applications, *JOM-Journal of the Minerals, Metals and Materials Society*, 33–39, Oct. 1989.

D.I. Bloemacher, Carbonyl Iron Powders: Its Production and New Developments, *Metal Powder Report*, **45**, 117–119, Feb. 1990.

INCO Ltd, Production and Properties of Carbonyl Nickel Powder, *Metal Powder Report*, **39**, 262–264, May 1984.

D.W. Johnson Jr and P.K. Gallagher, Reactive Powders from Solutions in *Ceramic Processing before Firing*, (eds. G.Y. Onoda, L.S. Hench,) 125–139, New York, Wiley, 1988.

E. Peissker, Production and Properties of Electrolytic Copper Powder, *Int. J. Powder Met. and Powder Techn.*, **20**, Nr.2, 87–101, 1984.

R. Kieffer and P. Schwarzkopf, Hartstoffe, Springer-Verlag, 1963.

K.A. Schwetz and A. Lipp, Boron Carbide, Boron Nitride and Metal Borides, in *Ullmann's Encyclopedia of Industrial Chemistry*, 197–307, Weinheim, VdCH-Verlagsgesellschaft mbH, 1985.

T.J. Carbone, Production Processes, Properties and Applications for Calcined and High-Purity Aluminas, 99–108, *Alumina Chemicals, Science and Technology Handbook,* (ed. Le Roy D. Hart), 1990.

H. Schmidt, Sol-Gel Derived Ceramics, in *Advanced Materials Processing, Vol 1,* (eds. H.E. Exner and V. Schumacher), 631–640, DGM Informationsgesellschaft, Oberursel, 1990.

S. Somiya *et al.*, Hydrothermal ZrO_2 Powder and its Sintering Behaviour, in *Science of Sintering* (eds. D.P. Uskokovic, H. Palmour III, R.M. Spriggs) Plenum Press, 507–518, 1989.

L.I. Trusov, V.I. Novikow and V.N. Lapavok, Powder Metallurgy and Nanocristals, *Science of Sintering* **23**, No. 2, 63–78, 1991.

3: Powder Properties and Characteristics

The properties of the starting powders to a great extent determine the properties of the final powder metallurgy parts. It is therefore essential to have a proper understanding of the role of powder properties in the powder metallurgy process and to apply suitable methods for a quantitative characterisation, at least of the most important powder properties.

3.1 TERMINOLOGY

Powders belonging to the so-called dispersed systems, which are arrangements of matter consisting of at least one disperse phase and a continuous surrounding medium. The dispersed phase itself consists of a number of individual elements, the particles. Dispersed systems are usually characterised by a qualitative description of the individual disperse elements and by the quantitative particle and structural characteristics of the system.

In metal or ceramic powders, a variety of disperse elements or particles can appear. The common powder metallurgy terminology for the description of these elements is given in Fig. 3.1. The term *powder particle* always stands for the primary particles, which do not develop by agglomeration of other particles, but are directly formed as individual particles during the production process. These powder particles can be amorphous or crystalline. Crystalline particles are either single crystals or consist of several crystals or *grains* (polycrystalline). Polycrystalline powders can be single- or multi-phased. The powder particles can form secondary particles or *agglomerates*, as shown in Fig. 3.1. In the powder metallurgy process, agglomeration often occurs unintentionally, but can also be intentionally produced by controlled stages. Agglomerates are defined as solid particles which are composed of other particles. This definition is independent of the nature of the attractive forces which cause the primary particles to form the agglomerate. Sometimes the term agglomerate is restricted to weak bonding between the primary particles. Hard agglomerates with strong bonding are then called *aggregates*.

The particle characteristics are measurable properties of disperse elements, which allow for a sequencing of the elements in accordance with these properties. The most important particle characteristic is its *fineness*. Besides the actual

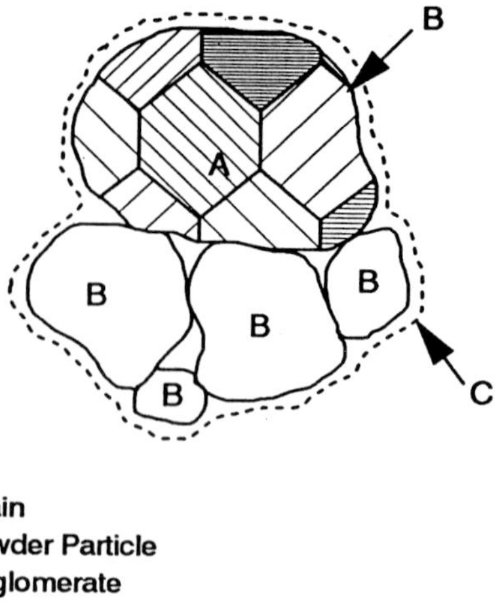

Figure 3.1 Powder elements in common powder metallurgy terminology

geometrical dimensions of the particle or of its projection, every physical property which is related to the dimensions by a single-valued function can be used, e.g. the particle mass, settling velocity, etc. The *particle size* is one such characteristic. It always has the dimension of length. The particle size of metal and ceramic powders ranges from less than one μm to several hundred μm. The range < 40 μm is sometimes called the *sub-sieve range,* as normal dry sieving is difficult to apply to such small particles. The range > 40 μm is therefore the *sieve range.* If dimension-related physical properties are measured, the particles are often regarded as spheres and the particle size is defined as the *equivalent diameter* of a sphere with the same physical properties as measured. Examples are the *volume diameter,* which represents the diameter of a sphere with the same volume as the particle under consideration, the *surface diameter* (sphere with same surface), the *Stokes' diameter* (sphere with same density and settling rate in a Stokes fluid field), and the *light scattering diameter* (sphere with same light scattering intensity). However, such a diameter ignores the shape.

The particle shape is therefore important. According to national and international standards, terms like *nodular, dendritic, acicular, fibrous, flaky, spheroidal, angular, irregular,* and *granular* are used for a simple qualitative characterisation of powder particle shape. Figure 3.2 shows some typical examples. For quantitative characterisation, shape factors are usually applied, representing relationships of characteristic linear, two- or three-dimensional parameters (i.e. the aspect ratio).

The structural characteristics describe the three-dimensional arrangements of the disperse elements in the system. They are not usually of importance for bulk powders but are of relevance for powder *suspensions,* which are dispersions of powders in a liquid.

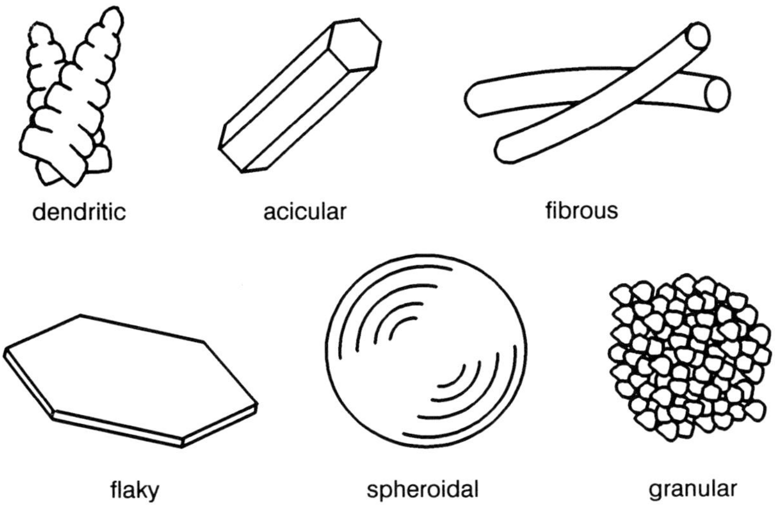

Figure 3.2 Qualitative characterisation of powder particle shape

3.2 PARTICLE SIZE DISTRIBUTION

3.2.1 Principal Aspects

For a number of equally sized spheres or cubes the particle size characteristics are simply given by one measurement. The individual particles of practical powders usually differ in shape and dimensions. The particle shape can deviate significantly from the ideal geometries mentioned above. The size characteristics of powders have therefore to be treated under two aspects: first, the real meaning of the determined size parameter, secondly the statistical distribution of this parameter, which has to represent the total number of particles in the powder.

The *size parameter* is identical with a real linear dimension of the particles only in some exceptional cases. Most methods for particle size measurement yield a size-dependent property of the particle. This property is then related to a linear dimension. The sieve diameter, for example, is the length of the minimum square aperture through which a particle will pass. The particle could be a sphere with the diameter of this aperture, or an elongated platelet with only one axis corresponding to the sieve diameter, a second axis much larger, and the third axis possibly smaller than the sieve diameter. Table 3.1 gives an overview of some important definitions of particle size. Most widely used are the equivalent spherical diameters. The equivalent spherical diameter is the diameter of a sphere with the same characteristics as a corresponding particle with regard to a given property.

Table 3.1 Particle size parameters

Symbol	Name	Definition
d_v	Volume diameter	Diameter of sphere with same volume
d_s	Surface diameter	Diameter of sphere with same surface
d_{sv}	Surface volume diameter	Diameter of sphere with same surface to volume ratio
d_c	Perimeter diameter	Diameter of circle with perimeter or particle projection area
d_f	Free-falling diameter	Diameter of sphere with the same density and free-falling speed in a given fluid
d_{St}	Stokes' diameter	Free-falling diameter at Reynold's number < 0.2
d_A	Sieve diameter	Width of minimum square aperture through which a particle will pass

In some cases the properties, which are estimated by the different methods of particle size analysis are directly related to a single geometrical parameter (length, surface, volume). Here, the length, surface or volume diameter are suitable size parameters. For a unit cube, d_s becomes 1.38 and d_v becomes 1.24 times the edge length. In other cases, the measured property depends on a combination of these geometrical parameters, which can also be exactly formulated. This is the case, e.g, for the Stokes' diameter, which represents the diameter of a sphere of the same density and the same free-falling velocity in a fluid at Reynolds numbers below 0.2 (see section 3.2.4.3). In many cases, an exact quantitative correlation is not possible, as already mentioned for the sieve diameter. The same holds for the free-falling diameter, if the Reynolds number exceeds 0.2, as the particle movement no longer occurs with random orientation under such conditions. 'Particle size' characteristics estimated by different analytical methods can therefore differ substantially, at least for such parameters.

In order to calculate the statistical distribution of the size parameter, it is necessary to characterise a statistically representative sample of the powder. The size characteristics can then be presented as *distribution functions.* These distribution functions represent the quantity proportion of particle belonging to a given size parameter z_i. Quantity proportions are formulated either as cumulative distribution $Q(z_i)$, or frequency distribution $q(z_i)$ (Fig. 3.3). $Q(z_i)$ gives the normalized quantity proportion of particles smaller than a size parameter z_i.

$$\frac{Q(z_i) = \text{quantity proportion}\,(z_{min} \ldots z_i)}{\text{total quantity}\,(z_{min} \ldots z_{max})} \tag{3.1}$$

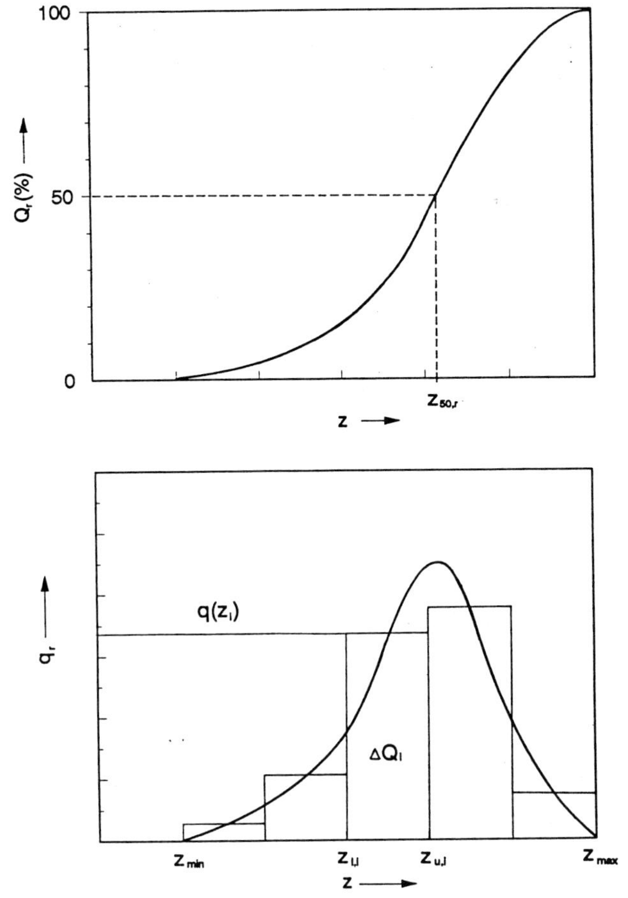

Figure 3.3 Particle size distribution functions. Cumulative distribution $Q(z)$ (upper diagram); frequency distribution $q(z)$ in differential and discrete (bar graph) form (lower diagram)

The frequency distribution $q(z_i)$ is defined as the fraction within a size class, divided by the class width.

$$q(z_{u,i} : z_{l,i}) = \frac{Q(z_{u,i}) - Q(z_{l,i})}{z_{u,i} - z_{l,i}} \tag{3.2}$$

If $Q(z)$ is differentiated, then

$$q(z) = \frac{\mathrm{d}Q(z)}{\mathrm{d}z} \tag{3.3}$$

Generally in statistics, only number distributions Q_0 or q_0 are used. In particle size analysis, quantity types are used besides numbers, depending on the measured data. Number distributions are measured by counting methods like those of microscopy. Three other types of distribution exist: length (Q_1,q_1) surface

(Q_2,q_2), and volume distributions (Q_3,q_3). In sieve analysis, for example, the quantity measured is the mass of particles within a given class, which is proportional to their volume. The results of sieving are therefore presented as mass distributions.

A mathematical transformation of one quantity into another is possible. For this purpose, the so called 'moments' $M_{k,r}$ of the distributions are required. The index k stands for the dimension of the size characteristic z. The index r corresponds to the chosen quantity type Q_r or q_r. $M_{k,r}$ is given by

$$M_{k,r} = \int_{z_{min}}^{z_{max}} z^k \times q_r(z)\,dz \qquad (3.4)$$

Any frequency distribution $q_r(z)$ can be derived from a distribution $q_t(z)$ by

$$q_r(z) = \frac{z^{r-t} \times q_t(z)}{M_{(r-t),t}} \qquad (3.5)$$

If, for example, the volume distribution q_3 has to be calculated from the number distribution q_0, equation (3.5) yields

$$q_3(z) = \frac{z^3 \times q_0(z)}{M_{3,0}} \qquad (3.6)$$

The moment of a distribution is also a suitable parameter for other mathematical operations. The average particle size $\bar{z}$, for example, can be simply calculated by

$$\bar{z} = \sqrt[k]{M_{k,r}} \qquad (3.7)$$

$\bar{z}$ represents the *arithmetic* average particle size. It is not identical with the so-called *median particle size* $z_{50,r}$, which is defined as the particle size, where Q_r reaches 50% (see Fig. 3.3)

Another operation often required is the calculation of the specific surface from a given particle size distribution. The specific surface is the surface/volume (S_v) or surface/mass ratio (S_m) of a powder

$$S_v = \varphi\left(\frac{M_{2,0}}{M_{3,0}}\right) = \frac{\varphi}{M_{1,2}} = \frac{\varphi}{z_{1,2}} \qquad (3.8)$$

φ is a shape factor, which depends on the relation of the real surface of an irregular particle to the surface of the equal volume sphere. S_m is derived by dividing S_v by the material's density ρ.

The graphic presentation of the size distributions is seen in Fig. 3.3. The frequency distribution is plotted either as a histogram or as the continuous derivation of the cumulative distribution. Various types of coordinate systems are utilised in plotting the distributions. The basic systems are linear (both axes) and linear/logarithmic. These systems are used for both the cumulative and the frequency distributions. Cumulative distributions are also plotted in special grids, in which the distribution functions become straight lines.

3.2.2 Common Types of Distribution Function

One of the most important functions is the so-called *RRSB distribution* (Rosin–Rammler–Sperling–Bennet), originally derived for pulverised coal and shown to apply to many other comminuted powder products.

If this distribution is applied in sieve analysis (quantity type $r = 3$), it can be written as

$$1 - Q_3(z) = R(z) = \exp\left(\frac{-z}{z'}\right)^n \tag{3.9}$$

Equation (3.9) can be transformed into equation (3.10), which represents a linear function of $\lg\lg(1/R)$ vs. $\lg z$

$$\begin{aligned}\lg\lg\frac{1}{R} &= n\lg z - n\lg z' + \lg\lg e \\ &= n\lg z + \text{const}\end{aligned} \tag{3.10}$$

The distribution is fully characterised by the parameters n and z', where n represents the slope of the line and z' is the particle size, for which the percentage retained (oversize) R takes the value of 36.8% or $1/e$ (Fig. 3.4).

The second important distribution function is the *logarithmic normal distribution.* While arithmetic normal distributions are based on the equal likelihood of differences of equal amounts in excess or deficit from a mean value, in log normal distributions ratios of equal amounts are equally likely. In such a distribution, the logarithm of the normalized particle size $\ln(z/z_{50})$ is normal distributed with a standard deviation of s. In a plot with a log-scale abscissa and a probability scale ordinate, the distribution function becomes a straight line (Fig. 3.4). The characteristic parameters of the distribution are the median particle size z_{50} and the standard deviation s. The latter is given by

$$s = \ln\left(\frac{z_{50,r}}{z_{16,r}}\right) \tag{3.11}$$

z_{50} and z_{16} can be derived from the plot as shown in Fig. 3.4. The log normal distribution is preferred when the actual type of distribution is not known. This is very convenient for the transformation of different quantity type distributions into one another. The standard deviation remains constant for every quantity type. The median size parameter is given by

$$z_{50,b} = z_{50,a} \times \exp[(b - a)s^2] \tag{3.12}$$

with b being the quantity type of the original, and a of the derived distribution. A volume distribution Q_3, for example, is derived from a number distribution Q_0 as

$$\begin{aligned} z_{50,3} &= z_{50,0} \times \exp(3s^2) \\ s(Q_3) &= s(Q_0) \end{aligned} \tag{3.13}$$

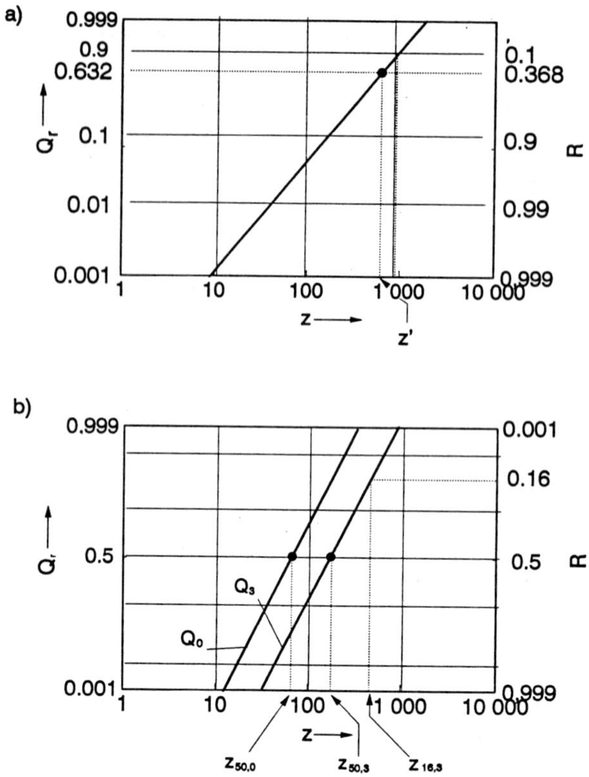

Figure 3.4 Plot of distribution functions in special coordinates. a) RRSB (Rosin–Rammler–Sperling–Bennet) distribution; b) log-normal distribution (see text for details)

The graphs of Q_3 and Q_0 are parallel lines (see Fig. 3.4). Other types of distribution functions are, e.g., *potential functions.* Logarithmic scaling of both axes results in a linear graph for this type of distribution.

3.2.3 Sampling of Powders

Powders are usually produced and processed in large quantities in the kilogram or tonne range. Particle size measurement is carried out with samples in the gram or even milligram range. The total error, f, in the measured cumulative size distribution is composed of the analytical error f_a and the sampling error f_s. If these errors are statistical in nature and independent from one another

$$f^2 = f_s^2 + f_a^2 \tag{3.14}$$

The sampling error should always be smaller than the analytical error, otherwise it will control the total error. If f_s, for example, is only half as large as f_a, it contributes only one fifth of f^2. In the usual range of analytical errors of about 2%, a sampling error of the order of only 1% is therefore tolerable.

It is by no means trivial to reach this error level. Even with great care, a statistical error is unavoidable. The real value $Q_0(z)$ of the total particle population in a powder is always a well defined fixed value. On the other hand, the real values $Q_0(z)$ of different samples of this powder follow a statistical distribution with a standard deviation σ. If the number of particles n in the sample is small compared to the total particle number, σ is given by

$$\sigma^2 = \frac{Q_0(1 - Q_0)}{n} \tag{3.15}$$

σ defines approximately the maximum error within a probability range of 68%. The maximum error within a probability range of 95% is about 4 times as high. Equation (3.15) can now be used to calculate the minimum numbers of particles for a given sampling error f_s, assuming ideal sampling. Table 3.2 gives an example of this minimum number for $Q_{0,50}$ at the probability range of 95%

For a sampling error of 1%, as mentioned before, the sample should consist of 10 000 particles. For practical counting methods, this is a fairly large number, which only can be handled by automatic systems. For other methods of particle size analysis the number of particles in the analysed sample is usually larger than this number, at least for fine powders. It is clear that the sampling error can be reduced to any minimum value by increasing the size of the sample, assuming ideal sampling.

If the particles of the parent population are randomly distributed, a single sample is already representative of the population. In the presence of segregations, representative sampling requires *splitting* or *homogenisation* of the parent population. Sample splitting is the preferred method for bulk powders, while homogenisation is preferred for suspensions. Sample splitting means the division of the bulk powder into a number of increments, some of which are put together again to make the sample. This can be done, for example, by coning and quartering. The material is poured into a conical heap, which is cut symmetrically into four identical increments. Two of these are put together to make a sub-sample, which is coned and quartered again and so on. Automatic devices such as chute splitters or spinning rifflers (Fig. 3.5) are generally used for this dividing procedure. The material is fed into a slowly rotating riffler. During each rotation, a number of z ($\approx$ 10) increments is created. After n rotations, the material has been divided into $z \times n$ increments, which form z sub-samples. Such devices can be used for subdivision down to a gram.

Table 3.2 Minimum number of particles required for a statistical sampling error f_s

f_s (%)	0.5	1	2	5	10
n_{min}	40 000	10 000	2 500	400	100

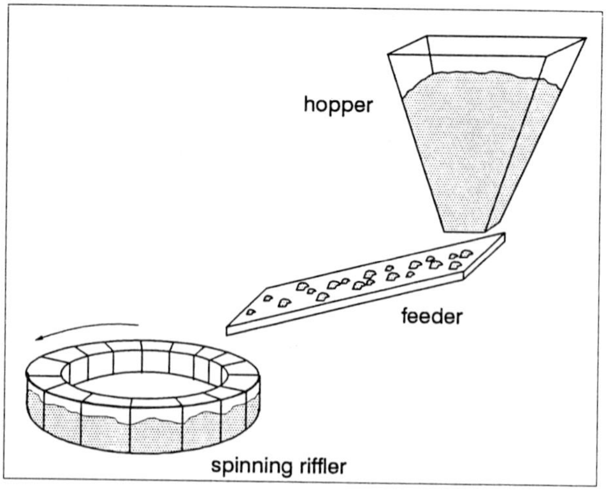

Figure 3.5 Sample splitting and homogenization (sampling) by a spinning riffler apparatus

For the majority of analytical techniques, smaller sample lots are required. A further subdivision can be undertaken by preparing a suspension of the subsample, and further sampling from the suspension. This can be done by withdrawing a part of the homogeneously stirred suspension with a syringe or pipette, or by using automatic suspension samplers.

3.2.4 Methods of Particle Size Analysis

In chemical engineering, there is a large variety of methods for particle size analysis, which can be applied to bulk powders, suspensions, slurries, aerosols, etc. Only some important methods with relevance for powder metallurgy or ceramics can be treated here.

These methods can be classified into: *counting methods, sedimentation methods, separation (classification) methods,* and *light scattering and diffraction methods.* The counting methods can be subdivided into direct and indirect methods, depending on whether the particles themselves are measured, or particle images or projections. Microscopy, for example, is an indirect counting method. With direct methods, the particles directly trigger the measurement. The Coulter principle (see section 3.2.4.5) is such a method. Sedimentation methods utilize the free-falling velocity of particles in a gravity or centrifugal field. Separation of the particles into different size classes and measurement of the fractions of every size class is realized in classical sieve analysis and in the fluid classification methods. The light scattering and diffraction methods analyse diffraction patterns of particles. Diffraction methods have become increasingly important in powder metallurgy and ceramics in recent years because of their versatility regarding measurable size range, and high speed of operation.

3.2.4.1 Sieve Analysis

Sieving is an obvious method of powder classification and one of the simplest and most widely used in particle size analysis. It covers the approximate size range 20 µm–125 mm using standard woven wire sieves. This range is extended down to 5 µm using micromesh sieves. The aperture sizes of test sieves are standardized (see section 3.6). They follow different geometrical progressions such as $\sqrt[4]{2}$ or the so-called Renard numbers $\sqrt[R]{10}$ ($R = 5,10,20$). Sieves are characterized by their aperture size in µm or mm, and are often also referred to by their mesh number, which is the number of wires per linear inch. A high mesh number thus implies a small aperture, for example 400 mesh corresponds to a 37 µm aperture with a wire thickness of 26.5 µm.

The sieves are made of either woven wire, punched or electroformed plates. The apertures of woven wire sieves are considered to be squares, although they deviate somewhat from this shape, due to the three-dimensional structure of the weave. Electroformed micromesh sieves are available with round and square apertures.

Dry sieving is applicable to relatively coarse non-cohesive powders. Fine and cohesive powders have to be analysed by wet sieving. The sieving operation is either performed by hand, or by various types of sieving machines. In machine sieving, the normal practice is to use a stack of sieves with an ascending order of aperture sizes, placing the powder on the top sieve (Fig. 3.6). The sieves are subjected to simple vibration or more complicated patterns of movement. Air-jet sieves or sonic sifters (Fig. 3.6) expand the size range for dry sieving down to about 10 µm. In air-jet sieves, the material on the sieve is fluidized by an air stream. Pressure from below forces the fine particles to pass through the sieve. In the sonic sifter, a vertical oscillating column of air sets the sample in a vertical motion. Most of the wet sieving methods use a stack of sieves filled with a liquid.

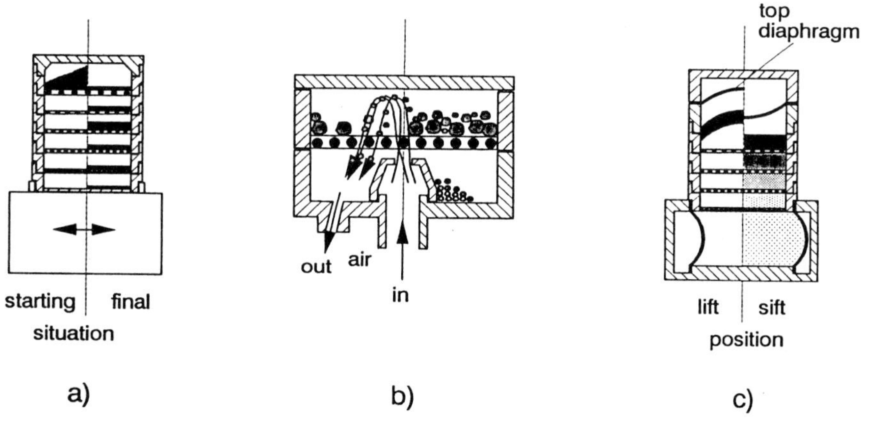

Figure 3.6 Sieving methods. a) vibrating sieving machine; b) air-jet sieving; c) sonic sifter

The sample is fed onto the top sieve. Sieving is accompanied by rinsing, vibration, ultrasonics, pressure differences and combinations of these measures. Wet sieving is essential for micromesh sieves and an additional step of drying the residual suspensions is necessary before the weight of the fractions can be measured.

The mechanisms of sieving are divided into two different categories (Fig. 3.7). The first relates to the passage of particles much finer than the sieve aperture, the second relates to the passage of the so called 'near mesh' particles. In category 1 the cumulative weight fraction, P, passing through a sieve in the time interval t is given by

$$P = at^b \tag{3.16}$$

where b is a constant nearly or equal to 1 and a is a function mainly of the two dimensionless sets of parameters

$$a = f(\frac{W}{\rho_s S A_0}, \frac{S}{d}) \tag{3.17}$$

W is the total load on the sieve, ρ_s the particle density, S the mesh opening, d the particle diameter and A_0 the sieve open area. a increases with S/d and decreases with increasing load/(density–aperture–open area) ratio. Effective sieving, i.e. short sieving times, requires low loading of the sieves. The tolerable load decreases rapidly with decreasing mesh aperture as both S and A_0 decrease. For woven wire screens, for example, the relative sieve open area is 41% for 1 mm mesh aperture, and only 25% for 20 µm mesh aperture. The open area decreases further for micromesh sieves down to 2.4% for a 5 µm aperture size. It is, therefore, essential to prevent 'blinding' of the sieves by particles being trapped in individual mesh apertures.

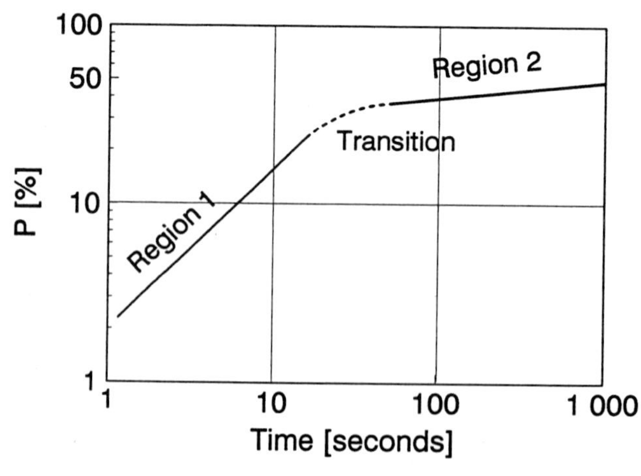

Figure 3.7 Time dependency of weight fraction passing through a sieve (T. Allen)

Practical sieve diameters therefore represent not the true linear dimensions of the particles at all, but only the ability of a majority of the particles, to pass through an arrangement of (non-ideal) apertures in a given time interval for a given motion of the sample. The main factors that influence the results are the shape and size of the apertures, sieve loading, intensity of agitation and time of sieving, the particle shape and the cohesiveness of the powder.

The quantity proportion measured in sieve analysis is the mass of the passing (P) or the residual particles (R) on a screen. The results are therefore usually presented as Q_3 or q_3 distributions, respectively.

3.2.4.2 Microscopy

Microscopy is one of the counting methods, in which each individual particle is observed and measured. In addition to *particle size*, microscopy can also give other information such as the *particle shape* and the *state of agglomeration*. The problem in the use of microscopy is the representativeness of the sample, as microscopical measurements are carried out on very small sample quantities and the sampling method can cause agglomeration. Microscopical methods can be treated under two aspects: methodology and image analysis.

Optical as well as electron microscopy are employed for particle size measurement. Optical microscopes can be used for particles down to about 0.8 µm. In this range of sizes the particle dimensions reach the wavelength of light which causes scattering and diffraction (see section 3.2.4.6). The reflection mode is applicable down to 5 µm; for particles less than 5 µm the transmission mode has to be used. A major problem is the small depth of focus, which is only about 0.5 µm at a magnification of 1000X.

Electron microscopy enables the measurement of particles down to the nm range. Transmission electron microscopy (TEM) with energies up to 200 keV and beam diameters of several µm is applicable for powders in the size range 0.001–5 µm. Sub-micrometre particles are often transparent to high voltage electrons, which enables additional information on the crystalline structure and structural defects of the particles. Clearly this approach is extremely time-consuming if statistically significant information is to be obtained. Generally it is employed to gather an idea about the size distribution and the variation in particle size..

Scanning electron microscopy (SEM) utilises medium energy electrons (5–50 keV) in a fine (< 0.01 µm) beam scanning the observation area. Secondary (SE) and backscattering electrons (BE) are counted by suitable detectors. Secondary electrons arise from the near surface regions of the particles and give high resolution images. Backscattered electrons arise from the somewhat larger volume of the particles near the electron beam, causing a slight reduction in resolution. They are dependent upon the atomic number of the element emitting them, resulting in a material contrast for regions of different atomic composition. This makes it possible to distinguish between different compositions in powder mixtures as long as there is a sufficiently large difference in atomic number between the elements. Also wavelength and energy disperse X-ray techniques can give more quantitative compositional information.

Sample preparation is a difficult problem in all microscopical methods. Each particle to be measured has to be isolated from the neighbouring particles. Most preparation techniques therefore use suspensions with low concentrations of the particles. These suspensions are applied to the microscope slide, electron microscope grid, or target stub and allowed to evaporate. There are various methods used for the fixing of the particles. TEM and SEM samples are coated with thin carbon or metal films to prevent charging effects.

The particles characterisation can be quantified directly using on-line procedures or from a number of micrographs taken from the samples. Image analysis can be carried out manually, or semi- or fully automatically, to give the number size distribution or shape factor distribution of the particles observed. In many cases, analyses have to be made from the projected images of the particles. This of course ignores internal porosity and concave parts of the surface. A more precise technique is therefore to examine sections from a random sectioning of the particles. This can be done for particles > 10 μm by the normal metallographic grinding and polishing methods. Estimates of particle size are then made from the projection of section images. Diameters commonly used are (see Fig. 3.8):

- *Feret's diameter*, the distance between two tangents on opposite sides of the particle. This is a one-dimensional projection of the perimeter of the particle outline.
- *Longest dimension*, the maximum value of Feret's diameter.
- *Maximum chord*, the maximum length of a line parallel to some fixed direction and limited by the contour of the particle.
- *Perimeter diameter*, the diameter of a circle with the same circumference as the particle perimeter.
- *Projected area diameter*, the diameter of a circle with the same area as the described projection.
- Diameters of *inscribed or escribed circles.*

A number of definitions exist for the particle shape factor Ψ, e.g.:

Ψ = Perimeter/(4π × projected area), or

Ψ = max. Feret's diameter/(Feret's diam. in a perpendicular direction)

Surface roughness is another particle characteristic. Particles of the same size and shape but different roughness can be distinguished by their fractal dimension, D (see Fig. 3.8). The fractal dimension accounts for the fact that the estimated parameter P_l tends to increase, as the step size l at which surface details are resolved increases. P_l is given by

$$P_l = \text{const} \times l^{1-D} \tag{3.18}$$

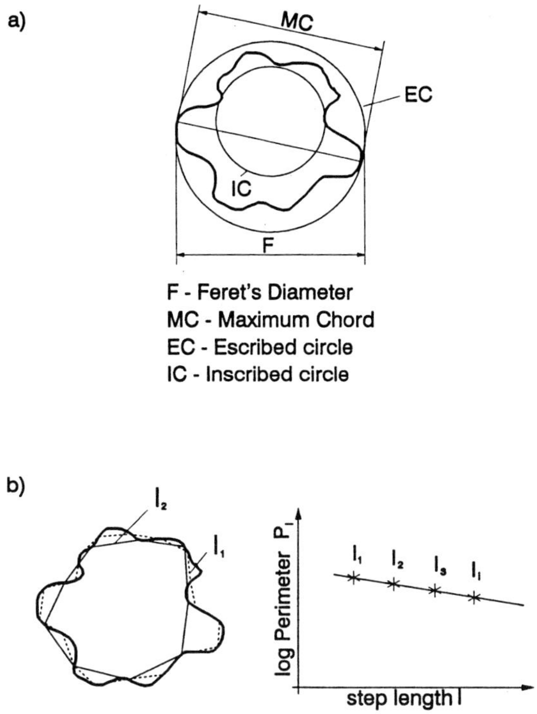

Figure 3.8 Particle characterisation by image analysis. a) particle size parameters; b) roughness characterisation by fractal dimensions

If the perimeter is measured by a polygonal line arrangement with varying step length l, and log P_l is plotted against l, the results form a straight line with the slope $1-D$. The steeper the line, the more rugged the outline of a particle. An ideally smooth surface has a fractal dimension equal to unity, i.e. it gives a horizontal line.

If the above parameters have to be measured manually, graticules are often used with linear scales or with grids inscribed with opaque and transparent circles. The former permit a direct comparison with linear dimensions, the latter with the projected area of the particles. Graticules were also used in some early semi-automatic systems. State-of-the-art image analysis is done on the basis of modern picture point logic. The picture is divided into a matrix of pixels which represent the image by different greyness levels in each pixel. In the first stage the picture has to be improved by processing functions to give sharp images, allowing touching particles to be separated, in order to select or reject certain particles and to improve the contrast between the particles and background debris. This can be performed automatically or manually by the operator. Modern software enables size and shape distributions based on selected parameters such as those mentioned previously to be determined. As the pixel matrix contains the complete image, even more complicated characteristics, such as orientation, texture etc., can be calculated.

3.2.4.3 Sedimentation Analysis

Gravitational Sedimentation Analysis

The fineness characteristic measured in sedimentation analysis is the constant free-falling or settling velocity v_f of a particle in an unbounded stationary fluid. Three main forces act on a particle settling in a gravity field with the acceleration, g, (Fig. 3.9): the gravitational force, F_G, the drag force F_D, and the buoyancy force F_B, which are in equilibrium once the particle has reached the velocity v_f. For a sphere of diameter D and density ρ_s falling in a fluid of density ρ_f, equilibration yields

$$F_D = \frac{\pi}{6}(\rho_S - \rho_f)\; g\; D^3 \tag{3.19}$$

For Reynolds numbers $Re < 0.2$, F_D can be expressed by equation (3.20), originally derived by Stokes for the behaviour of a sphere in a laminar Newtonian fluid of the viscosity η

$$F_D = 3\pi D\eta v_{St} \tag{3.20}$$

From equations (3.19) and (3.20) the Stokes' velocity v_{St} follows as

$$v_{St} = \frac{\rho_s - \rho_f g D^2}{18\eta} \tag{3.21}$$

The critical diameter $D_{St,c}$ beyond which equation (3.21) should not be used is given as $Re = 0.2$. It may be expressed by:

$$D^3_{St,c} = \frac{3.6\eta^2}{(\rho_s - \rho_f)\rho_f g} \tag{3.22}$$

The critical diameter is about 38 μm for iron and 27 μm for tungsten particles, respectively, settling in water. This size limitation can be extended by using more viscous liquids, e.g. glycerine/water mixtures.

A lower limit arises from Brownian movement of the particles as a consequence of collisions with the fluid molecules. In water, the displacement rate by Brownian movement exceeds that for gravitational sedimentation below a particle diameter of about 1 μm. In sedimentation practice, the assumptions of equation (3.21) are not exactly fulfilled. The real free-falling velocity is influenced by the particle shape, the finite extent of the fluid, and concentration and convection effects. The finite extent of the fluid, i.e. the walls of the suspension vessel, is of minor importance if the particle diameter is small compared with the vessel dimensions. A diameter ratio of 2×10^{-3} is recommended

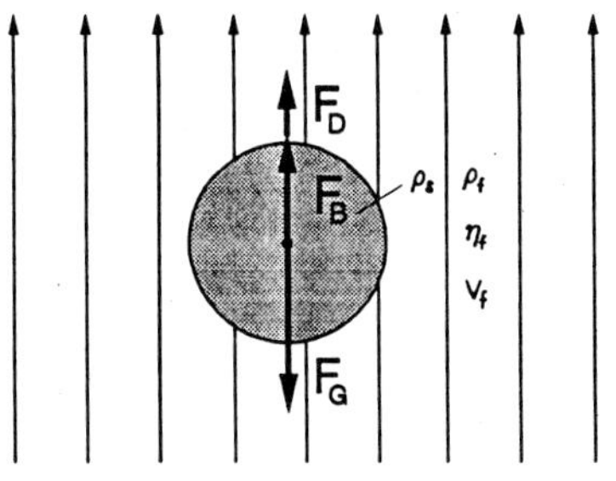

Figure 3.9 Principles of gravitational sedimentation

to allow for an error of < 1%. The volume concentration C_v of the particles in the liquid influences the sedimentation velocity, as the particles interfere with each other. At very high concentrations $C_v > 10^{-2}$, settling is hindered by increasing velocity differences between neighbouring particles at decreasing distances on the one hand, and an increasing upward flow of liquid to compensate for particle movement on the other hand. Under certain circumstances, settling is enhanced at low volume concentrations, e.g. by cluster formation, which increases the apparent free-falling diameter. Fortunately, this effect is limited to very narrow size distributions. In practice, volume concentrations of $< 10^{-3}$ are acceptable — sometimes even concentrations up to 10^{-2}. Convection currents can become a serious problem in gravitational sedimentation technique. The whole suspension must therefore be held at a stable temperature during the experiment. Density differences can also cause convection currents. This becomes critical if a zone of lower density exists below a higher density zone. Liquid then flows upwards at a velocity greatly exceeding the settling velocities of fine particles.

Within the Stokes' region, symmetrical particles such as spheres or cubes fall vertically and with the same velocity in any orientation. For less symmetrical particles an orienting force exists. For non-spherical particles, modification of equations (3.19) and (3.21) results in

$$3\pi d_d \eta v_{St} = \frac{\pi}{6}(\rho_s - \rho_f) g d_v^3 \tag{3.23}$$

$$d_{St}^2 = \frac{(\rho_s - \rho_f) g v_{St}}{18\eta} \tag{3.24}$$

$$d_{St}^2 = \frac{d_v^3}{d_d} \tag{3.25}$$

Within the Stokes region, the drag diameter d_d is usually assumed to be equal to the surface diameter d_s. This is known to hold for very low Reynolds numbers, while d_d exceeds d_s for increasing *Re*.

The methods of evaluation are commonly classified as suspension methods and line start or two-layer methods (Fig. 3.10). At the commencement of measurement by the suspension method, the particles are homogeneouosly distributed over the total suspension volume, while with the two-layer methods, they are injected at the top of the fluid. The measurement of quantity proportions is carried out either for a thin layer (incremental methods), or including all particles below or above a fixed measurement plane (cumulative methods). Various techniques result from these principles. The concentration of particles is estimated by means of gravimetric methods (pipette, sedimentation balance), absorption of electromagnetic radiation (photo and X-ray sedimentation), density (hydrometer) or from the hydrostatic pressure in the suspension.

As already shown in Fig. 3.10, the incremental two-layer methods yield directly the frequency distribution $q_r(v_{St})$ of the size characteristic 'sedimentation velocity'. The particle quantity observed during a small time interval at the observation level h after a sedimentation time t contains all particles (and only those particles) with a sedimentation velocity of $v_{St} = h/t$. $q_r(v_{St})$ can be transformed easily into the size distribution $q_r(d_{St})$ by equation (3.24).

In the incremental suspension methods, all particles whose sedimentation velocity is > h/t are below the control volume at a time, t. All other particles are present there in the same concentration as in the initial homogeneous suspension. The concentration $C(t)$ divided by the initial concentration $C(t = 0)$ therefore yields directly the cumulative oversize distribution $Q_r(v_{St})$.

Measuring the quantity proportions by cumulative methods simply means an experimental integration of the corresponding incremental results with respect to time. As can be seen from equation (3.3), the integral over the frequency distribution gives the cumulative distribution. The cumulative two-layer methods therefore yield the cumulative undersize distribution $1 - Q_r(v_{St})$.

The cumulative suspension methods are best understood by considering the weight fraction W_t found at the measuring level h after a time, t. W_t includes all

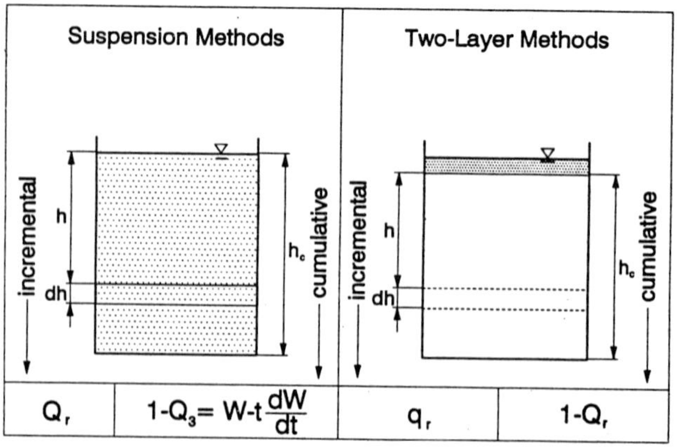

Figure 3.10 Evaluation methods in gravitational sedimentation size analysis (K. Leschonski)

particles whose Stokes's velocity is equal to or greater than h/t, and some particles with a smaller velocity which lie below the surface of the suspension.

$$W_t = \int_{v_{h,t}}^{v_{max}} q(v)\ dv + \int_{v_{min}}^{v_{h,t}} \frac{vt}{h} q(v)\ dv \tag{3.26}$$

Differentiation with respect to t, and multiplication by t yields:

$$t\ \frac{dW}{dt} = \int_{v_{min}}^{v_{h,t}} \frac{vt}{h}\ q(v)\ dv \tag{3.27}$$

(3.26) and (3.27) can be rewritten

$$W_t = 1 - Q(v_t) + t\ \frac{dW}{dt} \tag{3.28}$$

$Q(v_t)$ can be calculated from equation (3.28), since W, t, and dW/dt are known. The transformation into $Q(d_{St})$ again is performed by means of equation (3.24).

The simplest incremental method is with a pipette (Fig. 3.11). The changes in powder concentration are measured by drawing off definite volumes with a pipette introduced to a given depth h into the sedimentation vessel. These samples can be analysed by drying and weighing. However, this is a very time consuming process. Particle concentrations can be measured more easily from the absorption of light or X-ray radiation by the suspension, the method employed in incremental photosedimentation and X-ray sedimentation devices. The X-ray density is directly proportional to the mass of powder in the beam and is simply related to the absorbed transmitted X-ray intensity, making this a preferred method of gravitational sedimentation.

The most widely used cumulative device is the sedimentation balance. The principle is shown in Fig. 3.11. A balance pan at the bottom of the sedimentation vessel catches the falling particles. The increase of the sediment mass is automatically recorded with sedimentation time. From the resulting plot, the size distribution can be calculated according to equation (3.28). The less common manometric methods are based on a similar principle. The pressure near the bottom of the vessel is monitored. This is proportional to the mean density of the suspension between the surface and the measuring level, from which the mean concentration can be derived.

Centrifugal Sedimentation Analysis

In practice, most gravitational sedimentation techniques are of limited value in the size range < 5 µm, due to long settling times and increasing effects of convection. These difficulties can be reduced by centrifuging the suspensions, i.e. replacing the gravity field (Fig. 3.9) by a centrifugal force field. The same measurement principles can be used as with gravitational methods: suspension or two-layer, incremental or cumulative.

Two effective forces act on the particle: the centrifugal and the Coriolis force. The latter is of little importance in comparison with the centrifugal force. The

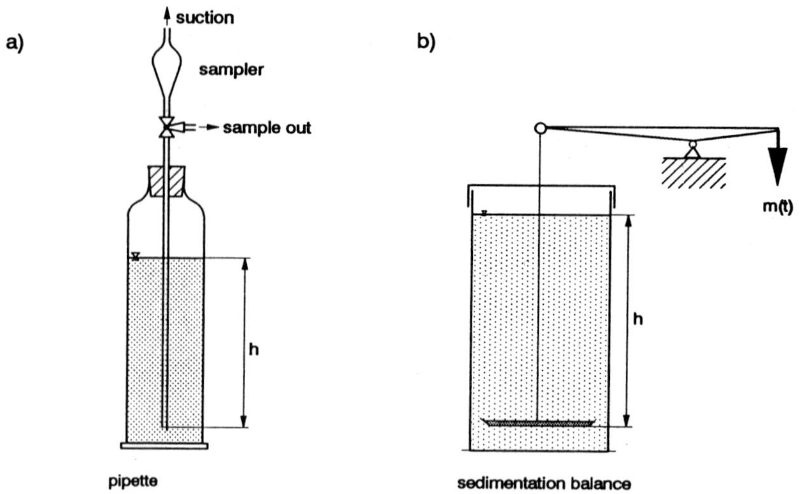

Figure 3.11 Gravitational sedimentation analysis by a) Andreasen pipette method (incremental sampling after time intervals $t_1.....t_n$) and b) sedimentation balance (cumulative sampling)

Stokes's velocity for a sphere can be derived in the same manner as from equations (3.19)–(3.21), replacing the gravity constant, g, by the centrifugal acceleration $\omega^2 x$, where x is the distance of the particle from the axis of rotation and ω the speed of rotation.

$$v_{St} = \frac{\mathrm{d}x}{\mathrm{d}t} = \frac{(\rho_s - \rho_f)\,\omega^2\,x\,D^2}{18\eta} \tag{3.29}$$

Thus, the settling velocity is not constant, but depends upon the radial position, x, of a particle. This makes the derivation of size distributions from the measured data much more complicated than in gravitational sedimentation. A way out of this problem is the two-layer or line start methods. The time, t, for a sphere of size D_m to settle from the surface of the liquid a distance S from the axis to the measurement zone, r, follows by integration of equation (3.29)

$$\ln\frac{r}{S} = \frac{(\rho_s - \rho_f)\,\omega^2\,D_m^2\,t}{18\eta} \tag{3.30}$$

With two-layer techniques, all particles at r will be of size D_m (incremental methods) or all particles greater than D_m will have settled out (cumulative methods) after a time, t. The size distribution can therefore be derived in a similar way to that in gravitational sedimentation, taking account of equation (3.30).

Another solution is to make the sedimentation height $(r - S)$ small by comparison with S and assume that the particles fall with a constant velocity. Data evaluation is then identical to that in gravitational sedimentation, except for the acceleration term.

No solution of the general settling equation exists for the suspension methods. Nevertheless, some special solutions are known and measuring techniques are based on them.

Some of the applied centrifugal techniques are schematically represented in Fig. 3.12. A number of these devices can also be used in the gravitational mode. Centrifugal sedimentation is widely used for very fine powders. The measurement ranges extend down to 0.01 µm. Computer assistance is provided with most modern systems for the derivation of the size distributions from the raw measurement data. Particle concentrations are mostly monitored by light or X-ray absorption for these automatic systems.

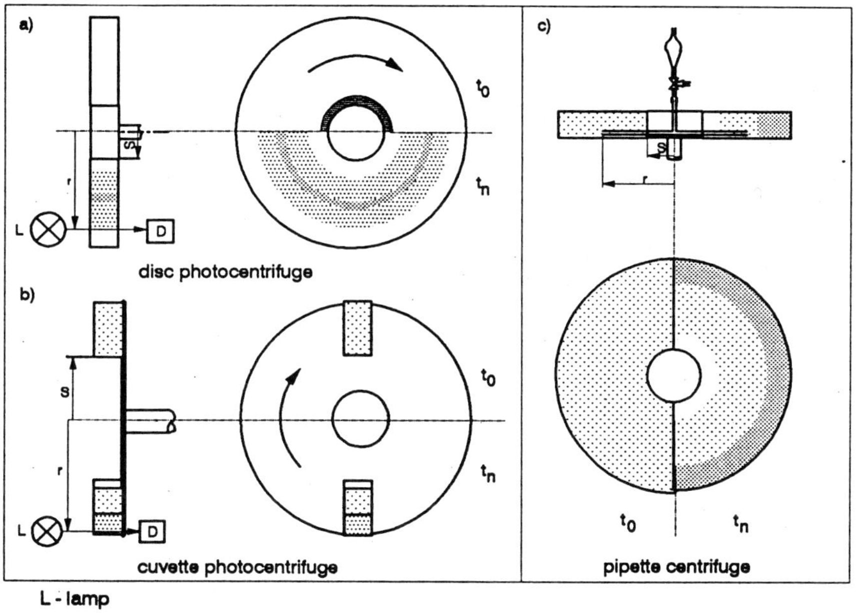

Figure 3.12 Centrifugal sedimentation by a) disc photocentrifuge (incremental two-layer method); b) cuvette photocentrifuge (incremental suspension method); c) pipette centrifuge (incremental suspension method)

3.2.4.4 Fluid Classification

Classification by liquids or gases covers methods which, like sieving, separate particles of different sizes. The proportions within every class can be measured to derive a particle size distribution. In powder metallurgy and ceramics, fluid classification is not used very often for particle size analysis, but is widely applied in powder production for process control or tailoring of the particle size distribution. Fluid classifiers are important components of continuously working powder mills, enabling particles to be removed from the process as soon as they have reached a desired size.

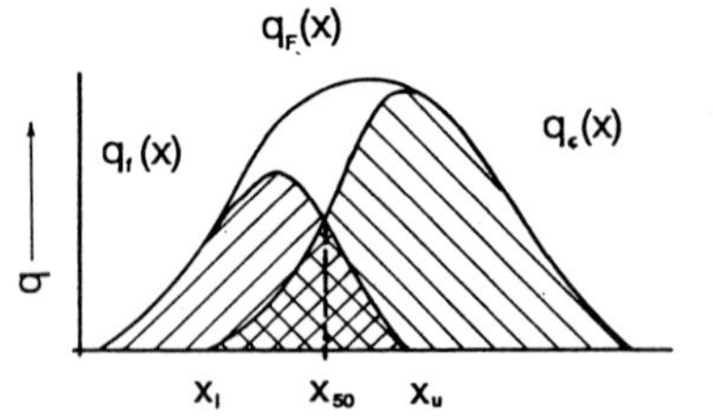

Figure 3.13 Classification of sample with frequency distribution q_F into a fine (q_f) and coarse (q_c) fraction (refer to text for classifier efficiency)

The efficiency of the classifier is very important (see Fig. 3.13). The coarse grade efficiency G_c of a separation is defined as the proportion of particles of size x, which occurs in the coarse fraction.

$$G_c = \frac{q_f(x)}{q_F(x)} \tag{3.31}$$

G_c is zero at $x = x_l$ and becomes 1 at $x = x_u$. The smaller the interval x_l to x_u, the sharper the separation. The cut size, which is usually defined as the particle size x_{50} corresponding to $G_c = 50\%$, gives no information on the efficiency of separation. It is therefore customary to use a ratio ϕ of other cut sizes, e.g.

$$\phi_{25/75} = \frac{x_{25}}{x_{75}} \tag{3.32}$$

$$\phi_{10/90} = \frac{x_{10}}{x_{90}} \tag{3.33}$$

For perfect classification, ϕ becomes 1. Values for $\phi_{25/75}$ above 3 are regarded as poor.

The separation principle is that at least two forces act on the particles causing the smaller and larger particles to separate. This occurs in the separation zone of the classifier from which the small and large fraction have to be removed without remixing. The forces which are usually balanced are gravity or centrifugal forces vs. drag forces or inertial forces. Different combinations are possible. All classifiers can be divided into two systems, the counterflow and the transverse flow (cross-flow) separators.

Counterflow classification in a gravitational field, also called elutriation, uses an upward current of flow, usually water or air (Fig. 3.14). This is the reverse of gravity sedimentation and ideally Stokes' law should apply. Particles with $v_{St} > v_{fluid}$ are sedimented, whilst finer particles are carried out of the separation zone. Most commercial classifiers use a cylindrical tube for the separation zone. Water elutriators are applied in the size range 10–200 µm, air elutriators in the range 1–100 µm.

For fine particles, centrifugal counterflow classifiers are generally used which have a flat cylindrical classification chamber (Fig. 3.14). The flow medium is withdrawn in the central axis of the chamber, resulting in a spiral flow field with

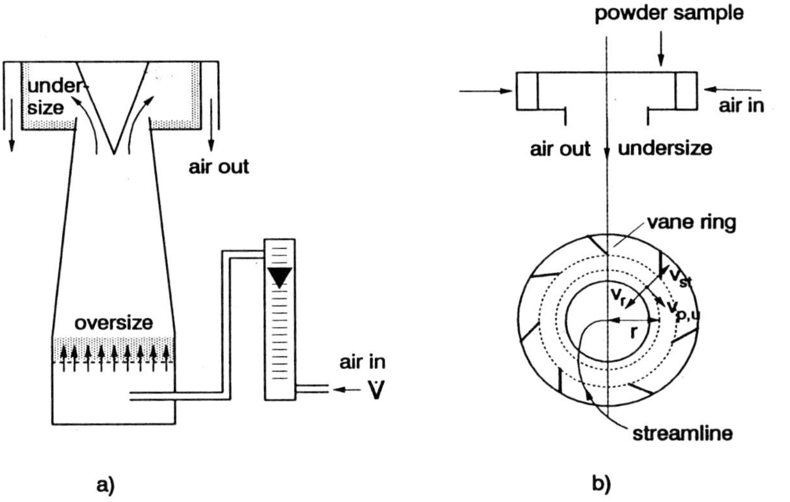

Figure 3.14 Counterflow classifiers (schematically); a) gravitational field; b) centrifugal field

a radial velocity component, v_r. Particles with a settling rate v_{St} in the centrifugal field equal to v_r rotate on a circular path of radius, r, with a velocity, $v_{r,u}$. Smaller particles are swept out with the fluid, whilst larger ones are sedimented to the circumference of the chamber.

$$v_{St} = \frac{(\rho_s - \rho_f)\, v_{r,u}^2\, D^2}{18\eta r} = v_r \tag{3.34}$$

v_r is related to the flow rate V' and the chamber height H as

$$v_r = \frac{V'}{2\pi r H} \tag{3.35}$$

If $v_{r,u}$ is known, the cut size, D, can be derived from these relations. In commercial classifiers, $v_{r,u}$ depends in a complicated manner on the dynamics of the spiral flow, and the cut size has to be derived by a calibration procedure. The cut size range of centrifugal air classifiers generally extends from 0.5–50 µm.

3.2.4.5 Electrical Sensing Zone Analysis (Coulter Principle)

Electrical sensing zone analysis belongs to the direct counting methods. The measuring principle is schematically explained in Fig. 3.15. The particles are suspended in an electrolyte into which a glass tube with a round orifice (diameter D) is immersed. A constant current is maintained between two electrodes, one of them situated outside, the other inside the tube. Defined volumes of the suspension are sucked through the orifice, e.g. by a mercury syphon. When a particle passes through the orifice, the current is changed and a voltage pulse occurs.

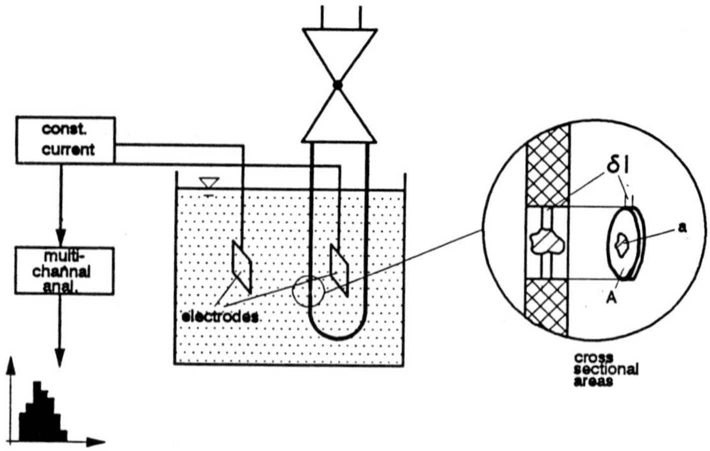

Figure 3.15 Principles of electrical sensing zone particle size analysis (Coulter counter)

The height of the pulse is commonly assumed to be proportional to the particle volume. A theoretical derivation of this relation can be made, taking the electrical resistance of a differential element of the orifice, through which a particle is just passing. This is a parallel arrangement of two resistors. If the resistivity of the particle is assumed to be very large compared to the resistivity, ρ_f, of the electrolyte, the differential resistivity increase δ (ΔR) due to the particle is obviously

$$\delta(\Delta R) = -\frac{\rho_f a \ \delta l}{A^2}/(1 - \frac{a}{A}) \quad (3.36)$$

From equation (3.36), the response of the instrument should not be expected to be proportional for rod shaped particles, but should be modified by the term a/A. Integration assuming spherical particles of a diameter d results in

$$\Delta R = \frac{\rho_f \ d^3}{A^2} \ F\left(\frac{d}{D}\right) \quad (3.37)$$

Hence, the pulse height is also not proportional to the volume of a sphere, but is modified by the function F, for which different solutions exist. Generally, the influence of F increases with increasing diameter ratio d/D. Thus oversizing of the large particles should be expected. However, experiments with model particles and real powders indicate that the assumption of a volume proportional response holds for shapes which are not too flaky or elongated. Therefore, the theoretical principle is still under discussion.

Perfect data is obtained if particles traverse the orifice singly. Two types of coincidence errors occur if this is not the case, 'horizontal' or 'vertical' interaction (Fig. 3.16). These have to be minimized by using extremely dilute suspensions. Coincidence corrections, based on statistical or empirical models are sometimes carried out, but are recommended to be kept as low as possible (usually in the range of a few %).

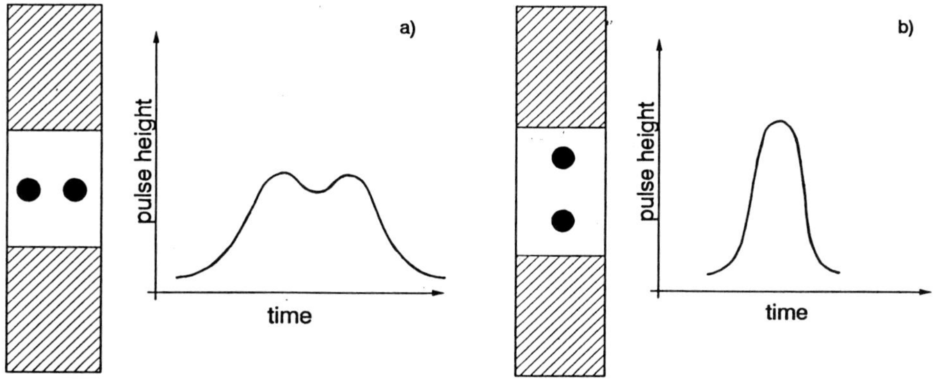

Figure 3.16 The coincidence problem in electrical sensing zone measurement; a) horizontal; b) vertical particle interaction

In practical application, the pulse height is calibrated by using small latex balls as standards. With simple instrumentation, a threshold for the voltage pulse is set according to the latex size. A defined suspension volume is sucked through the orifice, counting all pulses which exceed the threshold, i.e. all particles whose volume is equal or larger than the latex volume. Repeating this procedure for different sizes gives the number distribution of the particle volume. Normally, multi channel analysers are used. The pulses are classified according to their height, and simultaneously counted for the different classes observed. The number size distribution follows directly from the pulse height distribution. Calibration within one channel is usually sufficient. Calibration can also be made directly from the analysed sample in this case. This is done by comparing the measured powder volume in the metered sample to the actual volume, which is known from the concentration of the suspension.

The measuring range depends on the size of the orifice. The particle diameter should be in the range of 2–40% of the aperture diameter. The lower limit arises from background effects, the upper limit from non-linear response (see equation (3.37)) and blocking. Orifice diameters are available from 10–1000 μm and, hence, a size range of 0.2–400 μm can be covered. This method is most suitable in the sub-sieve range down to 1 μm with a size range ratio < 20:1.

3.2.4.6 Light Scattering and Diffraction Methods

Sizing methods which make use of the interaction of electromagnetic waves with the particles have already been mentioned in connection with x-ray and photo sedimentation. These methods apply radiation scattering not directly for sizing but for the estimation of the particle concentration in a suspension. However, scattered radiation also gives information on the size characteristics and can therefore be used in direct sizing methods. Although this has been known for a long time, the practical applications of such methods, mainly based on optical radiation, has only become common during the last decade. Three modern technologies, lasers, optic fibres and computers, have made this development possible.

Theoretical Background

The principles of these methods are based on the interaction of an incident optical beam with a single particle. This beam can either be *absorbed, scattered,* or *transmitted.* Scattering includes *reflection, refraction* and *diffraction.* The problem is theoretically treated by the Lorenz–Mie theory, which starts from Maxwell's electromagnetic equations on the scattering of a plane electromagnetic wave by a homogeneous sphere of diameter D (Fig. 3.17). The scattering intensity, I, in the distant field ($r \gg D$) is of the general form

$$I = I_0 \left[\frac{\lambda}{2\pi r}\right]^2 i(\theta, \phi, \alpha, m) \tag{3.38}$$

$$\alpha = \frac{\pi D}{\lambda} \tag{3.39}$$

where I_0 is the intensity of the incident beam, λ is the wavelength of the radiation in the surrounding medium, m the index of refraction of the particle relative to the medium, θ and ϕ are the angles of observation. The so-called Mie parameter α normalizes the diameter by the wavelength. The intensity function, i, derives in a complicated manner from the variables given in equation (3.39). Figure 3.18 gives an example of the dependence of the normalized scattered intensity I/I_0 from the size parameter α for fixed observation angles ($\theta = \phi = 0$: forward direction) and a given refractive index, m. Although there is no general analytical solution relating particle size to optical measurement, special solutions exist for two limiting cases of the problem: Rayleigh scattering for smaller, and Fraunhofer diffraction for diameters essentially greater than the wavelength λ.

size range	Model	size dependency
$D \ll \lambda$	Rayleigh scattering	$i \sim \alpha^6$
$D \approx \lambda$	Lorenz–Mie theory	
$D \gg \lambda$	Fraunhofer diffraction	$i \sim \alpha^4$

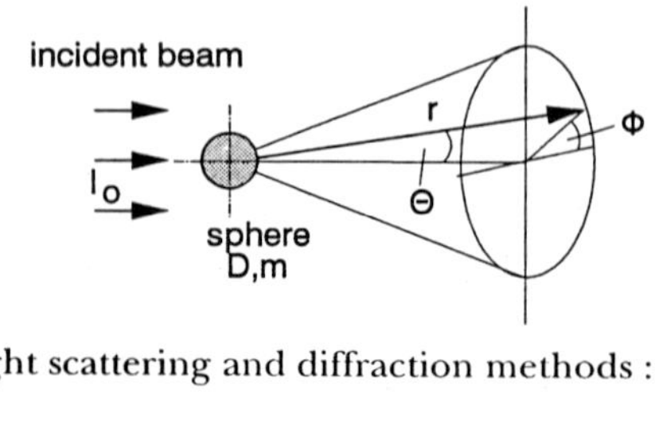

Figure 3.17 Light scattering and diffraction methods : geometrical situation

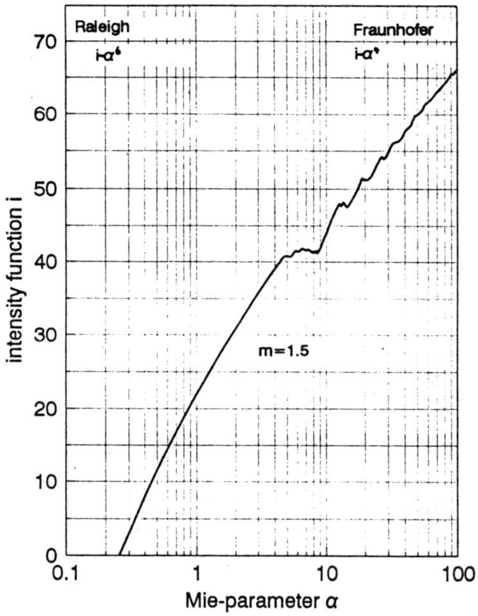

Figure 3.18 Dependency of intensity function i (equ. (3.38)) from Mie parameter α

In the Rayleigh region, the response of the field depends on the number of electrons, i.e. on the particle volume. The scattered-light intensity is proportional to the square of the particle volume. The scattering function, i, is given by

$$i = \alpha^6 \left[\frac{m^2 - 1}{m^2 + 2}\right]^2 (1 + \cos^2 \theta) \tag{3.40}$$

Equation (3.40) represents the sum of two terms, referring to the intensities of the vertically and horizontally polarized components, the former being independent of θ, the latter proportional to $\cos^2\theta$.

In the Fraunhofer region, the contribution of refraction diminishes and diffraction becomes dominant. The laws of geometrical optics apply. Within an angle $\theta < 8°$, the scattered light forms circular diffraction patterns. Their centre distance depends on the diameter of the sphere. The scattering function i is now given by

$$i = \frac{\alpha^4}{4} \left[\frac{2J_1(\alpha\, \theta)}{\alpha\, \theta}\right]^2 \tag{3.41}$$

The term within the brackets represents a so-called Airy function, characterised by a series of maxima of decreasing height with increasing value of the function (α θ) (Fig. 3.19), and thus corresponds to the diffraction patterns. α can therefore be derived from the measured intensity profile $I(\theta)$ of diffraction patterns using equation (3.41) in combination with equation (3.38). It should be pointed out that, contrary to the Rayleigh solution, equation (3.39) contains no materials' parameter (such as m), and therefore predicts diffraction behaviour independent of the material.

The above derivations for the Rayleigh and the Fraunhofer regions apply only to a single spherical particle. Multiparticle scattering can be treated under the assumption that there is no interaction of the radiation scattered from one particle with multiple scattering from the other particles. The total scattered intensity is the sum of the contributions of the single particles. As the particles are usually not monosized, I has to be calculated from

$$I = N \int_0^\infty I(D) q_0(D) \mathrm{d}D \tag{3.42}$$

where N is the number of contributing particles, which is usually not known. By substitution of N by the volume concentration C_v in the analysed suspension volume, V, transformation of q_0 into a volume distribution, and in combination with equation (3.38), equation (3.43) can be derived (see section 3.2.1, equation (3.6))

$$I = I_0 V \frac{3}{2\pi r \lambda} \int_0^\infty \frac{i}{\alpha^3} C_v \, q_3 \, (\alpha) \, \mathrm{d}\alpha \tag{3.43}$$

Equation (3.43) is of the type of a Fredholm's integral equation. It can not be solved for Rayleigh scattering but for the Fraunhofer region it assumes the form

(3.44)

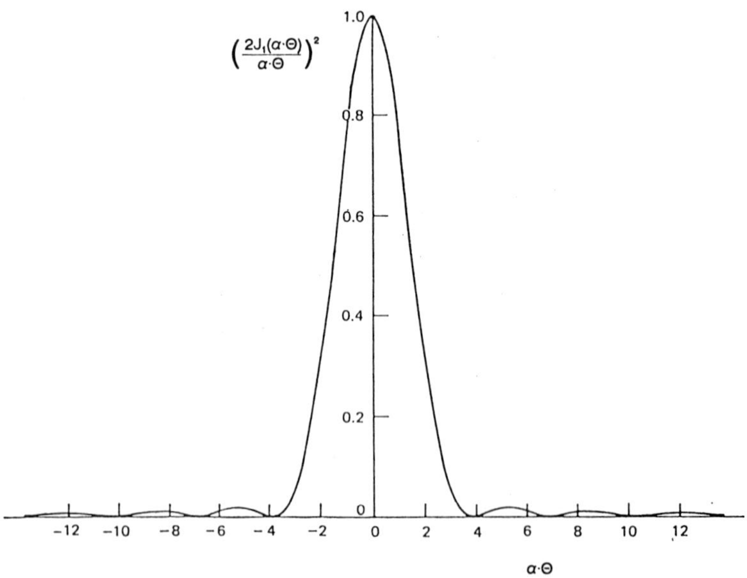

Figure 3.19 Airy function (intensity function for Fraunhofer diffraction, equ. (3.41))

Term III is the distribution function to be estimated. As term II is only a mathematical expression, it is possible in principle to derive this distribution from the measured intensity function $I(\theta)$.

Measurement techniques based on the above principles estimate the scattering behaviour either by monitoring the scattered intensity, I, or the *extinction* ΔS. The latter is the product of the incident intensity I_0 and the 'optical' cross section A_{ext} of the scattering obstacle, reduced by the scattered radiation S_{det} falling on to the detector element (Fig. 3.20).

$$\Delta S = I_0 \; A_{ext} \; - S_{det} \tag{3.45}$$

The *extinction coefficient* is defined as the relation between ΔS and the geometrical cross section. For a single sphere

$$K = \frac{4\Delta S}{I_0 \; \pi \; D^2} \tag{3.46}$$

Multiparticle extinction can be treated in a similar way to multiparticle scattering intensity.

$$\Delta S = -\frac{3}{2} \; I_0 \; A \; \Delta L \; C_v \int_0^\infty \frac{K}{D} q_3 \; (D) \; dD \tag{3.47}$$

ΔS is thus proportional to the concentration C_v and the width of the measuring cell ΔL. Both parameters have to be optimized to achieve an extinction of sufficient intensity without suffering problems caused by multiple scattering effects. An approximation of equation (3.47) is given by the Lambert–Beer equation.

$$\ln \frac{I}{I_0} = -\frac{3}{2} C_v \; L \int_0^\infty \frac{K}{D} \; q_3 \; (D) \; dD \tag{3.48}$$

Equation (3.48) is again of the type of a Fredholm's integral equation. General solutions for this extinction problem do not exist either for the Rayleigh, or for the Fraunhofer range and require the application of the full Lorentz-Mie

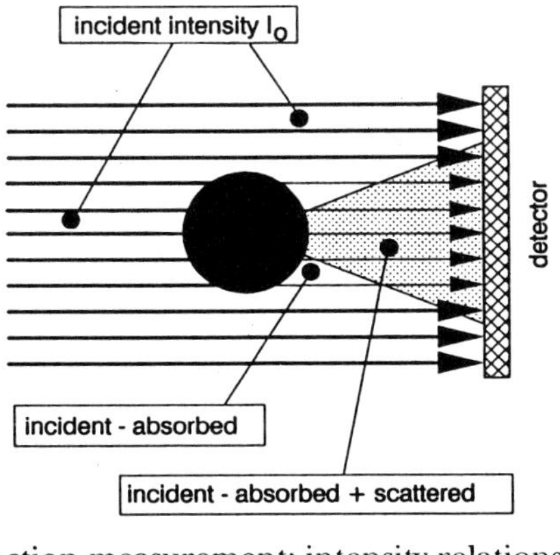

Figure 3.20 Extinction measurement: intensity relations in the detector plane

theory. Special sub-solutions are used in measurement techniques, in which not the complete distribution function, *q(D)*, is estimated, but average or integral parameters such as the particle concentration C_v or the specific surface S_v. Figure 3.21 shows the extinction coefficient K as a function of m and α. It is apparent that K becomes constant in the Fraunhofer range. In this case, integration of equation (3.48) is easily possible

$$S_v = 6 \int_0^{\infty} \frac{q_3(D)}{D} \, \mathrm{d}D \tag{3.49}$$

$$S_v = \frac{\ln(I/I_0)}{K \; C_v \; L} \tag{3.50}$$

For a known concentration, S_v can thus simply be determined from extinction measurements.

Measurement Techniques

In powder metallurgy the most important measuring techniques based on the above principles are optical counting techniques, laser diffraction, and turbidity measurement.

The principle of *optical counting methods* is shown schematically in Fig. 3.22. The particles either flow individually through an illuminated sensing zone, or a measured volume of suspension is scanned by a focused laser beam. Both methods require extremely diluted suspensions. The scattered intensity or the extinction is measured by a photo-electric detector. The height of the measured impulses corresponds to the size of individual particles according to the relations in equations (3.38) to (3.41) for the single particle problem. Usually, this relation is determined empirically by calibration covering a narrow size range.

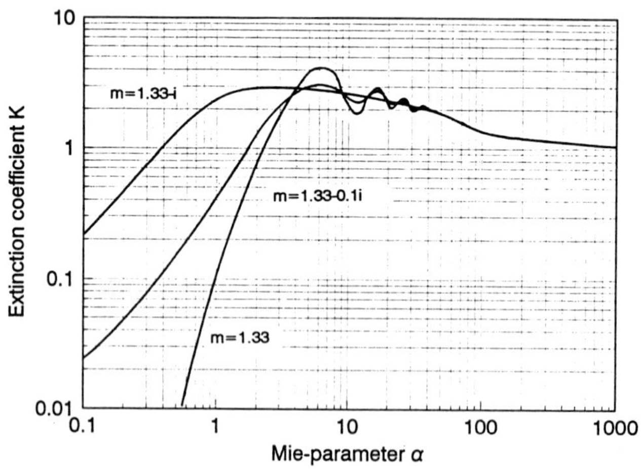

Figure 3.21 Dependency of the extinction coefficient K (equ. (3.44)) from the complex index of refraction m and the Mie-parameter α

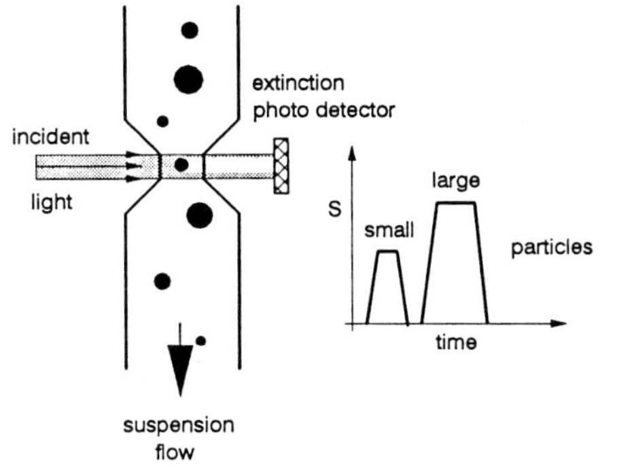

Figure 3.22 Particle size analysis by optical counting (schematically)

The impulse height distribution is monitored by a multi-channel analyser and converted into a size distribution on the basis of the calibration. For small particles, the refractive index is of influence and, therefore, has to be considered in the calibration, while this is not the case for large particles (see Fig. 3.21). These counting methods are applicable in the size range > 1 µm for particles in suspensions, and > 0.1 µm for airborne particles.

Laser diffraction has become by far the most important optical sizing method for metal and ceramic powders. As diffraction involves rays external to the particle, the intensity of the diffracted light is not dependent on properties of the materials (see section 3.2.4.6.1, equation (3.44)). Therefore, mixtures of powders of different materials can be measured without difficulty — which cannot be done with many other sizing methods. A diagram of the laser diffraction method is given in Fig. 3.23. The measuring zone is illuminated by an

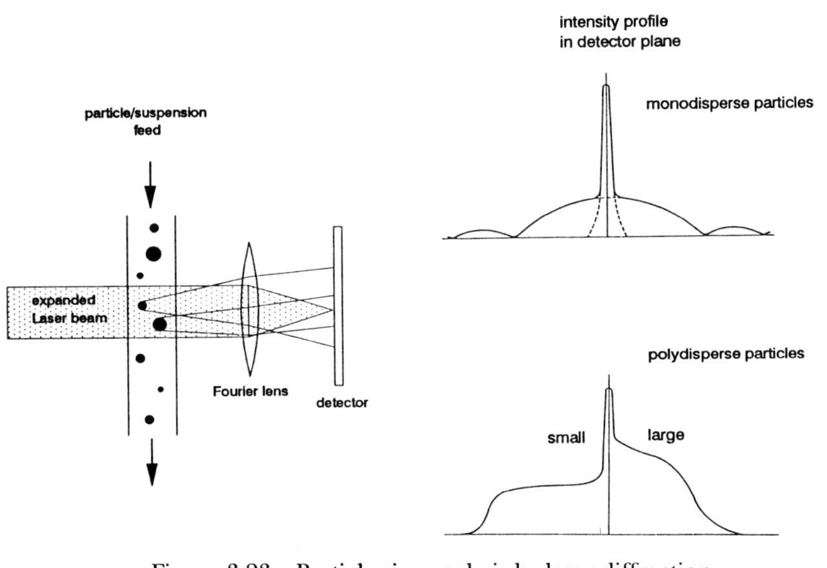

Figure 3.23 Particle size analysis by laser diffraction

expanded laser beam. The particles are fed through the measuring zone in a liquid or gas flow, by simple powder pouring or by stirring of a suspension in a closed measuring vessel. The diffracted and transmitted light are focused by a Fourier transformation lens on to the detector in the focal plane of the lens. By this optical arrangement, the diffraction pattern from a given particle within the detector plane becomes independent from the particle position in the measuring vessel, i.e. from particle movement. Multi-element photo-detectors are mainly used for measurement of the intensity distribution $I(I_0,\lambda,D,\theta)$. For monosized powders, a diffraction pattern with discrete minima and maxima appears, as for a single particle but at an enhanced intensity level. For powders with a varied size distribution, the intensity profile $I(\theta)$ becomes continuous, resulting from the superposition of the patterns of every particle within the measuring zone. This intensity profile is the basis (term I) for a numerical derivation of the frequency distribution $q_3(\alpha)$ from equation (3.44). He–Ne lasers with a wavelength of $\lambda = 0.633\,\mu m$ are generally used, limiting the applicability of the Fraunhofer approximation to a size of the order of $1\,\mu m$. Most modern instruments are stated to be suitable also for sizing in the submicrometre range. This has been made possible by the supply of different instruments with very different measurements, e.g., additional right angled scattered light measurement, application of Mie theory to the measured low angle intensity profile, or simple extension of the Fraunhofer solution into the sub-micrometre range. The first two variants, of course, require knowledge of the refractive index for the submicron particles. Substantial differences can arise from measurements by different instruments within this range. The measuring range of commercial instruments extends from 0.1 to $1000\,\mu m$. Because of the complex calculations, all instruments are fully computerised. The short measuring time required therefore makes laser technique very suitable for online particle size analysis, e.g. for process control in powder production, etc.

Turbidity measurement is a useful technique for average size determination. For large particles, it is based on a constant extinction coefficient (see section 3.2.4.6.1, equation (3.50)). Special solutions of the Lambert–Beer equation exist also for small light absorbing particles such as carbon-black, where the concentration C_v is approximately proportional to $\ln(I/I_0)$. Turbidity measurement is applied as a tool for concentration measurement in many other sizing methods, as already mentioned in the introduction to section 3.2.4.6.

3.2.4.7 Other Methods

Other optical sizing methods can potentially be applied, especially in the submicrometre range. One of these methods is *photon correlation spectroscopy (PCS)*. This makes use of the Brownian motion of small particles in a suspension, the frequency of which is inversely related to particle size. From the variation of scattered laser light patterns, the average particle size and an indication of the extent of the size distribution can be calculated (autocorrelation spectroscopy). The measurable size range is about 0.03–3 µm. The velocity of the particles can also be measured by the Laser-Doppler velocimetry principle.

Other methods, which should be mentioned, include *hydrodynamic chromatography (HDC)* and *field flow fractionation (FFF)*, which both belong to the fluid classification methods. HDC is very similar to the widely used chromatographic size analysis of molecules. A fluid is pumped through a column packed with impermeable spheres or through a long capillary. The particle suspension is injected into the flowing stream at the entrance of the column. Larger particles move with a greater velocity than smaller ones due to side wall effects. Monitoring the particle concentration at the column effluent vs. time yields the size characteristics. This method has been applied in the size range 0.01 to 200 µm. In FFF, a long narrow channel is used instead of a column or capillary. Fractionation occurs in a centrifugal or gravitational field. The size range is 0.01 to 1 µm. Both, HDC and FFF, are potential, but not yet commonly used, methods for powder metallurgy.

3.3 SPECIFIC SURFACE

As mentioned in the context of fractal dimension analysis (see section 3.2.4.2), the roughness and hence the apparent surface of a powder particle depends on the degree to which surface details are resolved by a measurement method. The actual surface of a powder should therefore be determined by methods whose size resolution corresponds to the surface dimensions relevant to the specific surface. In most cases, e.g. powder activity, driving forces in sintering etc., this is on an atomic or molecular scale. Gas adsorption methods and permeametry are suitable to approach this size level.

3.3.1 Surface Analysis by Permeametry

Permeametry is based on the flow of a viscous fluid through a closely packed bed of powder (Fig. 3.24). The pressure drop, Δp, increases with decreasing size of the pore channels in the powder bed. Two different types of flow have to be considered: viscous flow and molecular flow. Viscous flow is characterised by a mean free path, λ, of the fluid molecules, which is small compared with the pore size, d. The fluid–wall interaction is concentrated in a thin layer, and Δp is determined by the molecule–molecule interaction, or from this the fluid viscosity, η. The latter becomes negligible when the mean free path is large compared with pore size. This is the range of Knudsen diffusion, which is controlled by the mean molecular velocity and therefore depends on temperature, T, and molecular weight, M. With air as the measuring fluid, viscous flow is dominant at $d \geq 10\,\mu m$ at atmospheric pressure. Knudsen diffusion dominates at $d \leq 10\,\mu m$ at a pressure of 100 Pa.

Assuming viscous flow, the surface/volume ratio S_v can be derived by the Carman–Kozeny equation (see Fig. 3.24):

$$S_v^2 = \frac{1}{k}\frac{\epsilon}{(1-\epsilon)}\frac{1}{\eta}\frac{A}{L}\frac{\Delta p}{\dot{V}} \tag{3.51}$$

where ε is the porosity of the powder bed. Equation (3.51) can be derived by equating the surface of the capillary walls of the bed to the surface of the powder. The aspect factor, *k*, depends upon the size and shape distribution of the powder and takes values between 3 and 6. As *k* is usually not known, equation (3.51) is suitable mainly for comparative surface characterisation. Permeameters based on this principle can be divided into constant pressure and constant volume apparatus. Constancy of Δp is reached by maintaining a constant flow rate $\dot{V}'$. The constant volume instruments utilize a pressure difference created by a de-levelled U-tube system. Δp thus decreases with the system approaching the equilibrium level. The *Fisher sub-sieve sizer,* for example, is a constant pressure instrument. It is a modification of the early *Lea and Nurse apparatus,* which is still described in British Standards. The *Fisher sub-sieve size* is defined as the equivalent diameter of a sphere of specific surface S_v. Of the constant volume permeameters, the *Blaine apparatus* is the most common. It normally uses a powder of known surface area as a standard reference.

For Knudsen diffusion, the relation between S_v, Δp and the diffusional flow rate $\dot{N}'$ (mol/time interval) is given by:

$$S_v = \frac{24}{13}\sqrt{\frac{2}{\pi}}\epsilon^2 \frac{1}{\sqrt{MRT}}\frac{A}{L}\frac{\Delta p}{\dot{N}} \tag{3.52}$$

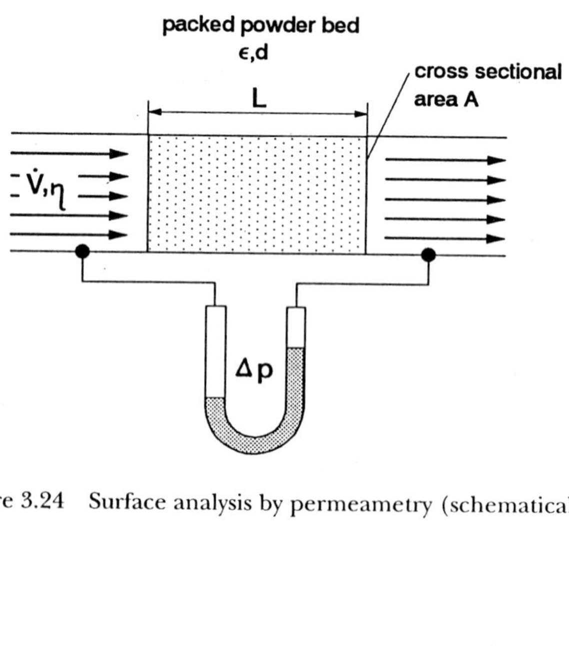

Figure 3.24 Surface analysis by permeametry (schematically)

The *Micromeritics Knudsen Flowmeter* is a commercial instrument. With it, the He flow rate through the powder bed is measured for different pressure levels below 2.5 kPa. S_v is estimated by extrapolation of the results to a pressure level of 0. The specific surface from Knudsen diffusion measurements corresponds to the surface as estimated by gas adsorption, reduced by the surface of open pores within the particles themselves.

3.3.2 Surface Analysis by Gas Adsorption

When a surface of a solid (*adsorbent*) is exposed to a gas (*adsorbate*), gas molecules are *adsorbed* for a finite time, after which they become *desorbed* and are replaced by other molecules. Depending on the interacting forces, adsorption is classified as *physisorption* (van der Waals forces), or *chemisorption* (chemical bonding). Physisorption may be compared to condensation of a vapour to a liquid and is a completely reversible process. It takes place at low temperatures and becomes significant below the critical temperature T_{crit} of the adsorbate. The equilibrium amount adsorbed (volume V) depends on the pressure, p, at which adsorption takes place, and on the nature of the adsorbate and adsorbent. Usually, this pressure is normalized by the saturated vapour pressure p_0 at the given temperature. The graph of V vs. p/p_0 can be measured and is called the *adsorption isotherm*. Adsorption theories deduce the monolayer capacity V_m from this isotherm, defined as the quantity of adsorbate needed to cover the surface with a molecular monolayer. The total surface of the adsorbent is then derived by dividing V_m by the area occupied by one molecule, σ.

The first theoretical approach relating V and p was the Langmuir equation, based on the assumption that adsorption is completed after a monolayer is formed. This assumption holds only for chemisorption, where bonding is limited to the gas molecules directly at the solid surface. In physisorption, second and further layers can be formed before a monolayer is completed. This is expressed by the *BET theory* (Brunauer, Emmet and Teller), which is derived by applying the Langmuir equation to subsequent layers and thus ending with a system of coupled equations. The solution results in the BET equation (3.53).

$$\underbrace{\frac{p}{V(p_0 - p)}}_{\text{I}} = \underbrace{\frac{1}{V_m c}}_{\text{II}} + \underbrace{\frac{c-1}{V_m c}\frac{p}{p_0}}_{\text{III}} \tag{3.53}$$

where c is a constant, related to the heat of adsorption. A plot of term I vs. p/p_0 should yield a straight line with an intercept and slope according to term II and term III (Fig. 3.25). The amount of an assumed complete monolayer can thus be calculated from the measured adsorption isotherm. As $c \gg 1$ in most systems, equation (3.53) can be approximated by

$$V_m = V(1 - \frac{p}{p_0}) \tag{3.54}$$

V_m thus can be deduced from only a single measured point of the adsorption isotherm.

In practice, nitrogen is usually chosen as the adsorbate. The area occupied by one molecule of N_2 is 0.162 nm² at the boiling temperature of liquid nitrogen (–195.6°C). The critical temperature T_{crit} is –146.9°C. This allows for complete desorption already at ambient temperatures. The measuring procedure is as follows (see Figure 3.26): The surface of the powder sample is cleaned from physisorbed molecules by evacuation and/or heating. Nitrogen gas is introduced into the measuring vessel which is then cooled in liquid nitrogen. The

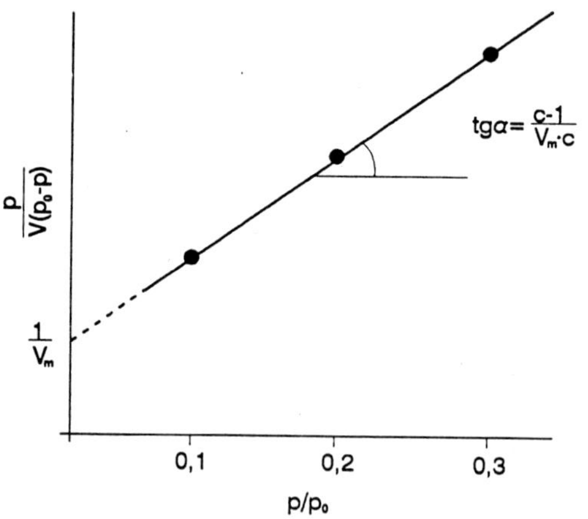

Figure 3.25 Extrapolation of monolayer volume V_m from measured adsorption isotherm (refer to equ. (3.53))

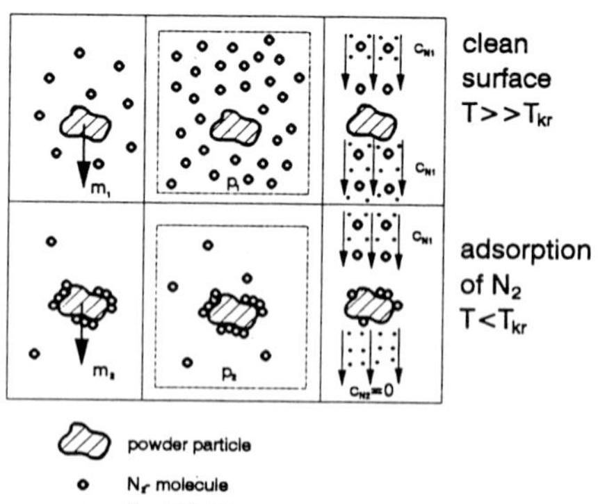

Figure 3.26 Specific surface measurement by nitrogen adsorption: common measuring principles (left to right: gravimetric, volumetric, gas chromatographic methods)

amount of adsorbate at this temperature is derived from the weight change of the sample (gravimetric methods), or via gas law from the pressure change in the measuring vessel (volumetric methods). Alternatively, *continuous flow gas chromatographic methods* can be used. A known mixture of He and N_2 is passed through the sample vessel. The nitrogen content in the mixture is monitored by thermal conductivity detectors situated at the inlet and outlet of the measuring vessel. During adsorption, the nitrogen concentration at the outlet decreases, until equilibrium conditions are reached. The amount adsorbed is found by integration of concentration in relation to time.

Measurements are carried out in the pressure range $0 < p/p_0 < 0.35$. Single point measurements are usually carried out at a fixed normalized pressure of 0.3. At higher pressure levels, *capillary condensation*, i.e. condensation of the gas inside pore channels within the particles, can take place. This phenomenon results in a hysteresis of the adsorption isotherm from which the pore size distribution can be derived by the Kelvin equation (see section 7.1.2). Capillary condensation is applicable in the size range of the so called *mesopores*, i.e. 1.5 to 200 nm.

3.4 TECHNOLOGICAL PROPERTIES

The behaviour of powders during powder metallurgy operations such as powder handling, conveying and compaction is related in a complex manner to the powder characteristics discussed in the previous chapters. A prediction of the powder behaviour on the basis of such properties is difficult. Test procedures have therefore been developed, which allow for a comparative characterisation of powders with respect to powder metallurgy processing. Standards cover flowability, apparent or tap density, and compaction behaviour of powders.

3.4.1 Powder Flowability

Flowability can be characterised in a simple way by the so called *Hall flowmeter test*, shown in Fig. 3.27. The measured parameter is the time required for 50 g of a powder to leave the flowmeter under the influence of gravity. The flow time depends in a complex manner on the internal friction between the powder particles, the friction between the powder and the funnel walls, the relationship between the funnel orifice and particle size, the density of the particles, and the funnel geometry. Obviously this test is only applicable for a comparative characterisation of free-flowing (non-adhesive) powders.

Cohesive powders can be treated according to the principles of flowability of bulk powders, which range between those of liquids and solids. Static liquids under gravity form a horizontal surface with hydrostatic pressure increasing linearly with the distance from the surface. Shear stresses cannot be transferred,

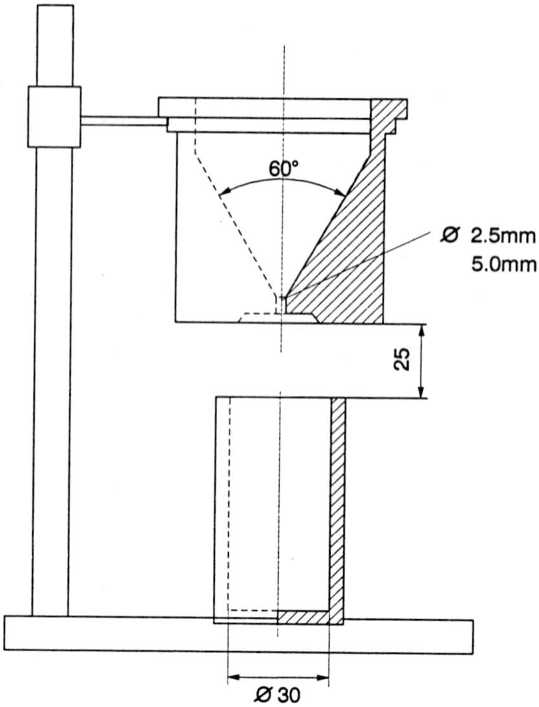

Figure 3.27 Hall-flowmeter for characterization of powder flowability and apparent density (according to ISO 4490 and ISO 3923)

e.g., even the lowest shear component causes movement (flow) of a liquid. The surface of bulk powders can be inclined to the horizontal to any desired amount within the so-called angle of repose. The pressure within a powder column does not increase in a linear manner with depth, but goes through a maximum. On the other hand, bulk powders can only transfer minimal stress and consequently behave quite differently from solids. A general description of the mechanical properties of a bulk powder is given by its *yield locus* (see Fig. 3.28), which provides information on the flow behaviour of a volume element of the bulk powder under a given state of stress. This is expressed in a Mohr's diagram, and shows the relation between compressed stresses σ and shear stresses τ necessary to cause the powder to flow. No flow is possible below the yield locus but a stress state at the yield locus causes flow; stress states beyond the yield locus cannot occur within the powder. Usually, a linear approximation is assumed, according to

$$\tau = \tau_c + \sigma \tan \phi \qquad (3.55)$$

τ_c represents the *cohesion* and φ the *angle of internal friction* of the powder. With increasing compressive stress, the yield locus comes to an end point beyond which compaction of the powder begins. In Fig. 3.28, this is the case for stresses according to the Mohr's circle with the maximum compressive stress σ_1. Th smaller circle represents the unconfined compressive strength σ_c. The powd behaviour can now be expressed in terms of τ_c and φ. Limiting cases are given in Table 3.3.

The cohesiveness of a powder can be expressed by the parameter σ_1/σ_c which is > 4 for cohesive and > 10 for (practically) free-flowing powders. Fluidised powders behave like liquids. This state can be reached with fine powders mixed with sufficient air to eliminate interparticle friction. The yield loci can be determined experimentally by various types of *shear cell apparatus* (see Fig. 3.28) in which a sample is sheared under increasing pressure. Automatic flowability testers based on this principle are available and have been proposed for powder metallurgy applications.

3.4.2 Apparent and Tap Density

The *apparent density* of a bulk powder is defined as the powder mass divided by the bulk powder volume. The latter is related to the particle packing, which is affected by the mode of filling, the container size, vibration during packing etc. These factors have to be kept constant. Standard test procedures are based on the *Hall flowmeter (funnel method)*, (Fig. 3.27) or the *Scott volumeter.* With both

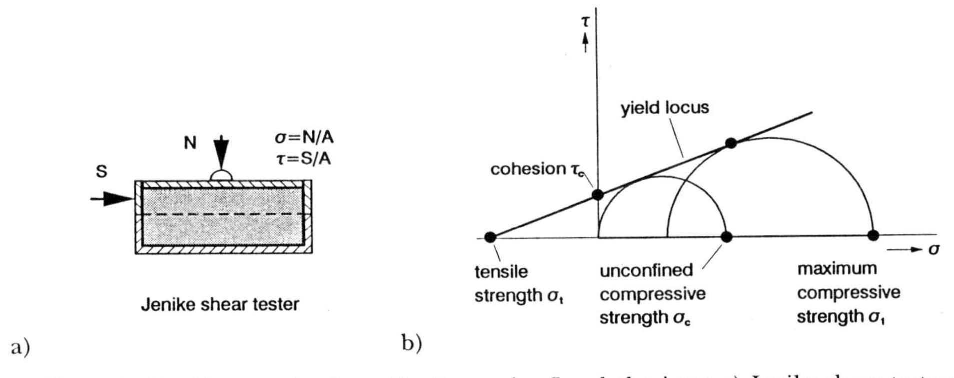

Figure 3.28 Characterization of bulk powder flow behaviour. a) Jenike shear tester; b) yield locus (normal/shear stress combainations for onset of flow)

Table 3.3 Flowability characteristics of different powders

Powder type	Cohesion τ_c	int. friction φ
ideal plastic	$\tau_c > 0$	$\varphi = 0$
cohesive	$\tau_c > 0$	$\varphi > 0$
free flowing	$\tau_c = 0$	$\varphi > 0$
fluidised	$\tau_c = 0$	$\varphi = 0$

methods, a cylindrical container of constant volume (25 cm^3) is completely filled with the powder. The measured mass of the powder is then divided by the container volume, yielding the apparent density. The funnel method is applied for powders which are free flowing through a Hall funnel with a 2.5 or 5.0 mm orifice. Cohesive powders are tested preferably with a horizontally oscillating funnel with an orifice of 7.5 mm. The Scott apparatus has a relatively large four sided orifice of 12.5 mm and is recommended for cohesive powders, which would change their properties under vibration.

The *tap density* is the apparent density of a powder packed vertically by vibration. The powder is poured into a calibrated glass measuring cylinder, which is tapped manually against a rubber plate, or by a standard tapping apparataus at a frequency of 1.7 to 5 Hz and an amplitude of 3 mm. Tapping is continued until no further densification is visible.

The variables which control powder packing are the particle characteristics (size distribution, shape, mass, resilience (i.e. the ability to resist conversion of kinetic energy to other forms of energy on impact), interparticle friction etc.), the container (dimensions, wall friction), deposition parameters (intensity of tapping, particle velocity), and the treatment after deposition (vibration).

With monosize spherical particles, the maximum packing density theoretically achievable is 74% (by close packing with a coordination number of 12). Random arrangements of monosize spheres reach 63% in dense and 60% in loose random packing. The packing density can be increased by the addition of much smaller spheres filling the interstices between the larger ones. The maximum packing density in close packing of binary mixtures is 86%, with about 73% of coarse spheres in the mixture. A 95% packing density is the limit in a ternary mixture with an optimum volume ratio of 66:25:9 for coarse, medium and fine components respectively, if the diameter ratios are large. With an increasing number of different sized particles, even higher values should be possible. The limiting densities can only be reached when all the particles fall into an ideal position. The vibration time required for such an arrangement is proportional to the number of particles, which is about 1:500:70,000 in the ternary mixture considered. Thus the time for ideal mixing will increase by a factor of 140 from the binary to the ternary ideal packing. It has also to be realised, that the size relation of the smallest to the largest particle is limited in practice, thus limiting the number of component particles in multi-size systems. Up to a mixture size ratio of 10, (a typical value in powder metallurgy) a binary system allows for a higher theoretical packing density than systems with more component particles. In general, packing density increases with the range of the particle size distribution and this tendency is more pronounced in densely packed arrangements than in loosely packed ones.

The packing efficiency decreases with increasing deviation from sphericity and decreasing particle size. The latter effect is at least partially due to increasing interparticle friction. Particle mass and resilience (for definition, see above) also play a role during particle deposition. Steel particles, for example, pack

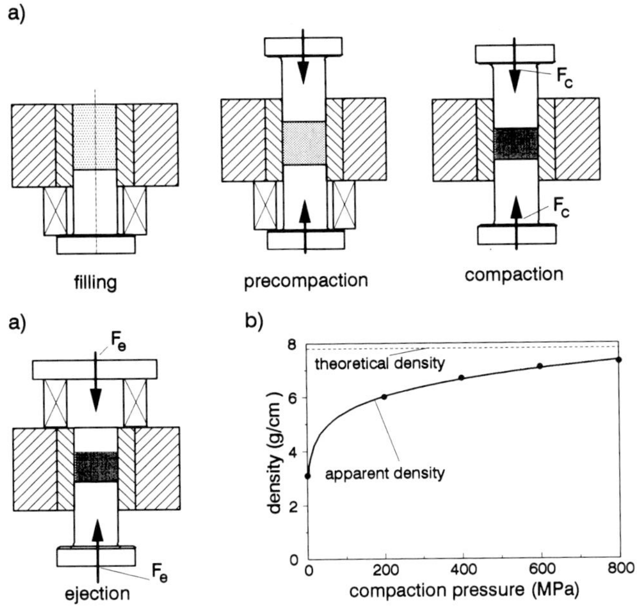

Figure 3.29 Compactability testing of metal powders. a) ISO 3927 floating die compaction tool; b) typical compactibility curve (water atomized iron powder)

more closely than lead shot under the same deposition conditions, due to their higher resilience. The relevant deposition parameters are the kinetic energy of the particles and the intensity of deposition (particles/second × area). Packing density increases with kinetic energy, but decreases when the intensity of deposition exceeds a critical value.

Near the container walls, the packing density is reduced by disorder in the particle arrangement. The critical range is within a distance of 10–50 particle diameters from the wall, as can be demonstrated from mono-size sphere experiments. This has to be taken into account, for example, in tool design for components with thin cross sections.

3.4.3 Compactibility

Compactibility or compressibility characterises the densification behaviour of bulk powders in terms of compact density and compaction pressure. The equilibrium density of a volume element within a powder compact depends on the stress components acting on that volume at any given moment during compaction. The equilibrium density of the whole compact at a given compaction pressure represents an average over many differential volume elements. It therefore depends not only on the stress distribution within the compact, but also on the deformation history of every differential volume element. Thus, the mean density of a macroscopic compact is not solely determined by an applied compaction pressure, but depends also on filling, compact dimensions, and the compact–die wall interactions. These problems are treated in section 6.1.1.1, together with other fundamentals of compaction. Practical test methods try to reduce the number of variables by utilizing (apparently) simple compact geometries and standard testing tools. Compactibility in uniaxial compression of metallic powders, excluding hardmetals, is standardized in ISO 3927 (1985). A floating die apparatus (Fig. 3.29) is used to produce cylindrical (diameter 20–26 mm, height/diameter ratio 0.8:1.0); or rectangular (30 × 12 × 5–7 mm) compacts at pressure levels up to 800 MPa. The compaction pressure is defined simply as the compaction force, divided by the cross section of the compact perpendicular to the pressure direction. After ejection, the compact density is estimated from its mass and dimensions. The density against compaction pressure graph (compactibilty curve) starts from the apparent density and, for most metal powders, increases asymptotically towards a limiting value, which is lower than the theoretical density of the system. In addition to compactibility, the ejection force, F_e, can also be measured in the test. Division of F_e by the mantle surface of the compact yields an average ejectional shear stress τ_e.

3.5 POWDER IMPURITIES

Chemical impurities are a common problem in all industrial materials. Because of the high specific surface of a powder, a large amount of the material is directly exposed to the surroundings during different stages of powder production and processing. Thus, impurities can easily be picked up from adsorption and chemisorption from the surrounding medium. The impurity content of powders, therefore, is often greater than that of solid materials. In many cases these impurities remain concentrated at the surface, giving rise to a second specifically powder metallurgy problem: the effect of the impurities on the final material properties. Even very small impurity concentrations can sometimes completely change these properties. To illustrate this, let us consider a high quality atomized titanium powder which becomes contaminated by a very small volume fraction of particles from another batch of powder with an unacceptably high oxygen content. The increase in total oxygen content is small, but by no means negligible. The surface oxide layers on the contaminated particles remain at their sites during HIP consolidation of the powder, resulting in defects in the component of the size of the original particle diameter. Another practical example is super-solidus sintering of high speed tool steels (see section 7.3.3.1), where the interaction between surface oxides and added carbon governs the final carbon content and thus the amount of liquid phase. Slight variations can therefore completely change the final microstructure.

Oxygen is the most important impurity in metal and non-oxide ceramic powders. ISO 4491 covers the determination of the oxygen content of metal powders by reduction methods: it distinguishes between *hydrogen loss, hydrogen-reducible oxygen*, and *total oxygen*. The *hydrogen loss* of a powder is defined as the mass loss divided by the total mass during hydrogen annealing of a sample. Annealing occurs under standardized conditions (temperature, reduction time, hydrogen dew point and flow rate, etc.). The hydrogen loss is easily determined but is not always correlated simply to the oxygen content. It includes other volatile substances, e.g., H_2O, N_2, P and hydrocarbons. It does not include oxygen in the form of stable oxides such as SiO_2, Al_2O_3, etc. Powders having a high oxygen affinity such as Cr can even be oxidized during the test, resulting in a negative value of hydrogen mass loss. Application of this test is limited to elemental, partially alloyed and pre-alloyed powders of low oxygen affinity without carbon. The *hydrogen-reducible oxygen* is determined by a similar procedure, including a separate drying step to remove humidity. Oxygen is determined from the H_2O in the reaction gas by analysis. The range of application is similar to that described above, but does not exclude carbon-containing powders. If carbon is involved, the carbon oxide in the reaction gas has to be catalytically converted into CH_4 and H_2O before analysis, and the oxygen content includes not only H_2-, but also CO-reducible oxides. *The total oxygen content* is measured by reduction–extraction. A dry powder sample placed in a graphite cruicible is heated under vacuum or inert atmosphere to very high temperatures ($\geq$ 2000°C). The oxygen content is determined from the reaction products CO or CO_2, for which different analytical techniques can be used (volumetric, chromatographic, infra-red absorption methods, etc.).

Another significant impurity, or dopant, is carbon. The *total carbon* content is usually determined by oxidation of a sample in an oxygen stream under quantitative measurement of the resulting reaction product, CO_2. For carbide powders especially the *free carbon* is of interest. Methods for free carbon analysis are based on the insolubility of free carbon in nitric and hydrofluoric acid. Free carbon can therefore be determined from the filtration residue remaining after dissolution of the powders. This residue is analysed by the same methods as are used for total carbon measurement.

Atomic absorption spectrometry is a standard method for the determination of the content of metallic impurities in hardmetals, but other common methods of wet chemical analysis can be applied.

Because of the problems resulting from impurity segregation, microanalytical methods are widely used in impurity control, but cannot yet be regarded as standard methods. Especially surface-sensitive methods such as Auger electron spectroscopy (AES), electron energy loss spectroscopy (EELS), etc., are suitable tools.

3.6 STANDARDS

Terminology, test procedures and evaluation of results in powder characterisation are standardized in national and international standards, including ASTM (American), BS (British), AFNOR (French), DIN (German), and ISO (International). Table 3.4 lists ISO standards, in accordance with the sequence of subjects in this chapter.

Table 3.4 ISO standards on characterisation of metal powders

ISO-No.	Year	Title/Content
2395	1972	Test sieves and test sieving; vocabulary
9276/1(D)	1987	Representation of results of particle size analysis
2591	1973	Test sieving
3954	1977	Powders for powder metallurgy purposes – sampling
4497	1983	Metallic powders: determination of particle size by dry sieving
4490	1978	Metallic powders: determination of flowability by means of a calibrated funnel (Hall flowmeter)
3923	1986	Metallic powders: determination of apparent density (3 parts, 1979–1986)
3953	1985	Metallic powders: determination of tap density
3927	1985	Metallic powders, excluding powders for hardmetals: determination of compactibility in uniaxial compression
4491	1989	Metallic powders: determination of oxygen content by reduction methods (4 parts)
3907	1985	Hardmetals: determination of total carbon content, gravimetric method
3908	1985	Hardmetals: determination of insoluble (free) carbon content, gravimetric method
4503	1978	Hardmetals: determination of contents of metallic elements by X-ray fluorescence, fusion method
7627	1985	Hardmetals: chemical analysis by flame atomic absorption spectrometry (6 parts, 1983 and 1985)

FURTHER READING

T. Allen, *Particle Size Measurement,* 4th Edn. (ed. B. Scarlett): Powder Technology Series, London, New York, Chapman and Hall, 1990.

H.G. Barth (ed.), *Modern Methods of Particle Size Analysis,* New York, Wiley, 1984.

D.J. Cumberland and R.J. Crawford, The Packing of Particles, in *Handbook of Powder Technology,* Vol. 6, (eds. J.C. Williams and T. Allen), Amsterdam, Elsevier, 1987.

S. Lowell and J.E. Shields, in *Powder Surface Area and Porosity,* 2nd Edn. (ed. B. Scarlett); Powder Technology Series, London, New York, Chapman and Hall, 1984.

L. Svarovski, *Powder Testing Guide: Methods of Measuring the Physical Properties of Bulk Powders,* London, New York, Elsevier, 1987.

DIN- Taschenbuch 133, *Partikelmeßtechnik,* 3rd Edn. Berlin, Köln, Beuth, 1990.

DIN- Taschenbuch 247, *Pulvermetallurgie,* Berlin, Köln, Beuth, 1991.

K. Leschonski *et al.* in *Grundzüge der mechanischen Verfahrenstechnik* (ed. K. Dialer, U. Onken and K. Leschonski), Grundzüge der Verfahrenstechnik und Reaktionstechnik 29–138, München, Wien, Carl Hanser Verlag, 1984.

Characterisation of Powders and Compacts, Princeton NJ, MPIF, 1992.

Standard Test Methods for Metal Powders and Powder Metallurgy Products, Princeton NJ, MPIF, 1993.

4: Powder Conditioning and Heat Treatment

Only in exceptional cases can metal or ceramic powders be introduced directly into a powder metallurgy shaping operation, because they originate from various different powder production processes. Usually, additional treatments are required to make a powder suitable for further processing. These operations are often summarized by the term 'powder conditioning'. They depend on the nature of the powder, the size and shape of the final part, and the subsequent operations to be carried out. This chapter covers some general aspects of these operations. Problems of special materials and processing methods will be treated in their corresponding chapters.

4.1 PRINCIPAL ASPECTS AND ALLOYING TECHNIQUES

During their processing, metal powders have to be handled, conveyed and filled into dies or moulds, and usually cold or hot compacted. The powders must be suitable for such operations, their characteristics providing the basis for the desired properties of the green (compacted) and the finally sintered component. The latter often requires additives, which are either temporarily present during some process steps and are removed in later processing, or are alloying additions, which remain in the material during the entire process. Therefore, the fundamental problems of powder conditioning are the introduction of these additives and the behaviour of the conditioned powder during its transport and compaction or forming behaviour.

Temporary additives have the function of lubrication, plasticizing and binding. *Lubricants and plasticizers* facilitate movement of the particles under externally applied forces by reducing interparticle and die wall friction forces. *Binders* enhance the strength properties of the green compacts, which have to be sufficient to withstand the stresses arising during subsequent handling. These stresses can be quite high, e.g. during ejection of a compact from an axial die press. The amount and nature of lubricants or plasticizers varies widely, depending on the forming operation. The consistency of powder composites ready for use ranges from free flowing solid powders to pastes or slurries. So even dispersant fluids are included as possible additives. Binders are necessary when the interparticle adhesion in the formed compact is insufficient. Interparticle adhesion depends on the size and quality of the contact planes between neighbouring particles, and is usually strong in parts produced by compacting from plastically deformable metal powders. The adhesion decreases with decreasing plastic deformation of the particles. Binders are therefore needed for hard powders and/or in forming operations with small interactions between particles e.g., pressureless shaping or injection moulding. They link neighbouring particles by bridging, which requires good adhesion between the particle surface and the binder. In many cases, temporary additive systems have two functions, plasticisation and binding. An example, again, is injection moulding, where the temperature-dependent viscosity of organic additives makes the powder thermoplastically formable at enhanced temperatures, and provides a rigid green part at ambient temperature.

Alloying additions are primarily required in order to achieve the desired material properties in the final component. In powder metallurgy processing, they often have additional functions to aid sintering. They can also influence the handling and formability of the powders. Various alloying techniques can be applied to powder systems, the most important of which are shown in Fig. 4.1.

Powder mixtures of elemental or alloy constituents are the easiest way to provide the desired concentration of alloying elements. The advantages of powder mixtures are their versatility and relatively good compactability. Master alloy powders can be used — not only elemental powders. Thus, for a particular composition of the final material, various types of starting powder mixtures can be chosen, which sometimes enable special problems related to the alloy system to be overcome. An example is the use of alloying elements of high oxygen affinity such as Cr, Ti, Al, etc., whose stable oxide films inhibit sintering. They can be used as master alloy powders, in which their activities are reduced by forming compounds or solid solutions with the base powder, or which form transient or even stable liquid phases during sintering. Thus, powder mixtures sometimes make it possible to optimize the final microstructure by 'tailoring' the path of a reactive sintering process (see section 7.3). The compactability of

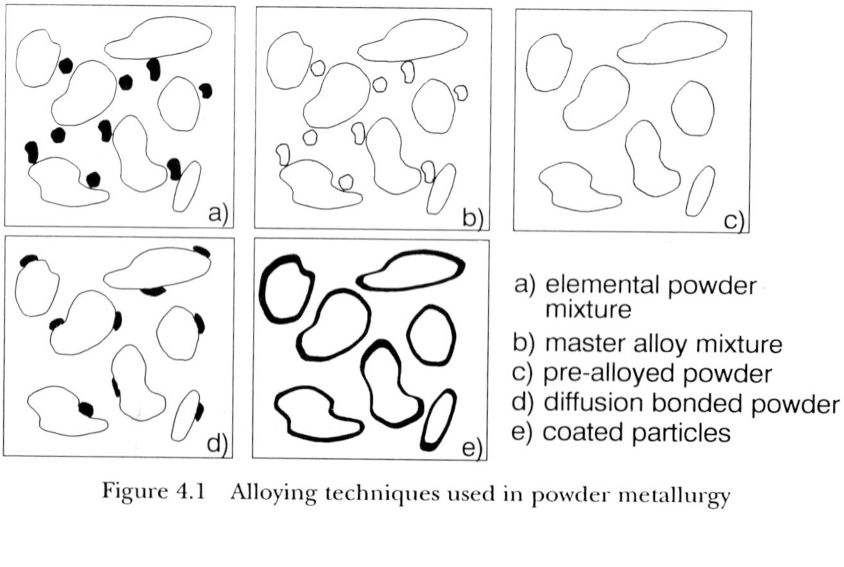

Figure 4.1 Alloying techniques used in powder metallurgy

powder mixtures is determined mainly by the base powder. The introduction of strengthening elements by simple powder mixing is therefore not as detrimental to compactability as pre-alloying of the individual powder particles. The disadvantages of powder mixtures are insufficient homogenisation during sintering, in many cases, and the danger of segregation during powder handling.

In *homogeneously alloyed (pre-alloyed) powders* every individual particle consists of a homogeneous alloy of the constitutents. This is achieved mainly by melt atomisation, but is also possible by mechanical alloying, coprecipitation or co-reduction and by mechanical comminution of the alloy. Alloying elements exhibit their strengthening effects within each particle, reducing its ability to undergo plastic deformation at a given compaction pressure, and thus decreasing compactability. The advantages of pre-alloyed powders are the homogeneity of the final microstructure, which is indispensible in many cases, and the elimination of segregation. When diffusional homogenisation leads to problems, e.g. due to diffusion porosity, pre-alloyed powders can be a solution. Pre-alloyed powders are used, for example, in the production of stainless steels, high speed tool steels, low alloy steels for sinter forging, and in many non-ferrous alloys such as those of copper, titanium, and cobalt.

Diffusion bonded (diffusion stabilized) powder mixtures are intended to combine the advantages of powder mixtures and pre-alloyed powders. They are produced by a special heat treatment of a mixture, whereby the particles of the alloying constitutents are diffusionally bonded (sintered) to the base powder particles. Thus, the danger of powder segregation can be completely eliminated. The pre-formed particle contacts in the powder mixture reduce the time required for final homogenization by diffusion. Additionally, it becomes possible to introduce the alloying elements in the form of very fine particles or even in the form of a solution of the metal salt, which forms a film on the surface of the base powder and is reduced during heat treatment. In this way, a further reduction of homogenisation time becomes possible. Diffusion bonded powders are mainly used in the production of sintered steel structural parts.

Particle coating with the alloying constituents is similar to diffusional alloying, but more effective for homogenization during sintering. It can be applied even in completely immiscible systems like metal–ceramics, where it provides the basis for ideal microstructures with a continuous skeleton of the metal phase at metal contents of only a few per cent. Coating of particles can be achieved via gas phase reactions or by precipitation from aqueous solutions.

As can be seen from Chapter 2, the range of particle sizes of powders covers several orders of magnitude, beginning at a few nanometres and extending up to some hundred micrometres. The handling and processing behaviour of such powders is completely different at the coarse and fine ends of this range. One major reason for this is the relation between surface attractive forces and inertial forces, which increases strongly with decreasing particle size. This causes problems in the handling and transportation of fine powders, which tend towards uncontrolled agglomeration and thus lose their flowability. Fine powders also tend to develop dust emissions and also cause severe damage to compaction tools by adhesive wear if they are entrapped in the gap between the die and the punch. Powders have therefore to be classified to remove dust or very fine size particles. For fine powders used in ceramics or hardmetal production the surface attractive forces have to be reduced or the inertial forces have to be increased. The former can be achieved by wet processing, i.e. *dispersing* of the powders, the latter by particle size enlargement, i.e. *controlled agglomeration (granulation)* of the primary particles.

Therefore, in powder conditioning, problems such as mixing or dispersing, segregation and agglomeration have to be considered very carefully.

4.2 MIXING, BLENDING, SEGREGATION

Blending is defined as the intermixing of powders of the same nominal composition. It is used to achieve a desired particle size distribution. *Mixing* implies intermingling powders of different chemical composition. Both can be mathematically described as an instationary change of concentration of the components along a local coordinate. The concentration change occurs by a convective or dispersive transport of the particles. The latter mode can be subdivided into diffusional transport and a stochastic particle motion caused by an energy input via stirring or other measures. The optimum dispersion which can be expected from a mixing process is a random distribution of the different types of particles, as compared with some other types of distribution (see Fig. 4.2). The quality of a mixture is usually described by the standard deviation s in concentration of a series of fixed volume samples. It has to be recognized, that s is sensitive to the sample volume itself. If the sample becomes large enough, s approaches zero and is independent of local variations in concentration. Thus, the sample volume necessary for a maximum tolerable standard deviation is another characteristic of mixture homogeneity.

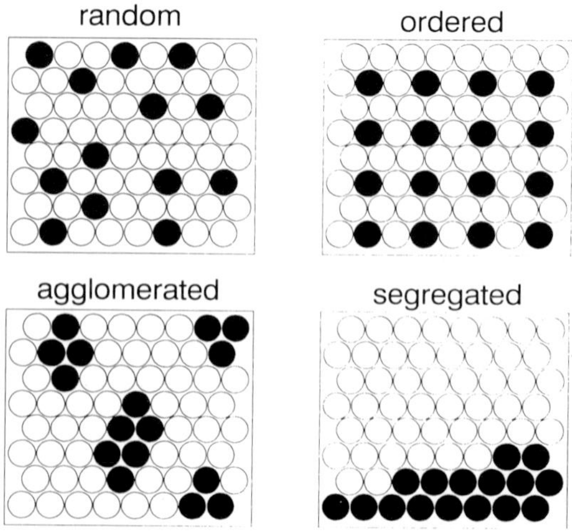

Figure 4.2 Dispersion of second phase particles in powder mixtures

Very varied equipment is available for blending, mixing and dispersing of powders. Some examples are schematically represented in Fig. 4.3. Tumble mixers utilize gravitational forces for producing particle motion. Shear-agitated mixers use paddles or other moving components in a stationary container for shearing planes within the bulk powder. Centrifugal mixers have a similar design, but operate in the range of rotational speeds, in which particle motion is controlled mainly by centrifugal forces. In fluidized bed blenders, the powder takes the character of a liquid and the contribution of convection and diffusion to particle motion and mixing increases compared with the former methods. For dry mixing and blending of metal powders, tumble and low-shear mixers are mainly used in practice, due to possible particle degradation in high shear equipment. Fluidized bed blending is satisfactory with respect to efficiency and lack of contamination by abrasion products, but it is sensitive to segregation caused by differences in particle size, shape or density. Agglomerated powders only attain homogeneity if the agglomerates are broken down during mixing. Wet mixing (co-milling of components) is, therefore, often applied for fine ceramic or hardmetal powders, using the same equipment for powder comminution as referred to in section 2.1.1.1 (ball mills, attritors etc.).

The *tendency to segregate* in a powder or powder mixture depends on the differences in particle size and shape, and especially on particle density. Segregation increases with increasing distance travelled by a free flowing powder. Size and density differences can balance out if the finer particles are more dense than the coarse particles. With coarse particles of higher density, fine particles floating on top of the powder mass increase the problem.

4.3 AGGLOMERATION

As already mentioned, agglomeration (granulation) allows enlargement of very fine powders to avert uncontrolled agglomeration and to make the powders

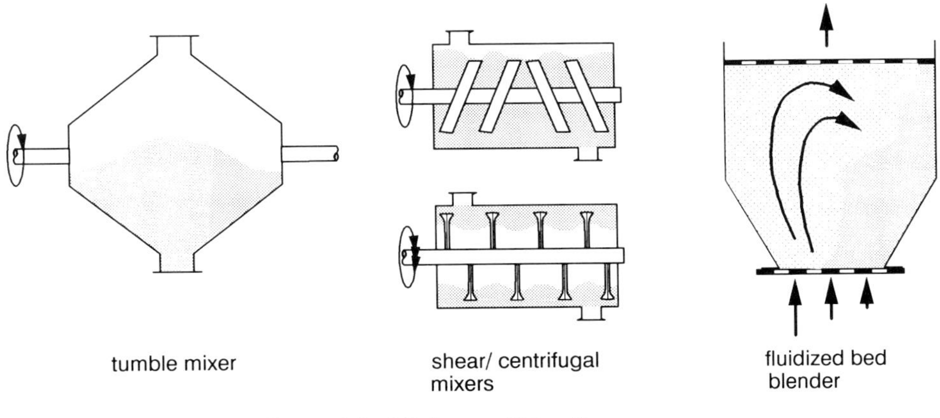

Figure 4.3 Mixing and blending systems

free flowing and safe. The bonding mechanisms for agglomeration (controlled as well as uncontrolled) can be:

- Solid bridges formed by crystallized salts or sintering contacts.
- Immobile liquid bridges formed by viscous binders, adhesives and adsorption layers.
- Mobile liquid bridges (capillary forces).
- Van der Waals and electrostatic forces.
- Mechanical interlocking of powders.

The adhesion forces decrease in the sequence given above and increase with decreasing size of the primary particles.

The methods of granulation are layering agglomeration, press agglomeration and spray drying. Some of the process variants are given in Fig. 4.4. In *layering agglomeration,* fragmentation and agglomeration occur simultaneously, the latter with a higher probability resulting in increasing agglomerate size with processing time. Layering agglomeration results in a relatively homogeneous arrangement of the primary particles and distribution of the binding agent added for granulation (as well as for later processing). Granules of up to several mm in diameter can be produced by layering methods. *Press agglomeration* occurs under high pressure and results in high strength, low porosity agglomerates of well defined size and shape. In *spray drying,* powder suspensions are atomized into droplets, which are dried by evaporation of the liquid during the free fall of the droplets. Atomisation of the suspensions is undertaken by centrifugal (rotating disc) atomisation or by single-fluid or two-fluid nozzles, similar to melt atomisation. The granule size is determined by the size of the droplets and the solid content of the suspension. The largest agglomerate sizes achieved are 600 μm with a single-fluid nozzle, and 300 μm with rotating discs, which also yield the narrowest size distribution. In the specific droplet drying process, spray dried agglomerates are often inhomogeneously packed, and sometimes even take the form of hollow spheres.

Granulation becomes necessary in powder metallurgy practice only in the sub-sieve range, depending on the interparticle attractive forces and the density of the material from about 10–5 μm downwards. Thus, granulation has to be applied mainly for systems which are dispersed in a suspension in prior processing steps (e.g. sol–gel methods). Starting from a suspension, spray drying is, therefore, by far the most important practical granulation method. It is used for most ceramic powders, for hardmetal powder mixtures and other fine metal powders such as molybdenum. Processing of these metals requires closed systems suitable for non-aqueous suspensions. Layering agglomeration by tumbling or fluidized bed spray agglomeration are also used in commercial practice. All temporary additives needed during further processing are generally introduced during the agglomeration procedure. In addition to size enlargement and introduction of additives, granulation also has the effect of suppressing the segregation tendency in powder mixtures.

Agglomerate strength is an important parameter, as well as agglomerate shape, size distribution and porosity, which control flowability and tap density.

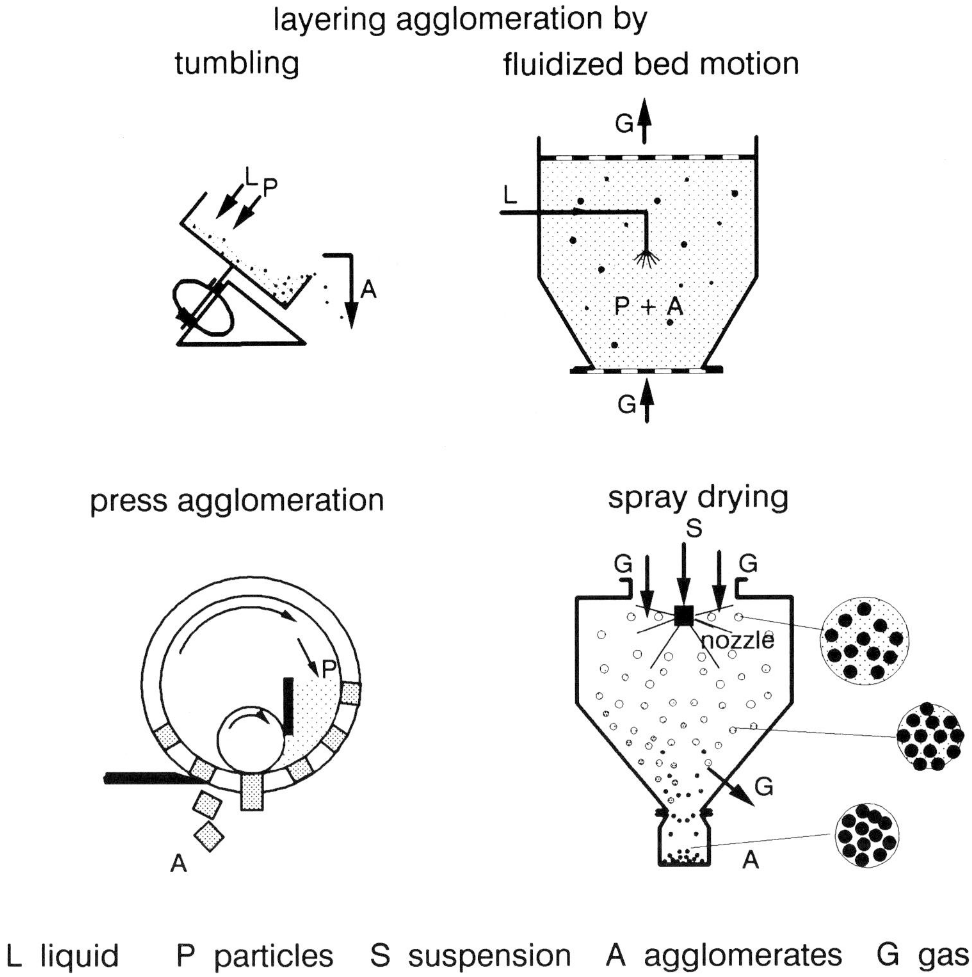

Figure 4.4 Granulation (agglomeration) methods

The granules must withstand handling and transportation, but have to be weak enough to be completely deformed during shaping. Otherwise, inhomogeneities in the green compact cause agglomerate-related defects by differential sintering (see section 7.2.3.2).

4.4 HEAT TREATMENT

Heat treatment of powders is performed for various purposes: chemical reduction, decarburisation, annealing, degassing, size enlargement. The primary reduction and decarburisation treatments have already been dealt with in Chapter 2. Raw powders resulting from various production processes often exhibit a reduced compressibility, due to rapid cooling during atomisation, work hardening from mechanical comminution, residual interstitial impurities such as oxygen, carbon or nitrogen, and to oxide layers formed during production or storage. *Annealing* is therefore common for many metal powders, when a good subsequent compressibility is required. During annealing, residual stresses are relieved by recovery and recrystallisation. Alloying elements in supersaturation can be precipitated in the form of relatively coarse particles, thus reducing the work hardening tendency during powder compaction. Apart from these microstructural effects, annealing can reduce the impurity levels of nitrogen, carbon and oxygen. In many cases, decarburisation is accomplished simultaneously with deoxidation by reaction of carbon and oxygen in the powder to form carbon monoxide. The additional use of reducing atmospheres such as hydrogen or dissociated ammonia further lowers the oxygen content and prevents re-oxidation. Practical examples include carbonyl iron powder, in which about 0.8% carbon and 2% oxygen in the initial powder are reduced to 0.02%, and 0.15% respectively by hydrogen annealing. Electrolytic iron powders are heavily strain-hardened by milling and contain 0.2% C and about 1% O_2 before annealing. This is reduced to 0.02% C and < 0.6% O_2 by an annealing treatment in hydrogen-containing atmospheres. Martensitic and ferritic stainless steel powders are annealed after atomisation to enhance compactibility, while austenitic stainless steels are generally used in the 'as atomized' condition. Water atomized high speed tool steel powders are annealed in vacuum or hydrogen to transform martensite and retained austenite to ferrite and spheroidal carbides and also to reduce the oxygen content. Only after such annealing procedures can the powders successfuly be compacted by die pressing.

Superalloy or high performance spherical powders such as titanium or gas atomized tool steel powders are often consolidated by hot compaction in sealed claddings. Absorbed gases from atomisation and subsequent handling, such as argon, nitrogen or oxygen have to be removed, otherwise they become part of the consolidated product and can form unwanted pores. This is done by evacuation *degassing* methods, either at ambient or enhanced temperatures.

Size enlargement by heat treatment results from sintering of powder particles. In most cases this is an unintended consequence of the above heat treatments, necessitating a subsequent crushing of the annealed product agglomerates. The heat treatments are therefore performed at temperatures as low as possible, to minimize sintering and to allow for a subsequent mild comminution without significant strain hardening. In principle, size enlargement can also be carried out intentionally for grain coarsening, i.e. production of monocrystalline particles and coarser powder particles, e.g., from chemically derived powders.

FURTHER READING

C.E. Capes, Particle Size Enlargement, in *Handbook of Powder Technology*, Vol. 1, (ed. J.C. Williams and T. Allen), Amsterdam, Oxford, New York, Elsevier, 1980.

ASM Powder Metallurgy Committee, *Metals Handbook* Vol. 7, Powder Metallurgy, 9th Edn. Metals Park, OH, American Society for Metals, 1984.

K. Leschonski *et al.*, in *Grundzüge der mechanischen Verfahrenstechnik* (ed. K. Dialer, U. Onken, and K. Leschonski), Grundzüge der Verfahrenstechnik und Reaktionstechnik, 29–138, München, Wien, Carl Hanser Verlag, 1984.

5: Health and Safety

This chapter deals with the health and safety aspects related to powder properties and powder handling. Safety aspects in other fields of processing like powder shaping, sintering, secondary operations etc. are outside the scope of this book.

5.1 TOXICITY OF POWDERS

Although there are no excessive health risks associated with the majority of metal powders manufactured and used in powder metallurgy, there are some acknowledged problems with specific metal and alloy powders. To ensure the health of personnel exposed to those powders, knowledge of their toxic effects is necessary. Toxicity, which is a material's capacity to produce harmful effects to the body, is determined by the material's biochemical characteristics, the route through which it enters the body, and its dosage. The usual routes of a material to the body are inhalation, oral ingestion or absorption through the skin. Metal powders usually enter the body as airborne particulates being breathed in the workplace, because the processing and handling of metal powders usually generate dust.

The behaviour of the particulates is influenced by their size and density. Larger particles (> 12 μm) mostly lodge in the upper respiratory tract, which is cleared in a relatively short time, so that the half-life of the material deposited in this area is less than 24 h. The particles are expelled by coughing or are swallowed and either reabsorbed in the gastrointestinal tract or excreted in the faecies. Smaller particles tend to penetrate further down into the alveolar region of the lungs, with half-lives measured in months or years. If small particles of low density are inhaled, they do not remain in the lungs, because they stay suspended in the air and are, for the most part, exhaled again.

The presence of particulates in the respiratory tract may cause allergic, fibrogenic and immunological responses. The particles may also be absorbed and transported to another organ by the bloodstream.

The permissible exposure limits are regulated in the USA by the Occupational Safety and Health Administration (OSHA) in the form of threshold limit values (TLV). In the United Kingdom, the *Control of Substances Hazardous to Health Regulations* (COSHH) defined in 1988 maximum exposure limits (MEL),

Table 5.1 Threshold exposure limit values (US-standards) for metal powders

	TLV [mg m^{-3}]		STEL [mg m^{-3}]	
Beryllium (Be)	0.002	metal and compound dust	0.025	metal and compound dust
Aluminium (Al)	10.0	metal dust and aluminium oxide		
	5.0	pyro powders		
	5.0	welding fumes		
	2.0	soluble salts		
Copper (Cu)	1.0	metal dust	2.0	metal dust
Iron (Fe)	5.0	iron oxide		
	1.0	soluble salts		
Nickel (Ni)	1.0	metal dust*)		
	0.007	nickel carbonyl		
	0.1	soluble salts	0.3	soluble salts
Cobalt (Co)	0.05	metal dust	0.1	metal dust
	0.1	cobalt carbonyl		
Lead (Pb)	0.030	inorganic lead		
	0.075	organic lead		
Molybdenum (Mo)	5.0	soluble salts	10.0	soluble salts
	15.0	insoluble compounds	20.0	insoluble compounds
Tungsten (W)	1.0	soluble salts	3.0	soluble salts
	5.0	insoluble compounds	10.0	insoluble salts

*) MEL reduced in 1992 from 1.0 to 0.5 mg m^{-3}

being equivalent to TLV. These are the time-weighted average of airborne concentration for an eight hour workday. For various materials, a short-term exposure limit (STEL) is defined as a 15 minute or 30 minute time-weighted average that should not be exceeded at any time during a workday. If no short-term limit is specified, it is recommended that a figure of three times of the MEL should be used as a guide to control the short term excursions in exposure. The TLV and MEL values are recommendations and should be used as guidelines for good practice. Table 5.1 shows TLV values for several metal powders and metal compounds.

In the European Community, a *Guide to Legislation on Health and Safety in the European PM Industry* was completed in 1991. The purpose of this guide is to inform manufacturers, importers and users of metal and alloy powders operating in the EC of their statual obligations under new EC legislation, as applicable in the national regulations of the Member States. This first EPMA guide presents guidelines for the preparation of metal safety data sheets and a labelling guide for metal and alloy powders, which fall within the scope of materials classified as dangerous substances and dangerous preparations. The guide contains TLVs for a few metal powders, being the same as those already existent in USA. The EC is expected to adopt TLVs for other substances and it is planned to update the guide regularly.

For conventional iron and steel powder metallurgy the toxicity of Ni powder by inhalation is of some significance and the workers should shield themselves by masks during handling of Ni powder. Ni carbonyl is extremely toxic, indeed, being important mainly for the producer of powders. The toxicity of Be on the skin is caused exclusively by soluble salts, but chronic Be disease (berylliosis) may be caused by many types of dust, affecting 1–2% of persons with special sensitivity. The high toxicity of Pb and Cd is another health risk in powder metallurgy (e.g. during the manufacture of bearings), especially when fine powders are used. The toxicity of Co is significant in the hard metal industry. Cr, which is available nowadays as fine elemental powder but is not included in Table 5.1, may be classified as toxic similar to Ni.

The high toxicity of SiC whiskers, especially in the diameter range below 3 μm has been discovered only fairly recently. They have been found to be carcinogenic after inhalation. Great care has to be taken during processing of whisker-reinforced materials, which should take place in vented boxes.

5.2 PYROPHORICITY AND EXPLOSIVITY OF POWDERS

A hazard with many finely divided metal powders is their thermal instability in ambient air. Powders may become pyrophoric (self-igniting at normal ambient temperature) when the heat of reaction is generated more rapidly than it can be dissipated into the environment. The *pyrophoricity* of a metal powder depends directly on particle size, surface area and the heat of formation of the oxide. The most common metals in powder metallurgy which are known to be pyrophoric are: Al, Cr, Co, Cu, Fe, Ni, Ta and Ti. Pyrophoricity is not limited to elemental powders, but occurs also with compounds, like TiB_2. During the ignition process a large amount of airborne oxide particles may occur.

The danger of pyrophoricity increases with the reduction of the metal powder size. For example, aluminium powder of about 0.03 μm size, or zirconium powder of 3 μm size, are extremely pyrophoric. Porous powders of much larger particle size, like iron, nickel, or copper when prepared from their oxides at low reduction temperatures, may also be pyrophoric if the specific surface area approaches about $1\,m^2\,g^{-1}$. The presence of moisture tends to increase the pyrophoricity because the hydrogen developed during the reaction often results in lowering the ignition temperature. Chemical protective layers, like stearate or thin oxide layers mitigate the pyrophoric character of metal powders.

Pyrophoricity should be distinguished from *autogeneous ignition*, which refers to the lowest temperature at which combustion begins and continues in a metal powder when it is heated in air.

Another even more important hazard during the processing and handling of metal powders is the *explosivity* of dust clouds which may be generated in pulverizers or mixers, by powder pouring, stirring or in powder elevators. The typical ignition sources for dust clouds are flames and mechanically or electrically generated sparks. The explosivity of a metal powder depends on different physical and chemical factors as well as on atmospheric conditions.

Physical factors increasing the explosivity are the same as those that increase pyrophoricity: small particle size and high specific surface area. Fine or porous powders are usually more likely to generate homogeneously suspended dust clouds and are easier to ignite than coarser powders. Protective layers (stearate) or thin oxide films reduce the explosivity. Important for the explosivity of most metal powder dust clouds is the oxygen content of the ambient atmosphere. If the explosivity conditions are fulfilled, the destruction potential of powder dust explosions is very high. In areas where a high probability of dust explosions exists, venting systems should be installed. Critical tube or vessel systems should be protected by bursting disks or designed strong enough to bear the explosion pressure. The classification of metal powders due to their explosivity is shown in Table 5.2. The light metals and especially Zr powders are dangerous with this respect.

Table 5.2 Conditions for Explosivity of Selected Metal Powders (Dahn)

	High explosivity	Moderate explosivity	Low explosivity
Powders	Mg, Al, Zr	Cu, Fe, Mn, Zn, Sn, Si	Co, Pb, Mo
Oxygen content in ambient gas	< 3%	> 3%	> 10%
Ignition temperature	< 600°C	300 to 800°C	> 700°C and strong ignition source
Explosive limit	20 to 50 g/m³	100 to 500 g/m³	very high powder concentration

In case of fire, dry powder extinguishers must be used; the use of water is prohibited. The best prevention of metal powder explosions is a knowledge of the explosivity characteristics, the avoidance of dust cloud generation as well as of dust accumulation.

FURTHER READING

Metals Handbook 9th Edition, Vol. 7, Powder Metallurgy, 194–208, Metals Park, OH, American Society for Metals, 1984.

M. Rhodes (ed.), *Principles of Powder Technology*, Chichester, Wiley, 1990.

V.A. Tracey: 'Health and Safety in the Powder Metal Industry', *Powder Metallurgy* **35**, 1992, 2, 93–36.

European Powder Metallurgy Association: *Guide to EC Legislation on Health and Safety in the European PM Industry*, Shrewsbury, 1992.

Powder Metallurgy Environmental Regulation Guide, MPIF, Princeton, NJ, 1991.

6: Compaction and Shaping

Bulk powders are transformed into preforms of a desired shape and density by compaction or shaping. One usually distinguishes between processes with and without the application of an external pressure. Some process variants combine the shaping operation with the sintering step and end up with the sintered part instead of a preform.

6.1 PRESSURE-ASSISTED SHAPING

The pressure-assisted forming operations can be subdivided into cold and hot compaction methods. From the material's point of view, cold compaction takes place in a temperature range within which high temperature deformation mechanisms like dislocation or diffusional creep can be neglected. In most practical cases, cold compaction occurs at ambient temperature, while hot compaction is carried out at enhanced temperatures. Low melting materials like lead, however, undergo hot compaction even at room temperature. Processes, where powders are dispersed in a low melting carrier medium, e.g., injection moulding, could be associated with both cold and hot compaction.

6.1.1 Cold Compaction

6.1.1.1 Fundamental Aspects

In most applications, high densities of the green compacts are desired. Higher compact densities usually result in better green strength properties and smaller dimensional changes during sintering to the final density level, which exceeds 85–90% of the theoretical density of all powder metallurgy products, except products with functional porosity, such as filters and bearings. In a large variety of materials, even full density has to be achieved to meet the desired properties. This density level is well beyond the apparent density of normal bulk powders, which is controlled by the particle arrangement after feeding the powder into a cavity (Fig. 6.1). The apparent density is usually much lower than in a random,

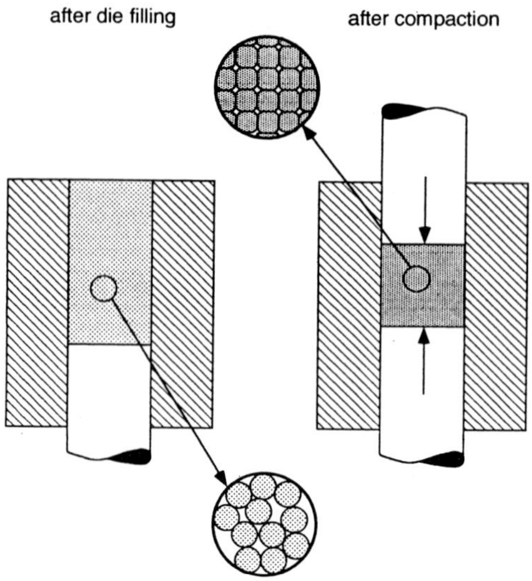

Figure 6.1 Densification during compaction by changes in particle arrangement and shape

dense particle arrangement (see section 3.4.2), as packing is hindered by particle bridging. Densification is therefore principally possible solely by rearrangement of the particles without changing their shape until a random dense packing is reached. Further densification definitely requires a shape change by plastic deformation or fragmentation of the particles.

Bridging can be overcome either by the application of a load to the particle contacts or by the reduction of the frictional forces in the particle–particle and particle–cavity wall contacts, which is possible by lubrication, e.g. through an interfacial liquid. The volume fraction of pore fluid is an important characteristic of such particle–liquid systems. It is characterised by the degree of pore saturation (DPS) which is defined as the fractional volume of pores filled by the liquid. Figure 6.2 represents the limiting cases. In pressure assisted compaction

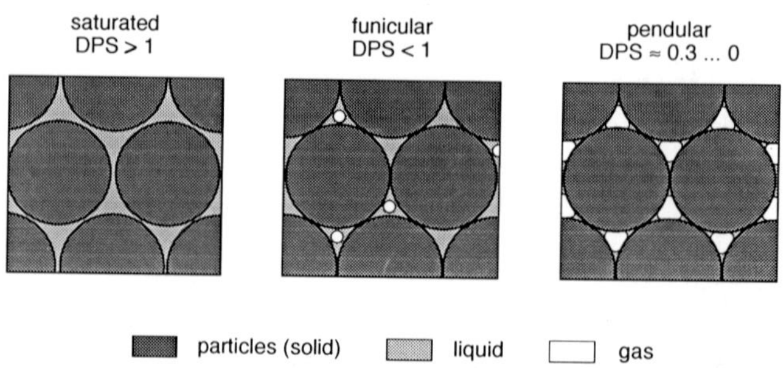

Figure 6.2 Powder systems in presence of a liquid phase (DPS: degree of pore saturation)

and shaping of powder metallurgy parts, the cases of DPS≈0 (bulk only very small amounts of lubricants), and DPS≥1 (saturated sta tion moulding feedstocks) are of relevance.

With increasing volume fraction of the liquid, the consistency o system changes from bulky over plastic and paste-like to slurry. A applied pressure contributes in a different manner to the state of stress in the liquid and the particle skeleton. Shear stresses can be resisted by the skeleton only. The normal stress is divided between the liquid and the skeleton, depending on their volume fractions.

In saturated suspensions the applied pressure is carried only by the liquid. No densification is possible until the liquid is allowed to drain. The shear components of the stress result in deformation (flow) and allow for shaping of the compact and for a particle arrangement close to the random dense packing. Material flow takes place, as long as loading results in a shear stress component.

Bulk powders exhibit a compressible behaviour. When a pressure is applied to the loose powder filling, some of the bridging particles enter a different state of aggregation. With increasing pressure, the fraction of large voids or packing defects is reduced by such restacking events. However, the strength of remaining particle bridges increases, as the particle contacts become flattened by elastic and plastic deformation, and the frictional forces are increased by cold welding and interlocking of rough particle surfaces. Thus, rearrangement and shape change of particles occur simultaneously and are interacting mechanisms. The dominance of rearrangement is limited to the beginning of densification, depending on the apparent density, particle shape, size, and deformability. Its contribution to total densification increases with increasing rigidity of the particles. It is in the order of 10% for typical metal powders, while it makes the major contribution in compaction of fine ceramic powders. The remaining interstices between the particles can be filled by plastic flow of material into the voids, or by fragmentation of the particles, allowing void filling by a secondary type of rearrangement. The required compaction pressure increases steadily with density, resulting in a state of equilibrium between external loading and the compact compliance.

For a quantitative description of the behaviour of the powder system under an externally applied pressure, different approaches exist. The first one is the classical way of regarding the compact as a homogeneous sample and looking to macroscopic measurable characteristics like pressure–density relationships, radial pressure coefficient etc. The second approach is to use continuum mechanics, often in combination with computer simulation, to predict the stress and density distribution in the compact as a response to the external forces. The third approach concerns the micromechanics of compaction, i.e. the behaviour of individual particles under the external load.

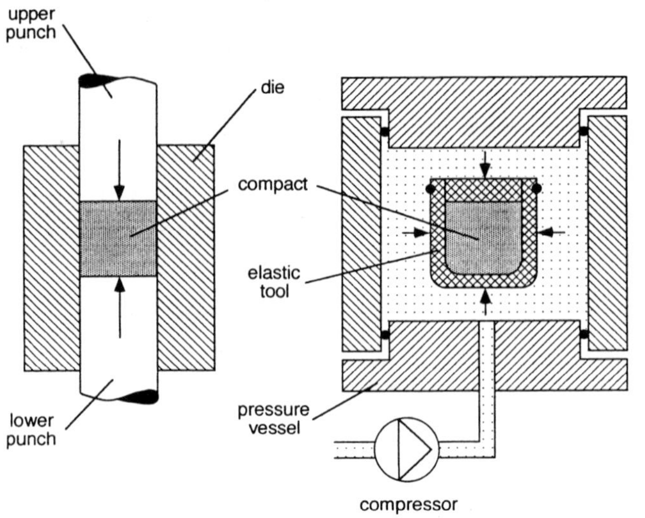

Figure 6.3 Cold pressing principles (left: axial- resp. die-pressing; right: isostatic pressing)

6.1.1.2 *Cold Pressing*

6.1.1.2.1 Principal Aspects

Cold pressing is the most important compaction method in powder metallurgy. It starts from bulk powders containing very small amounts and sometimes even no lubricant or binder additions. One usually distinguishes between axial (die) and isostatic pressing (Fig. 6.3). In axial pressing, the powder is compacted in rigid dies by axially loaded punches. The axial compaction pressure P_{ax}, which is defined as the punch load divided by the punch face area is the main process variable. In isostatic pressing, the powder is sealed in an elastic mould and exerted to the hydrostatic pressure P of a liquid pressure medium.

The compaction behaviour of powders, expressed by their overall pressure-density relations is shown in Fig. 6.4. The controlling parameters are mainly the particle size and the ability for plastic deformation. Densification starts from the apparent density, which is similar for the coarse iron and alumina powder, and which is not too far away from random dense packing for both of them. The fine powders exhibit a significantly lower starting density, due to hindered packing. With increasing compaction pressure, the average density of the compacts increases. The slope of the curves differs significantly for the ductile metal powder and the non ductile alumina. The density increase for a given pressure increment is always higher for the ductile material. In all cases, the curves asymptotically approach a final density level, which is below the theoretical density of the materials. The density difference between the fine and the

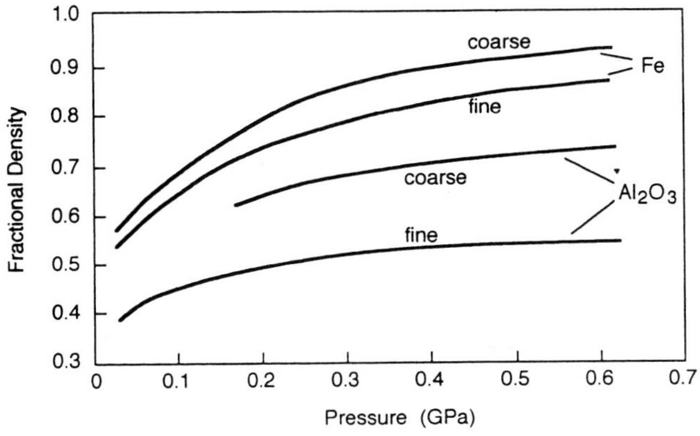

Figure 6.4 Typical pressure-density relations in compaction of metal and ceramic powders (after Fischmeister)

corresponding coarse powder is not reduced with increasing pressure. Much higher pressures are necessary for the fine powders to reach the same density as the coarse ones. Figure 6.4 exactly represents the microscopic situation: the brittle alumina powder can be compacted only to a density relative to that of its random dense packing, while the maximum densities of the ductile iron powder are much higher, which can only be explained by extensive filling of interparticle voids through plastic deformation. Interparticle friction and bridging effects increase with decreasing particle size. With increasing density, the resistance to densification increases rapidly.

It is common, to correlate this behaviour with a three-stage compaction process with different controlling mechanisms. These are: particle rearrangement in the first, elastic-plastic deformation in the second, and work hardening (ductile powders) or fragmentation (brittle powders) in the final stage. As already explained, these stages are overlapping with strong interactions between the different mechanisms.

A large number of mathematical expressions for the description of pressure–density relations have been published over the years. Most of them are based on fitting of experimentally derived curves. An example is the well known Konopicky equation, which gives a linear interdependence of compaction pressure P and the logarithm of the fractional porosity U, as it is found for most metal powders over a more or less extended pressure interval (see Fig. 6.5):

$$P = \frac{1}{C_1} \ln \left(\frac{U_0}{U} \right) \tag{6.1}$$

C_1 is a constant and U_0 is the extrapolated fractional porosity for $P = 0$.

In some cases, an attempt has been made to correlate the parameters of the resulting equations with powder and materials properties, e.g. the yield

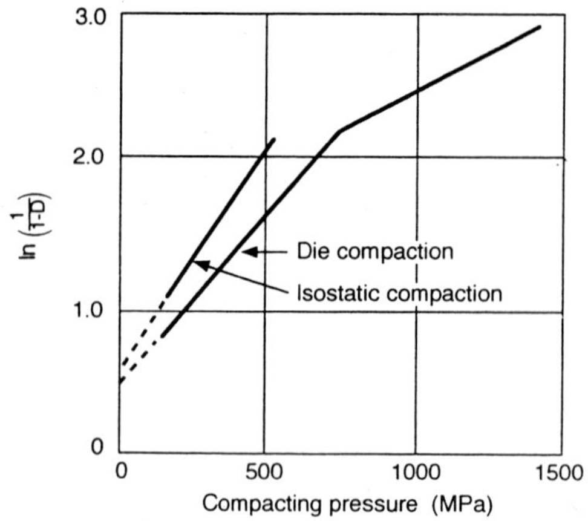

Figure 6.5 Pressure-density relationship of iron powders, ordinate adapted to equ. (6.1) (U = 1 – D; after Morgan and Sands)

strength, and thus to give them a physical meaning. Its significance, however is doubtful, as each parameter may depend on several powder properties.

Another approach is to put a true physical model behind the equations. Equation (6.4), for example, is of the same type as solutions for plastic yielding of a hollow sphere of radius R_o with a central void of radius R_i under a hydrostatic pressure P, as derived by Torre:

$$P = 2\,\sigma_y \ln\left(\frac{R_o}{R_i}\right) \tag{6.2}$$

σ_y is the yield strength of the material. If the compact is regarded as a hollow sphere, the void representing the total pore space, and the shell the solid fraction, U equals $(R_i/R_o)^3$ and equations (6.2) comes to:

$$P = \frac{2}{3}\,\sigma_y \ln\left(\frac{1}{U}\right) \tag{6.3}$$

Equation (6.3) is obviously invalid for $P = 0$, where it predicts a porosity of 100% instead of the practically observed apparent density. Taking into account U_0, Torre derived a modified solution with:

$$P = \frac{2}{3}\,C_2\,\sigma_y \ln\left(\frac{U_0}{U}\right) \tag{6.4}$$

Comparison of equations (6.1) and (6.4) indicates a linear relationship between $(1/C_1)$ and the yield strength σy. This proportionality was shown by Heckel to be:

$$\frac{1}{C_1} = 3\,\sigma_y \tag{6.5}$$

A semi-logarithmic plot of porosity versus compaction pressure should result in a straight line with a slope according to equations (6.1) and (6.5). Figure 6.5 gives an example for the behaviour of iron powder under isostatic and axial compaction. Two linear relationships are visible. This behaviour was also observed for a variety of other metals. The slope of the first straight line depends on the yield strength, approximately as predicted. The decrease in slope in the high density range was attributed to work hardening, which is not included in the model.

The present state of physical modelling is represented by a microscopic model developed by Fischmeister and Arzt. This model regards an arrangement of monosize spheres under an external hydrostatic pressure (Fig. 6.6). Compaction starts from a random dense packing (D = 0.63) of these spheres. This arrangement excludes restacking of particles and allows the calculation of the average particle contact number (coordination number) Z_0 at the beginning of densification. From the so called random distribution function Z_0 follows to be 7.3. With increasing pressure, the particle contacts undergo plastic deformation. The distancies between the particle centres decrease, resulting in increasing co-ordination numbers and density. When reaching theoretical density, the original spherical particles are deformed to polyhedra, as represented in Fig. 6.6. The average contact number, Z, and the average contact area, a, are parameters, which only depend on the actual and initial density, D and D_0 respectively, and on the particle radius, R. Fig. 6.7 shows the dependence of the average fractional contact area $(Z\,a)/(4\,\pi\,R^2)$ per particle on the density as derived by different authors in comparison with experimental measurements. Up to 90% of TD, the predictions are in fairly good agreement with the measured values. The external pressure, P, is assumed to be balanced by the contact forces, F. With the knowledge of $(Z\,a)$, the multiparticle problem can then be reduced to a local problem, regarding the deformation of one representative contact plane (Fig. 6.6). Assuming a homogeneous contact stress P_{eff} and a constant yield strength σ_y, F comes approximately to:

$$F = 3\;a\,\sigma_y \tag{6.6}$$

Taking into account work hardening by a hardening function $H(D)$, Fischmeister and Arzt derived the following pressure–density relation:

$$P(D) = 3\,\sigma_y\,\{1 + H(D)\}\,\frac{a(D)\;Z(D)}{4\,\pi\,R^2}\;D \tag{6.7}$$

The increasing resistance of a compact against further densification is thus separated into two terms: the material's deformation behaviour, which can simply be measured in a uniaxial compression test, and a (known) geometrical hardening function.

Pressure–density relations provide information obtained from the integral volume of a compact. They give a good description for isostatic compaction of simple shaped parts, where a homogeneous state of stress and density can be

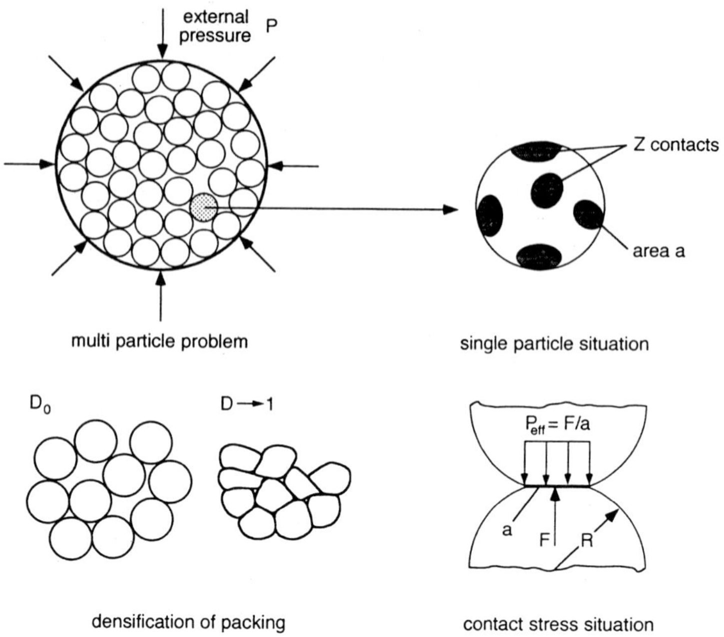

Figure 6.6 Microscopic modelling of powder compaction (Fischmeister and Arzt)

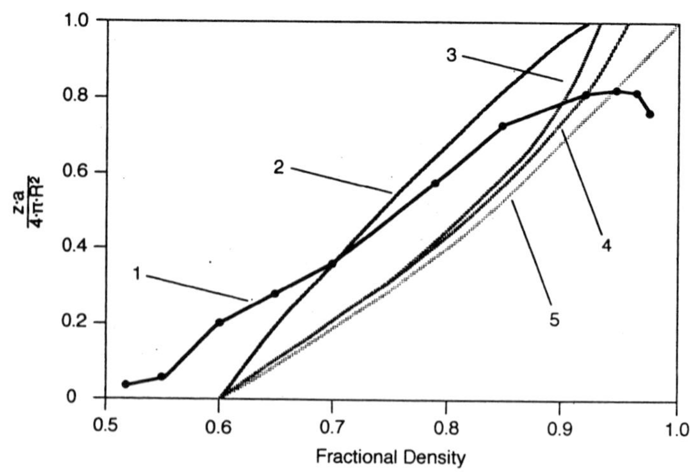

Figure 6.7 Normalised contact area. 1, experiment (Pb spheres, Agniel); 2, theory (short distance material transport, relevant in HIP-modelling, see section 6.1.2.1; Arzt); 3,4, theory (long distance material transport, different approximations; Arzt and Fischmeister); 5, theory (approximation, Helle et al)

expected. In axial pressing, stress and density vary locally, as powders differ fundamentally from liquids in transmitting pressure. Fig. 6.8 illustrates the situation for a cylindrical compact. An external load, F_u, at the upper punch is balanced by a force, F_l at the lower punch and a force F_d from the die, which results from powder-die wall friction. Assuming the axial pressure to be constant over the cylindrical cross section, a differential volume element of height dz is kept in equilibrium by the forces $a\{P_{ax}(z)-P_{ax}(z+dz)\}$ and $2\pi R dz \tau(z)$, as shown in Fig. 6.8. The frictional shear stress τ is the product of the friction coefficient μ and the radial pressure P_{rad}. The latter is not externally applied like in isostatic compaction, but depends on the powder fluidity. For a first approximation, P_{rad} can be taken as a constant fraction $k\,P_{ax}$ of the axial pressure. Integration along the axial coordinate z results in:

$$P_{ax}(z) = P_u \exp\{(-2\mu\ k\ z)/R\} \tag{6.8}$$

$$P_{ax}(H) = P_u \exp\{(-2\mu\ k\ H)/R\} \tag{6.9}$$

These relations indicate a decrease in axial and radial pressure with increasing distance to the face of the loaded upper punch, which should result in a decreasing compact density in the axial direction, as is always observed in practice. They also explain the importance of the height/diameter ratio $H/2R$ of the compact, which controls the minimum local pressure. The pressure ratio factor k depends on the actual compact properties and takes always a value of <1. It increases with increasing pressure up to a typical level of 0.5 for metal powders.

Thus, in axial compaction the radial pressure is substantially lower than the axial pressure. The stress tensor can therefore not simply be described by one single invariant as in isostatic pressing, where the hydrostatic pressure $P=\sigma_1=\sigma_2=\sigma_3=\sigma_m$ completely determines the volume strain (density change). In axial pressing, at least one second invariant is required to characterise the stress tensor, which is responsible for the shear strain (shape change). This second invariant is related to the so called deviatoric shear stress τ_d. The density of any volume element of the compact depends on both invariants σ_m and τ_d of its stress tensor. In terms of continuum mechanics, compaction is

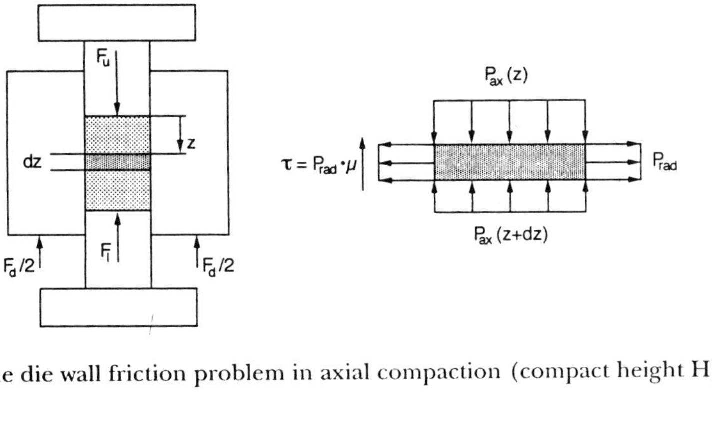

Figure 6.8 The die wall friction problem in axial compaction (compact height H)

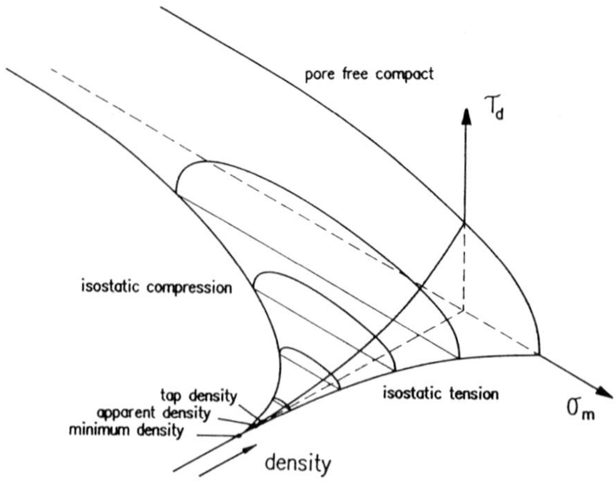

Figure 6.9 The yield-(flow) surface of metal powder compacts in the σm–τd-density space (after Ernst)

characterised by the yield locus and a flow rule. The yield locus is the state of stress σ_m, τ_d, at which a volume element begins to undergo irreversible (plastic) deformation. Figure 6.9 shows a qualitative illustration of the yield locus of a metal powder. The yield loci form asymmetric elliptical curves with increasing radii for increasing compact density. This means increasing resistance against irreversible deformation under all types of stress, hydrostatic tension, and compression, as well as shear. Approaching theoretical density, the elliptical character changes to a parabolic one, taking into account that dense metals will no longer be plastically deformed under pure hydrostatic compressive stress. Compaction starts at the tip of the three dimensional yield surface, representing the minimum density, where the particles are in contact without any external forces. Gravity and vibration already result in some density increase (apparent and tap density respectively). This is the state, where pressure assisted compaction has its onset.

If a porous compact with ideal elastic–plastic material behaviour is loaded by a combined stress σ_m, τ_d, it is elastically deformed until the stress reaches the yield surface, where the stress remains during further deformation. The corresponding density change can be positive or negative, depending on the state of stress. Densification requires hydrostatic compressive stress larger than the yield limit at the actual density. For strain hardening materials, the yield surface can only predict the start of irreversible deformation. Plastic deformation takes place in a yield 'space' beyond the yield surface. If the yield surface, resp., space is known (at least in the interesting σ_m, τ_d range), finite element simulations of compaction under any state of stress become possible. However, this requires comprehensive experimental measurements, as no physical model which could describe the behaviour of powder compacts under all types of multiaxial stresses is available at present.

6.1.1.2.2 Practice of Axial Cold Pressing

Compaction Sequence and Tooling

Axial pressing is by far the most important practical forming method. The powder is compacted between the punch faces and the die walls, which undergo only very limited elastic deformation. The compacts can, therefore, be fabricated to very close geometrical tolerances. The compaction sequence consisting of die filling, compaction, and ejection of the compact can be carried out in both mechanical or hydraulic presses or as mixed mode presses at high production rates. Axial powder pressing is therefore a very economic method for the mass production of precision parts.

The density gradients explained above increase with increasing height/diameter ratio or, more generally, with increasing ratio of the mantle surface and the cross section of the compact. The problem can be significantly reduced if the compaction pressure is allowed to act on both punches of the system. This so-called double-action pressing is schematically compared with single-action pressing in Fig. 6.10. In single-action pressing, the density decreases from the top to the bottom of the compact according to the pressure gradient derived in equation 6.8. In double-action pressing, the axial pressure is symmetrical to the central cross section of the compact, the so-called neutral axis or plane. The minimum pressure and density are found in this plane by:

$$P_{ax} = P_u \exp \{(-\mu k\ H)/R\} \qquad (6.10)$$

Besides the effect of formally reducing the height/diameter ratio by a factor of 2 and thus increasing significantly the minimum pressure, this process has another advantage. The symmetrical density distribution prevents distortion of the part, which can occur as a result of differential sintering (see section 7.2.3.2) of more and less dense layers in compacts with unsymmetrical density distribution. Double-action pressing is therefore applied for the majority of parts, even if the height/diameter ratio is not very large.

Double-action pressing can be realised by several tooling systems as shown in Fig. 6.11. With the double-action or ejection tooling system, the compaction force is applied to the upper and lower punch simultaneously, with a stationary die. The double-action effect can also be achieved with a floating die tooling system. The die is mounted on a yielding mechanism, consisting of springs, or hydraulic or pneumatic auxiliary cylinders. The yielding mechanism balances the weight of the die and allows for free floating, when additional forces are acting on the die. As the upper (loaded) punch enters the die, frictional forces develop and cause the die to move downwards. The force on the lower stationary punch increases until the frictional forces from both punches come into equilibrium. The withdrawal tooling system utilizes a loaded upper and a stationary lower punch in combination with a moving die. The die movement is hydraulically controlled in contrast to the friction controlled movement of the floating die system. The ejection of the pressed compact is carried out by

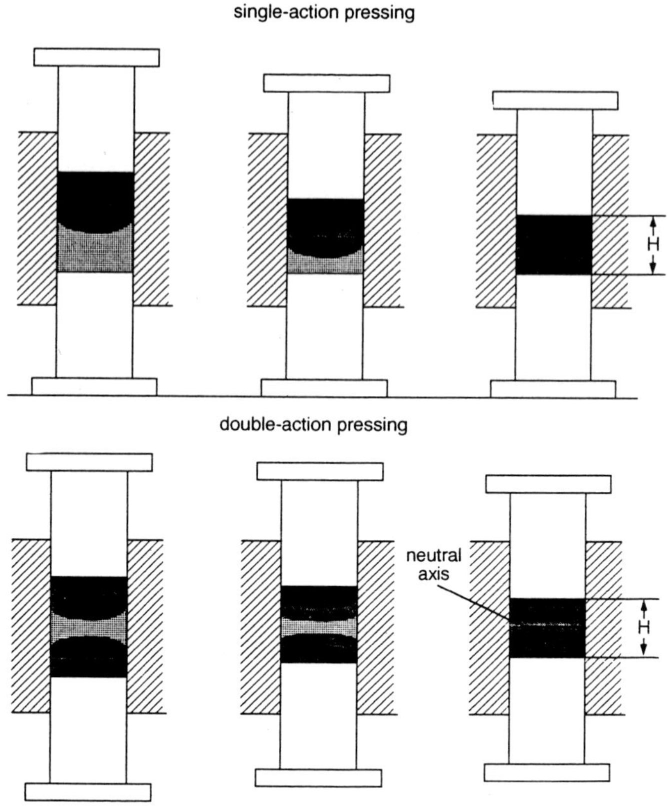

Figure 6.10 Axial cold pressing: single-action vs. double-action pressing

upward movement of the lower punch in the double action and floating die systems, and by withdrawing the die in the withdrawal system.

Practical reasons for selecting one or another tooling system depend on the capacity of the production press and the compact geometry. An obvious example is the load capability of the moving press rams, which both, equally, have to provide the full compaction force in the double-action system. Whereas in the floating die and withdrawal systems, the load capacity of the lower ram has only to be sufficient to provide the ejection and withdrawing force respectively.

However, the decision to use a specific tooling system is governed mainly by the complexity of the compact geometry. This problem is not included in the simple schematics of Fig. 6.11, which only covers single-level parts without any through-holes. Because of interparticle friction, material transfer perpendicular to the pressing direction is hardly possible. This results in the necessity for a constant compression ratio in each part of the compact, as shown schematically in Fig. 6.12. For compacts with more than one thickness level, this requirement can only be satisfied by dividing the lower punch into independently moving

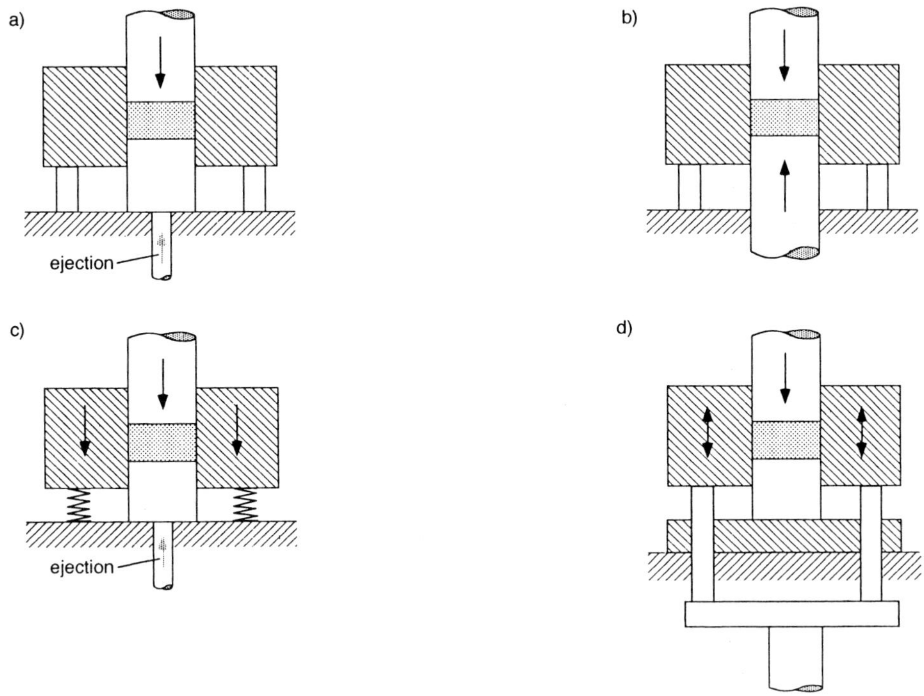

Figure 6.11 Axial cold pressing -tooling principles-. a) single action pressing, b) ejection tooling system; c) floating die tooling system; d) withdrawal tooling system

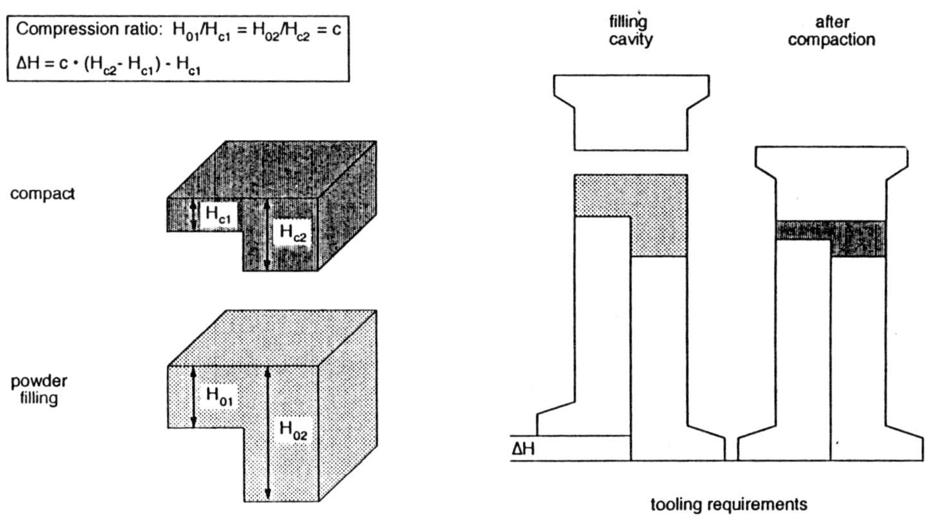

Figure 6.12 Tooling requirements for class III and IV parts: splitting of lower punches

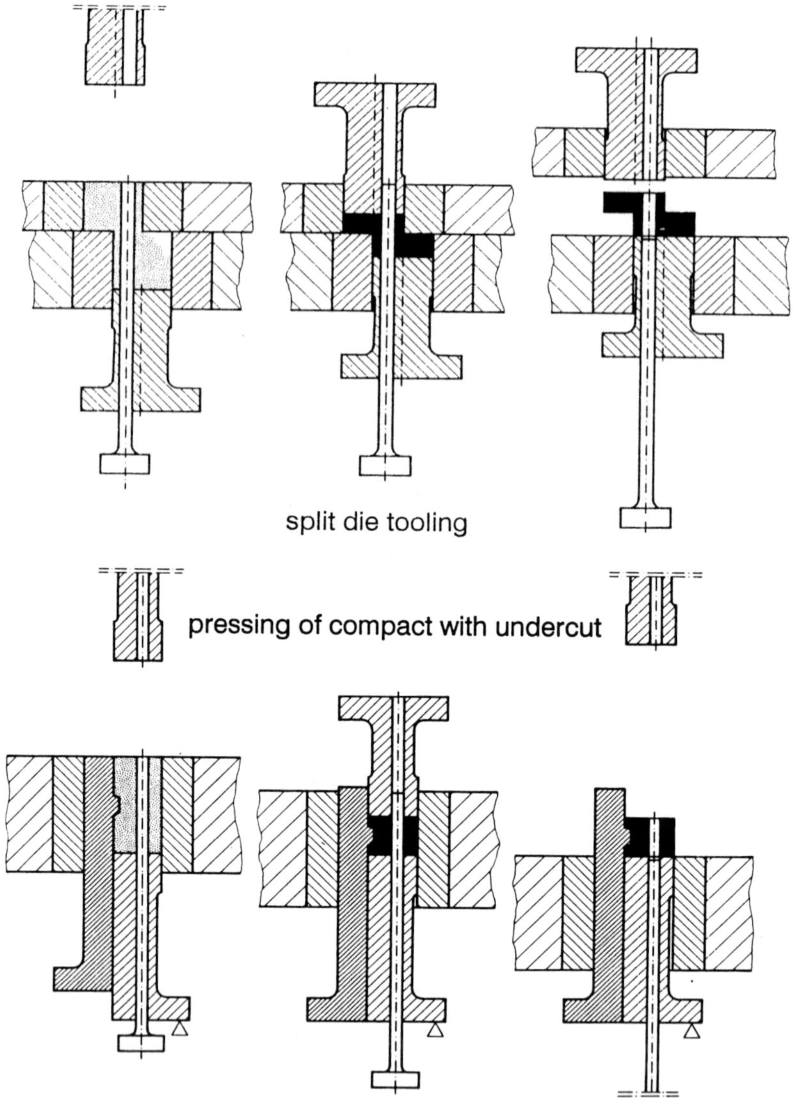

parts for every thickness level. Therefore the tooling for multi-level parts can become rather complicated and expensive, limiting the choice for a tooling system.

This has resulted in a common part classification. According to the Metal Powder Industries Federation, parts fabricated by axial pressing are classified with respect to their complexity:

Class I parts are single-level parts fabricated by single action pressing up to a thickness of 7.5mm.

Class II parts are single-level parts of any thickness fabricated by double-action pressing.

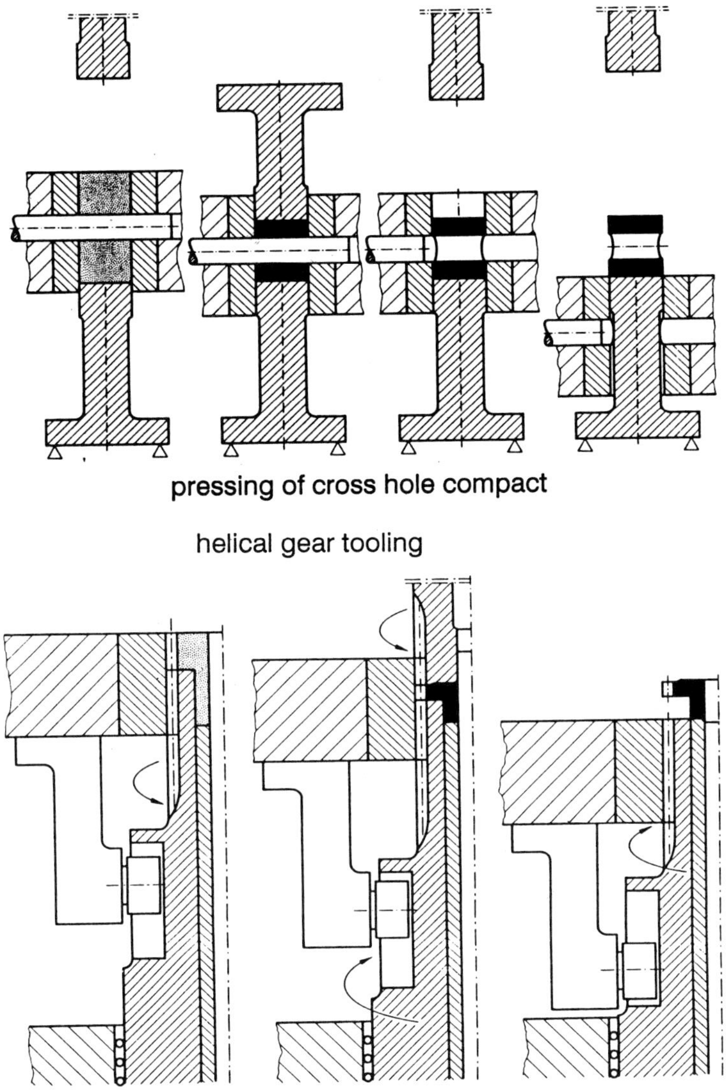

Figure 6.13 (above and opposite) Fabrication of critical geometries by axial pressing (for explanation see text) (Beiss)

Class III parts are two-level parts of any thickness fabricated by double-action pressing.
Class IV parts are multi-level parts of any thickness fabricated by double-action pressing.

Single-action tooling systems are limited to production of Class I, double-action tooling systems to Class I and II parts. Class III and IV parts are usually produced by withdrawal tooling systems.

A common limitation in part complexity is that features perpendicular to the pressing direction cannot be compacted and stripped, e.g. cross holes and threads. With very sophisticated tooling even these limitations can partially be overcome. Some examples are shown in Fig. 6.13: The first is the split die system, which enables the compaction of parts with asymmetric upper and lower sections in the pressing direction. Such parts could not be ejected from a closed die tooling system. The second example shows the pressing of a part with an undercut. Even cross holes can be made if they are located in the neutral axis, as shown in the third example. The fourth example shows a tool set for axial pressing of helical gears, which can be produced up to helix angles of 25°.

A complete pressing tool system can consist of a variety of upper and lower punches, core rods for forming through-holes, the die, and a tooling system assembly, on which the components are mounted. The use of such assemblies increases production flexibility, because the time for changing complete tooling systems is relatively short.

Compaction Presses

Operation of the tools requires special mechanical or hydraulic powder presses. Cam-driven mechanical presses are limited to capacities of about 1 MN. The capacity of eccentric-drive mechanical presses extends from ~ 7–7.5 MN. Standard hydraulic compacting presses are available in a capacity range from 0.3–20 MN. The production rate of mechanical presses is generally higher than that of hydraulic presses by a factor of 1.5–5. Their maximum stroke rates are ~ 3000 h^{-1} compared to 650 h^{-1} for hydraulic presses. Hydraulic systems are advantageous for long components. The maximum depth of powder fill is ~ 200 mm in standard mechanical and 400 mm in hydraulic presses.

Compaction in mechanical presses is always controlled by part length. Variations in filling density result in variations of compaction pressure and compact density. Overfilling can lead to severe tool damage. With hydraulic presses, part-length control as well as compaction pressure control is possible. In the latter mode, variations in filling density result in variations of part length, while compaction pressure and density remain uneffected. Modern systems use computer controlled servo-hydraulics, resulting in very precise punch movements and the possibility of controlling the pressing sequence of multiple punch tooling systems very effectively. Only by this development has it become possible to overcome common shape limits, as shown for example in Fig. 6.13.

Limiting Factors in Axial Pressing

Besides geometry, limitations in axial die pressing arise from the filling behaviour of common metal powders, the load capacity of the presses, and from the overall process economics. Usually filling becomes non-reproducible if the width of the die cavities falls below 2 mm. This results in increasing density gradients in thin sectioned compacts. The minimum weight, e.g. of iron and steel parts, is ~ 0.2 g for the same reason. At a common compaction pressure of 600 MPa, steel parts with a cross-section of up to ~ 300 cm^2 can be produced in the standard range of press capacity. The maximum weight of these compacts is ~ 2–5 kg. For bigger parts, other fabrication technologies are often more competitive than powder metallurgy. Compaction pressures beyond 600 MPa are feasible and sometimes used in practise, but give rise to significantly increased tooling costs. Depending on the shape complexity, axial pressing requires large quantity output to compensate for the tooling costs and thus becomes competitive mainly in mass production.

6.1.1.2.3 Practice of Isostatic Cold Pressing

Pressing Equipment

Isostatic cold pressing is performed in practice either by the so-called 'wet bag' or the 'dry bag' method. With wet bag tooling, the flexible bag of the mould is filled outside the pressure vessel, sealed and finally placed within the vessel. The bag is completely surrounded by the pressure-exerting fluid. With dry bag tooling the bag is permanently sealed within the pressure vessel and powder filling occurs without removing it from the vessel. Wet bag tooling is therefore more versatile but requires (mostly manual) handling of the filled bag. Dry bag tooling is more suitable for mass production, but is more limited in shape flexibility.

Usually, liquids are used for exerting pressure. Criteria for selecting a pressure medium are compatibility with the tool, vessel and pumping system, and its compressibility, price and availability. In practice, special oils or water with anti-corrosive and lubricating additives are generally used. With increasing compaction pressure these fluids deviate significantly from ideal incompressible behaviour. Even water, often thought to behave incompressibly, has a stored energy of 4.5 kJ/kg at a compaction pressure of 200 MPa. Under such conditions, the energy content in a 1 m^3 vessel is equivalent to 1 kg of high explosive, such as TNT. This is not very much, if compared, for example with the energy content of a standard 50 litre container of hydrogen, which is equivalent to 15 kg TNT. Nevertheless, safety aspects play an important role in vessel design and operation, because the high pressure results in extreme endloads on the covers of the vessels. The type of loading is cyclic, according to the pressurisation–depressurisation sequences of press operation. Various vessel designs are in practical use, including simple forged monolithic vessels, double- or multiple-layer forged vessels and prestressed designs, e.g., wire wound vessels. The end-load problem is solved by threaded closures or threadless methods, like pin and yoke closures. Large high pressure units are often installed within restricted areas or bunkers, or are covered by special energy absorption protection shields. All systems have to be inspected periodically by non-destructive testing methods to prevent failure caused by fatigue-initiated crack nucleation.

Tooling

Tooling for isostatic pressing (Fig. 6.14) differs considerably from axial pressing tools. In the simplest case, a tool or mould consists of a flexible (elastomeric) mould with the shape of the desired compact, the dimensions of which are enlarged by the compaction ratio which is defined as the relation between the compact density and the fill density of the powder. When the sealed mould is exposed to pressure the complete free surface of the compact is under hydrostatic loading which causes isotropic densification of the compact. In 'dry bag' systems and, more generally, for the fabrication of hollow parts like tubes, tooling can also include rigid components, such as inside mandrels or an outside form support. Depending on the tool design, local deviations from isotropic densification occur. In dry bag compaction, for example, often only the radial pressure component is allowed to act on compacts with high height/diameter ratios. The compact surfaces formed by the rigid parts of the tooling can be produced with surface qualities and geometrical tolerances comparable to those reached in axial die pressing. The surfaces formed by flexible moulds exhibit larger tolerances and poor surface qualities.

Elastomers such as natural and synthetic rubber, silicone rubber, PVC and polyurethane are suitable bag materials. Criteria for material selection are compatibility with the pressure fluid, the required stability of the mould (thick or thin wall) and the quantity of parts to be produced within the mould lifetime. PVC and rubber moulds, which can be produced by simple dipping methods, are often preferred for single-use tooling. Polyurethane is the most common

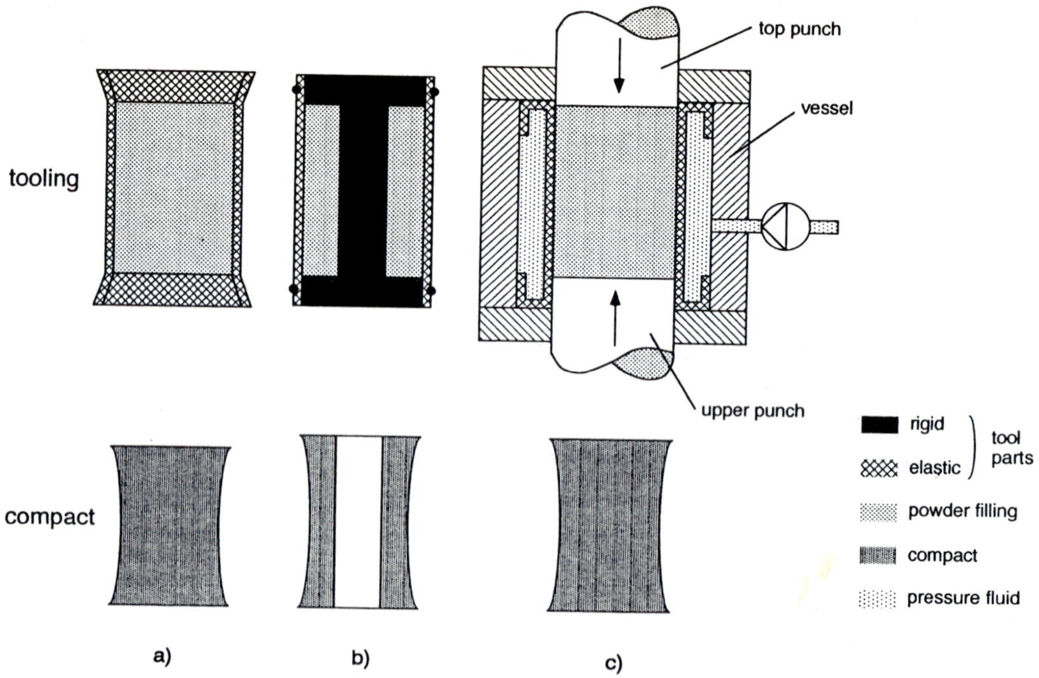

Figure 6.14 Examples for tooling in isostatic cold pressing: a), b) wet bag; c) dry bag tooling

bag material for multi-use toolings. Polyurethane moulds are relatively expensive, because of the complicated casting method required for mould production.

Advantages, Disadvantages, Practical Applications

Isostatic pressing offers several advantages compared with axial die pressing. The first is the shape capability, which is much less restricted in isostatic pressing. Because of the absence (or at least strong reduction) of die wall friction effects, thin walled compacts with large height/diameter ratios can be produced. As the hydrostatic compaction pressure is acting homogeneously within the pressure fluid, the only size limitations for the compact are the dimensions of the pressure vessel. The largest systems in operation have dimensions of 2 m in diameter and 3–4 m in height and a pressure capacity of 250 MPa. The maximum length of isostatically compacted parts published so far is 5 m for ceramic tubes.

The second advantage of isostatic pressing is the very homogeneous density distribution within the compacts, which again is related to the absence of wall friction. This homogeneous density distribution explains the fact that in most cases the green densities obtained by isostatic pressing exceed the overall density obtained by axial die pressing having the same nominal compaction pressure. It is also responsible for a greater green strength and a more reproducible sintering shrinkage. However, very small density gradients may occur towards the centre of compacts due to interparticle friction. This tendency is more pronounced at low compaction pressures. The third main advantage in isostatic pressing is the absence of an ejection step. This is important mainly for compacts of low green strength, e.g. ceramics. The compaction pressure for such materials in die pressing is limited by ejection shear stresses, which cause failure when the compacts leave the die. These shear stresses are proportional to the residual radial strain of the die after deloading, which increases with increasing compaction pressure. Thus, the maximum safe compaction pressure in die pressing of such materials is relatively low — of the order of 100 MPa for ceramic powders, while no principal limitation exists in isostatic pressing. Other advantages are the possibility of reducing or even avoiding the use of lubricants, as a consequence of the absence of die wall friction, and the relatively low tooling costs, which allow for an economic production of small numbers of parts or even of single parts.

Disadvantages of isostatic pressing are the greater dimensional tolerances, at least of the part sections which are formed by the flexible mould, and the lower production rates, which are governed by the necessity of closing and opening the pressure vessel during every pressing cycle. This problem has been overcome by the dry bag method, which is applied, for example, in the production of spark plug insulators. For this application, a battery of isostatic press cavities is operated and controlled by one pressurising system. Production rates of 2800 pieces per hour can be reached in an eight cavity unit at compaction pressures between 30 and 150 MPa. The production rate for ceramic grinding balls up to 40 mm in diameter can even reach 10,000 pieces per hour with such a system. However, the shape capability of the dry bag method is not as versatile as with the wet bag method.

Cold isostatic pressing is also used for the manufacture of semi-finished products as well as net or near-net shaped parts from metal and ceramic powders. Semi-finished parts include blocks and tubes made from hardmetal and ceramic powders, which are subsequently shaped by green machining (indirect shaping — see Table 10.1). Other semi-finished products are blocks of high-speed steels or aluminium alloys, which are subsequently consolidated by hot isostatic pressing or hot extrusion respectively. Shaped articles include for example, filter elements, tubes and crucibles, balls for application in bearings or as milling media, spark plug insulators, high-speed steel tool preforms, and special structural parts with threads or undercuts, which cannot be produced by die compaction.

6.1.1.3 Powder Injection Moulding (PIM)

Powder injection moulding is a developing shaping technology in powder metallurgy and ceramics, even though first applications in ceramics date back to the late 1920s. In powder metallurgy the term: metal injection moulding (MIM) is often used. The interest in this technology has accelerated over the last 15 years, as the demands of parts design approached the shape and property limits of conventional powder processing routes. The principles of the process are well known from the manufacture of parts from polymeric (thermoplastic) materials, where injection moulding is one of the most important forming methods which is applied in large scale production.

6.1.1.3.1 Principal Aspects

Processing Principles

A flow chart with the main steps of the complete process is given in Fig. 6.15. The moulding operation requires a preconditioning step in which the powder is transformed into a mouldable state, the so-called feedstock. This is done by mixing the powder with (mostly) organic binders, which make the powder flow under moulding conditions, but also provide good shape stability and green strength of the moulded parts. These requirements are satisfied by changing the rheological behaviour during the moulding operation.

Feedstock Rheology

The forming operation itself is controlled by the rheological behaviour of the feedstock at the moulding temperature which depends on the rheology of the binder and the volume content and characteristics of the powder particles. The rheological behaviour can be characterised by shear tests (Fig. 6.16), where the system is sheared between two parallel plates with a relative velocity v under a normal load F_n.

At the moulding temperature feedstocks exhibit only a slight dependence on the normal load F_n. Such systems exhibit a linear velocity profile $v(y)$ in a shear test.

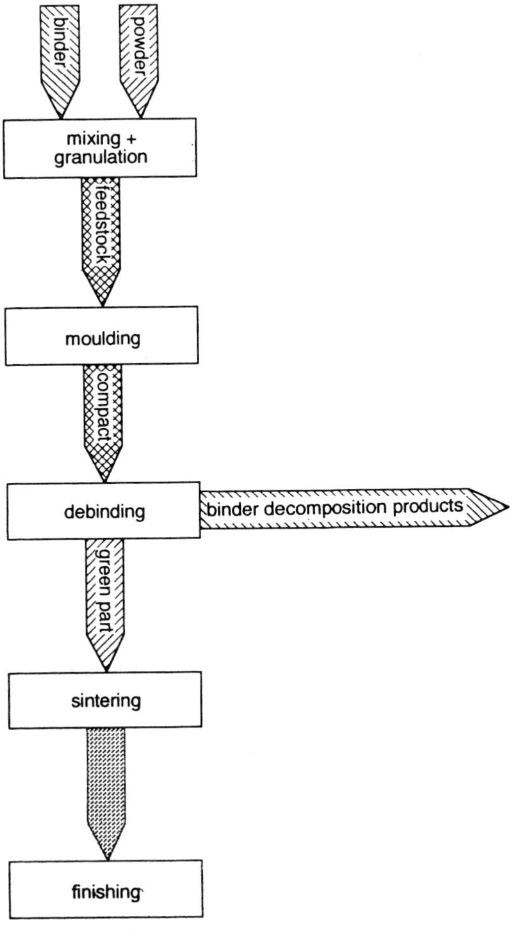

Figure 6.15 The powder injection moulding (PIM) process

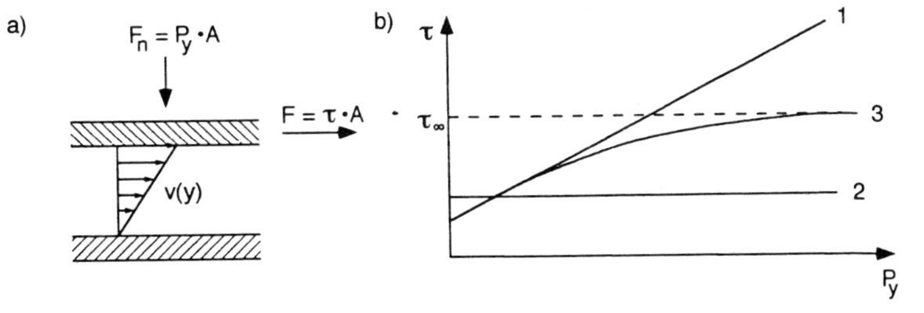

Figure 6.16 Rheological behaviour of powder systems: a) shear test; b) influence of normal pressure P_y on shear stress τ for bulk powders (1), incompressible fluids (2), and pastes (3) (after Buggisch)

The shear stress τ increases monotonically with the shear strain rate dv/dy resp. $\dot{\gamma}$, as given in Fig. 6.17. τ and $\dot{\gamma}$ are related by the viscosity η, with

$$\tau = \eta \, \dot{\gamma} \tag{6.11}$$

η is constant for a Newtonian fluid like water. For pseudoplastic or shear-thinning fluids, the viscosity decreases with increasing shear strain rate. Fluids with increasing viscosity are called dilatant or shear thickening fluids. Shear-thinning and thickening results from shear-induced structural changes in the fluid. Most polymer melts, polymer solutions and suspensions belong to the pseudoplastic fluids. Some fluids, e.g., slurries containing attracting molecules, require a yield stress τ_y to initiate flow (see Fig. 6.17). Such fluids are called Bingham Fluids, if the plastic viscosity η_p is independent of the shear rate

$$\tau - \tau_y = \eta_p \, \dot{\gamma} \tag{6.12}$$

Fluids with a yield point may also exhibit pseudoplastic or dilatant behaviour.

Non-Newtonian behaviour has several consequences in respect of important technical applications, e.g. the velocity profiles in circular tubes, shown in Fig. 6.18. Pseudoplastic fluids exhibit small velocity gradients in the centre of a tube, but very large velocity gradients near the tube wall. The viscosity close to the wall can be some orders of magnitude lower than in the tube centre in a pronounced pseudoplastic behaviour. In a Bingham fluid, the velocity even becomes constant in the centre plane, where the axial pressure gradient is not high enought to result in a shear stress $\tau > \tau y$ (plug flow). Thus, shearing or relative movement of different fluid layers occurs mainly or exclusively in a thin near-wall layer. In highly concentrated suspensions and with smooth tube walls the wall adhesion is no longer strong enough to keep the suspension velocity

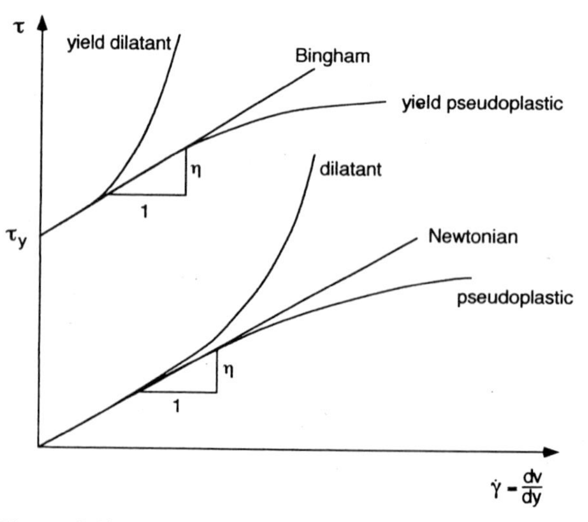

Figure 6.17 Rheology of powder suspensions (see text)

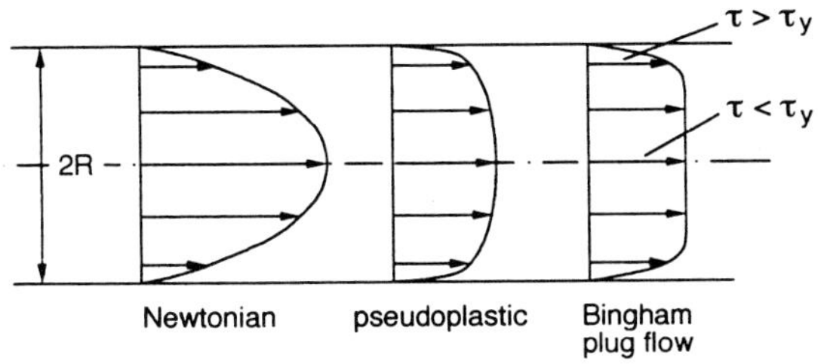

Figure 6.18 Velocity profiles of Newtonian and non-Newtonian fluids in circular tubes (after Buggisch)

zero at the wall. The fluid therefore slips along the wall with a velocity v_s (slippage flow), enhancing the volume flow $\dot{V}$ by an additional term $\dot{V}_s$ when

$$\dot{V}_s = \pi R^2 v_s \tag{6.13}$$

Such differences in friction and shear stresses along tool or runner cross sections can be a source of demixing effects in moulding feedstocks.

The viscosity of a suspension is controlled by the dispersant, the solid volume fraction of particles, and the particle characteristics. Even with a Newtonian behaviour of the dispersant it is normally highly dependent on the shear rate. The relative viscosity η_r, which is defined as the viscosity of the suspension η_s divided by the viscosity of the dispersant η_d, is predicted by a variety of models.

Powder injection moulding (PIM) feedstocks have been modelled successfully on the basis of the concept of an immobile liquid — assuming the liquid required to fill the space between the particles to be immobile. The viscosity of the suspension is then determined only by the mobile fraction of liquid. This leads to a maximum solids loading $c_{v,max}$, where all of the liquid becomes immobile and the viscosity approaches infinity. The relative viscosity η_r of a concentrated suspension like a PIM feedstock therefore depends on the relative solids loading $c_{v,rel} = (c_v / c_{v,max})$. An appropriate expression is given in equation 6.14

$$\eta_r = K(1 - c_{v,rel})^{-n} \tag{6.14}$$

K is a coefficient containing the particle characteristics and the shear rate. It is near unity for coarse particles and increases with decreasing particle size and decreasing shear rate whilst n is an exponent near a value of two.

The maximum solids loading is found to be directly correlated to the packing density of the particles. Therefore, the packing density (see section 3.4.2) of the powder is a key factor in tailoring of a PIM feedstock. The higher the packing density, the less the amount of binder required to achieve a suitable viscosity for the moulding operation. For coarser powders, equation (6.14) predicts no influence of the mean particle size on the suspension viscosity as long as the size distribution and hence the packing density remain unchanged. An irregular particle shape results in a higher viscosity, because of decreased packing density, and maximum solids loading. The same effect occurs with agglomerated powders. For fine powders, more binder is required for a uniform surface coating. Depending on the film thickness, this effect usually becomes pronounced at particle sizes below ~ 1 μm. The fractional tap density D_{tap} can be taken as a qualitative indicator for the maximum solids loading. The quantitative value of $c_{v,max}$, however, exceeds D_{tap} by 10–20%, as packing in the powder – binder mixture is much more effective than in the dry powder. Optimum solids loading is recommended at ~ 0.95 $c_{v,max}$. The solids loading c_v of practical feedstocks ranges from ~ 50–65 vol.%. The lower limit is controlled by debinding and sintering problems and is stated to be about 45 vol.%.

Maximum feedstock viscosities during moulding are in the range of $< 10^2$Pa s at shear rates in the order of 10^2–10^5 s^{-1}. As the powder increases the viscosity of the binder by a factor of 10–10^4, suitable binders should have viscosities no higher than 10 Pa s.

Base Powders

The particle size of PIM base powders is usually chosen to be < 25 μm. Fine powders decrease the size of moulding defects and increase sinterability. The lower size limit is determined by the solids volume fraction and the acceptable debinding time. Problems have to be expected for particle sizes < 1 μm. Nevertheless, injection moulding of submicrometer powders belongs to the 'state of the art' in the production of hardmetal and advanced ceramic components. Spherical, rounded, or at least equiaxed particle shapes are preferred for a high packing density even if such powders limit the green strength of the debonded compacts. The particle surface should be smooth and clean to enable thin binder films. Fine gas and water atomised powders or carbonyl powders are well within this requirement. The particle size distribution is often optimised for a high packing density, i.e. bimodal or wide size distributions are chosen. The freedom to tailor the size distribution is, however, limited by subsequent sintering problems such as exaggerated grain growth resulting from wide size distributions. There are attempts to use coarser cheaper powders.

Binder Systems

The most widely used binder systems are thermoplastics such as polymers and waxes. A small molecular chain length is desirable for a low viscosity and short debinding time. Apart from binder rheology, several other parameters have to be considered in binder selection, e.g. the wetting behaviour and thermal expansion mismatch with the powder particles, or the change in volume on melting. Good wetting increases the maximum solids loading and supports capillary transport of the binder in debinding. High thermal expansion and melting volume change can cause cracking during cooling or debinding. Another important selection criterion is the debinding behaviour. Common binders are usually multicomponent systems — consisting of components with differing melting and decomposition temperatures, which enables selective removal during debinding.

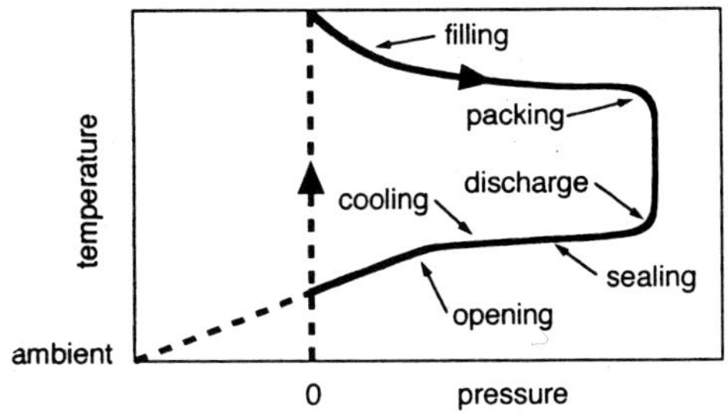

Figure 6.19 Temperature-pressure sequence in powder injection moulding (after German)

Moulding

Moulding of a thermoplastic feedstock consists of heating it above the melting temperature of the binder, forcing the melt under pressure through a sprue and runner channel system into the mould cavity, and subsequently cooling the mould and solidifying the powder–binder mixture. The interaction of the main processing variables, temperature and pressure is shown schematically in Fig. 6.19.

During moulding, the mould cavities have to be filled uniformly with the feedstock, avoiding density gradients, demixing or sedimentation of the powder. This requires a progressive filling of the cavity. Excessive moulding rates at high pressure–low viscosity conditions result in the so-called jetting phenomenon, illustrated in Fig. 6.20. Jetting causes defects such as weld lines and entrapped air. Too slow moulding leads to incomplete filling, due to premature freezing of the feedstock. This is called a short shot. The mould filling rate is determined by mould geometry, feedstock viscosity and moulding pressure. The moulding pressure is limited (apart from by jetting) by flashing along the mould parting line and by adhesion, which cause ejection problems. Proper conditions result in a very homogeneous density distribution.

The mouldability of a given feedstock can be characterised by the shot length reached during moulding into a spiral flow channel, which, to a first approximation depends on the inverse of the viscosity. Demixing and sedimentation are promoted by changes in the flow direction and low viscosities. For moulding feedstocks with a low softening temperature, high yield strength and subsequent pseudoplastic behaviour are desirable.

Cooling and freezing of the compact are governed by its surface to volume ratio and the heat capacity and thermal conductivity of the mixture. The hardening time varies with the square of the compact thickness.

Modelling of the complex conditions in moulding, requires mathematical methods, such as finite element or finite difference analysis. With the wide application of injection moulding in the polymer field, advanced computer programs are available, including viscosity models, heat flow calculations, mould filling and shrinkage programs and even programs for tool design and cost calculations.

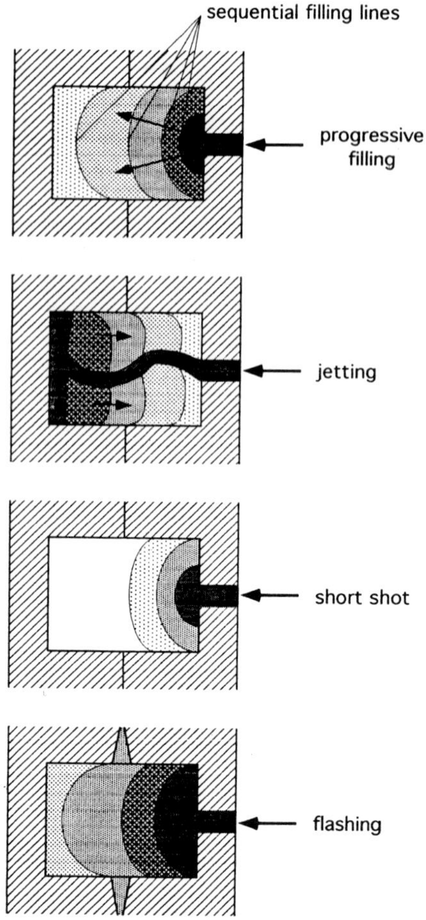

Figure 6.20 Defect creation in powder injection moulding through unsuitable process parameters

Debinding

Before sintering, the binder has to be removed from the compact without particle disruption and without contaminating the compact. The compact shape must be retained during this operation which requires some replacement of the binder–particle interactions by particle–particle interactions. Two major debinding techniques are used in practice: thermal debinding and solvent extraction of the binder. Thermal debinding can be carried out by evaporation, thermal decomposition or by extraction of the liquid binder by a wick substrate or powder bed. Solvent extraction is done by immersing the compact in a solvent, which partially dissolves the binder.

In thermal debinding the transport mechanisms are capillary flow of the liquid binder, and diffusion or permeation controlled flow of the gaseous evaporation or reaction products through the pore channels of the compact.

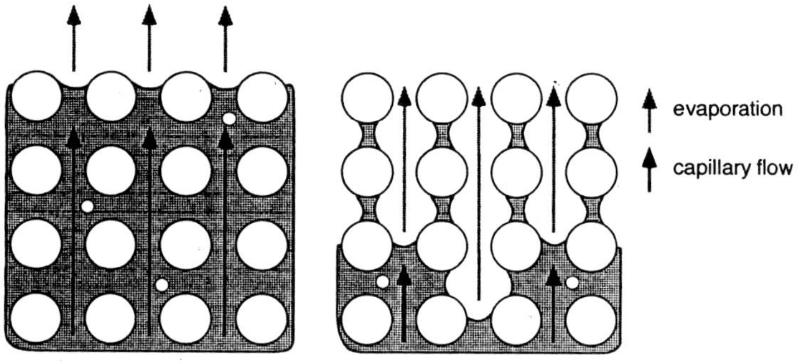

Figure 6.21 The role of capillary flow of the liquid binder in thermal debinding. Initial state (left): evaporation rate controlled; later state (right): no capillary flow to surface, permeation controlled

Capillary flow is important during the first stage of binder evaporation/decomposition and for wick extraction. When the binder begins to evaporate from the surface of the compact, the liquid binder is forced by the capillary pressure to flow from the interior to the surface, as long as a saturated state (Fig. 6.21) exists at the surface. Usually the evaporation rate is the slower and therefore the rate-controlling factor during this stage, resulting in a constant debinding rate. When the surface dries, capillary flow is no longer possible. The remaining binder, mostly concentrated in the inter-particle bonds has to be removed by internal evaporation and permeation or diffusion to the surface. This is accompanied by a steadily decreasing debinding rate.

Solvent debinding is applied mainly as a first debinding step, creating an open pore structure for further debinding by evaporation. It can be used with binder systems consisting of at least two components from which one is extracted by a liquid or vapour solvent. Sufficient soluble component is required to provide a continuous network to the surface. The other components have again to be removed by thermal debinding; 80% binder reduction by solvent extraction is said to be a realistic proportion.

Depth and amount of binder removal are proportional to the square root of the debinding time, i.e. debinding time is dependent on the square of the component thickness. The particle packing density in the compact should be low for binder removal but this is detrimental to shape stability and can result in enhanced sintering shrinkage. Increasing debinding problems are predicted for packing densities over ~ 75%. Coarse particle sizes allow for easier binder removal, but are detrimental to sintering. Higher temperatures generally increase the debinding rate, but are very dangerous, as vaporisation of the binder in the compact interior can disrupt or even explode the compact.

6.1.1.3.2 Practice of Powder Injection Moulding

Feedstock preparation

The first step in feedstock preparation is mixing. Mixing requires a

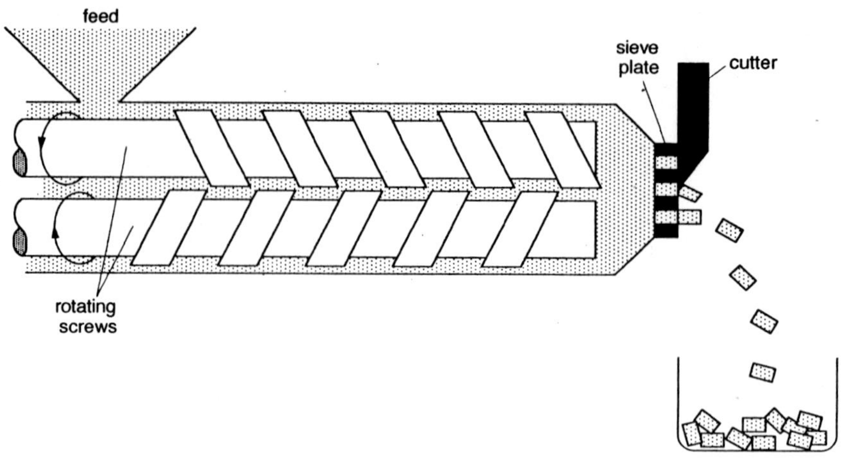

Figure 6.22 Twin screw extruder/granulator for PIM feedstock preparation (schematic)

homogeneous dispersion of relatively small particles in a dispersant of relative high viscosity. Complete de-agglomeration of the particles is essential. For thermoplastic binders, enhanced temperatures and shearing are applied in mixing. Several high shear mixers are used in feedstock preparation. The most effective system is the twin screw extruder, schematically represented in Fig. 6.22.

The feedstock has to be stored and later fed into the moulding stage in the form of a free-flowing granulate. It is therefore cooled at the end of the mixing stage and made into pellets. The pellets can be fed back into the extruder several times to homogenise them. A tolerable fraction of granulated moulding scrap is usually added with the last pass. Recycling of this scrap consisting of sprues, runners and imperfectly moulded parts is important for process economics, especially for very small items, when the finished product may even be the lesser fraction of the moulding shot.

Injection Moulding

Figure 6.23 shows the main components of the moulding equipment. The mould consists of two major parts, the rear and front cavity plates, which are clamped during moulding and can be opened for ejecting the compact. The feedstock is injected from the nozzle through the sprue, the runner and the gate into the mould cavity, which has the desired compact shape. A tool can contain multiple cavities. The runners should be short to minimise cooling during mould filling. Large runner diameters aid mould filling, but increase the amount of waste feedstock which has to be recycled; a diameter of 6 mm is considered typical. The gates should freeze before the compact and form a notch for trimming the runner from the compact after ejection; they therefore usually have small cross sections. The mould also contains channels at suitable locations for the heating and cooling media respectively.

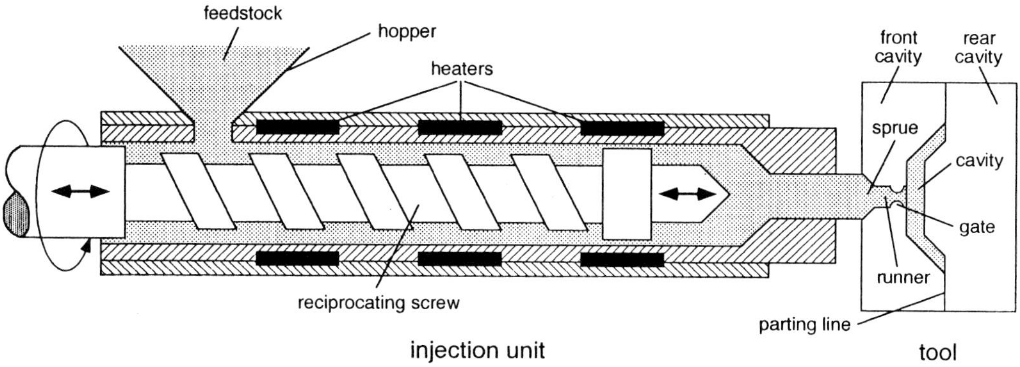

Figure 6.23 Key components of injection moulding equipment (schematic)

The injection moulding machine consists of a clamping and an injection unit. The first unit operates the mould movement, i.e. opening, compact ejection, closing and clamping. The second unit affects the feedstock melting and injection. The reciprocating screw unit is the most common injection system. The screw is designed for rotational and reciprocating movement inside a heated barrel. Prior to injection, the screw is in a backward position and transports the feedstock by rotation to the front of the barrel (Fig. 6.23). After this plasticizing step, the screw moves like a plunger and forces the feedstock into the mould. This movement creates the injection pressure which is held until the gate has solidified. Modern moulding equipment provides programmable closed-loop control systems for the various processing variables and the moulding sequence. Controlled parameters include temperatures, forces and pressures, and the injected feedstock volume.

The time for one shot varies between 5 s and 1 min. Multiple cavities in excess of ten in one tool set are used in PIM. The main features of injection moulding machines are the clamping force, injection pressure, shot rate and the shot volume. The clamping force must exceed the injection force. Practical moulding temperatures are in the range of 100–180°C; practical peak injection pressures are in the range of 15–30 MPa.

Debinding and Sintering

Debinding is a very complex operation. It is often the slowest stage in the PIM process. Reported debinding times vary from 1 hour to two weeks with initial heating rates as low as 1 K h^{-1}. Thermal debinding by itself is the method generally applied in practice. Solvent debinding techniques are shorter, even when the process has had to be completed by thermal debinding of the solvent-irreducible binder fraction. The overall advantage of solvent techniques is also supported by the fact, that the residual binder can be removed directly in the sintering furnace, thus avoiding handling the completely debindered part.

Thermal debinding is carried out by heating the compact from room temperature up to about 600°C. The process atmospheres are air, inert and reducing gases, or vacuum, often with changing atmospheres over different temperature ranges. Air is preferred at the start of the debinding process, but has to be replaced at increasing temperature by inert or reducing gases for metal powder compacts. The furnace atmosphere is often circulated within the furnace to minimize local variations in temperature and partial pressure of the evaporation products. Solvent extraction is carried out either with liquid or vapour solvents. With liquid solvents, the compacts are immersed into a heated solvent bath for several hours. Vapour solvent debinding is performed in a partial vacuum with a low partial pressure of a high vapour-pressure solvent such as freon (Cl_2CF_2 or $Cl_2CF\ CF_3$).

Substantially reduced debinding times are claimed for a novel 'catalytically' controlled debinding technique, which is based on polyacetal as a main binder component. This polymer decomposes directly from the solid state into its gaseous monomer formaldehyde after reacting with nitric acid. The polyacetal can therefore be removed from the compacts at a temperature below its melting point. Heating to this temperature of, typically, 110°C is relatively uncritical, as the catalytic decomposition is limited to the binder–atmosphere interface, preventing any pressure buildup in the interior of the compact. Thus, an open pore structure is achieved quite quickly, through which the residual binder components can be easily removed at higher temperatures by permeation.

Sintering of the debindered compacts is usually carried out in the same way as compacts from other forming processes. Care has to be taken over carbon control, which, if debinding is not carried out under adequate conditions can be influenced by residual carbon resulting from pyrolysis of the binder.

Application Aspects

As already mentioned, PIM enlarges the shape capability of the powder metallurgy process and is used mainly for parts with very complex shapes. Because of limiting factors such as debinding time or powder costs the maximum size or cross-sections, respectively, of parts which can be fabricated by PIM is somewhat smaller than with conventional shaping methods. Figures of 100 mm and 100 cm^3 are mentioned for the largest part's dimension and volume respectively of the largest part. Taking only shape capability, the PIM process competes favourably with forming techniques such as precision casting or machining of powder metallurgy preforms. Its main advantage results from the possibility of net shape production of complex parts via the powder route, as with typcial powder materials like cemented carbides, tungsten heavy alloys, superalloys and ceramics. Another potential field is short fibre or whisker reinforced metal matrix composites (MMC), where the tendency not to segregate and the possibility for fibre alignment make the PIM process superior to other forming techniques.

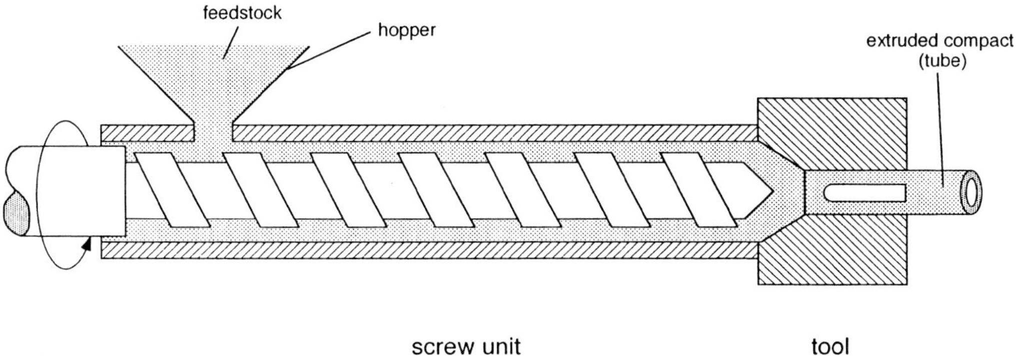

Figure 6.24 Powder extrusion equipment (schematic)

6.1.1.4 Cold Powder Extrusion

Powder extrusion is a long established technology in the manufacture of traditional ceramic products, such as bricks and tiles, stoneware tubes, etc. In powder metallurgy, this technology is applied, for example, in the production of rod shaped preforms for hardmetal drilling tools. Similar to injection moulding, it is a plastic forming method, where a plasticized powder system is forced by a piston or screw unit into a forming die (Fig. 6.24). A main difference is in tooling, which limits extrusion to products of constant cross section, such as rods and tubes. Additionally, the possibility of changing the rheology of the extrudate during the process by heating or cooling is very limited, because of the semi-continuous process. Thus the powder-binder system must provide flow under the extrusion pressure, but also shape stability of the extruded compact under gravity and handling stresses.

The typical behaviour of a paste, has been explained in section 6.1.1.3.1 and Fig. 6.16. At high normal stresses, the flow behaviour may be approximated by the Bingham equation (6.12). The velocity profiles in the extrusion barrel, die and finishing tubes depend on the relations between the extrudate yield strength τ_y and the wall friction τ_{wall}. Flow takes place by the mechanisms of slippage, differential flow and plug flow (see Fig. 6.18). The slip flow component can be increased by die wall lubrication. The permeability within the extrudate has to be kept very low, to prevent separation of the particle–binder mixture by migration of the binder through the particle skeleton.

Industrial extrusion pressures range from 4–15 MPa. An extrusion velocity of 1 m/min is a common figure. Tooling can be very sophisticated, as seen in ceramic honeycomb catalyst substrates with up to 200 channels cm^{-2} and channel walls as thin as 0.2 mm.

Close-packing particle size distributions containing large fractions of particles < 1 μm are preferred in the formulation of suitable powder systems. Organic binders with medium to high viscosity are used as plasticizer. In oxide ceramics, where extrusion is widely applied, aqueous binder solutions are chosen. Non-aqueous extrusion, as is needed in the case of metal powders, is based on solvents such as liquid wax, petroleum, alcohol, etc.

6.1.1.5 *Metal Powder Rolling (Roll Compaction)*

The compaction of metal powders in the gap between two rollers is an alternative method of producing semifinished products such as sheets and strips. Although this method was patented as early as 1902 (granted to Siemens and Halske for making tantalum strip), the work by Naeser and Zirm (1950) on making iron strip stimulated the development of roller compaction on a wider basis. This process, which looks simple at first glance, consists of the following steps:

- Rolling the green strip from metal powder.
- Sintering (continuous or batch) as coils.
- Rolling (hot or cold) of the sintered strip for densification and coiling.
- Final cold rolling for adjusting specifications.
- Annealing.

After the first step the powder is mechanically bonded, being porous and brittle but nevertheless self-supporting. The principle, with typical roll positions, is shown in Fig. 6.25. The mill is usually operated vertically (i.e. the roller positions are horizontal) because this avoids difficulties in feeding loose powder into the gap. The 2-roller system is used in most cases, but 4-roller mills with

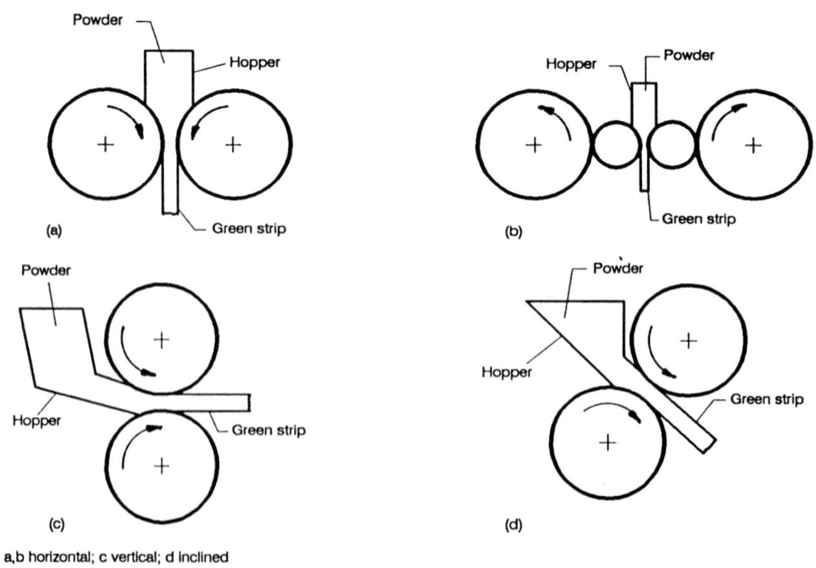

Figure 6.25 Principle of powder rolling with typical roll positions: a), b) horizontal; c) vertical; d) inclined

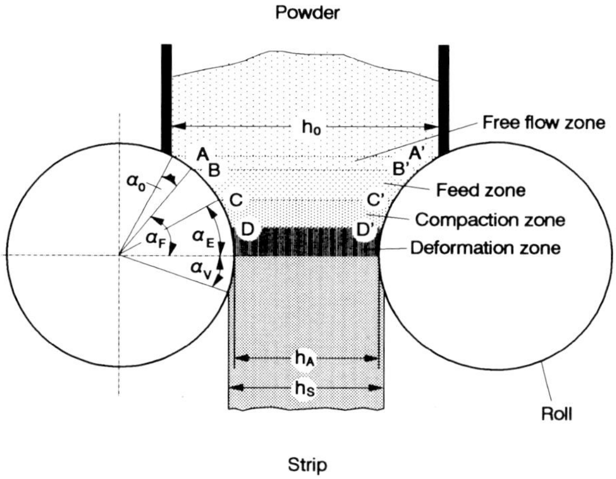

Figure 6.26 Powder/roll contact area in powder rolling (refer to text)

inner rollers of small diameter can be used for preparation of thin strips in the range of several tenths of a millimetre. The 'inclined' arrangement (Fig. 6.25d) is not of practical interest. The amount of powder passing the mill gap is controlled by its flow properties as well as by the frictional forces between powder and roller surface and between the powder particles themselves. Another possibility for powder supply is the forced feeding by additional mechanical means, such as a screw, providing a somewhat higher rate than free flow.

The contact area between powder and rollers can be divided into four zones (Fig. 6.26), namely:

- the free flowing zone (*A–B* or *A′–B′*, respectively), in which only gravity is acting, defined by the angle α_0. This angle is determined by the feed width h_o above the rollers.
- the feed zone (*B–C* or *B′–C′*, respectively), in which friction forces between powder and roller come in action. The densification mechanism is particle rearrangement. The boundary of this zone is defined by the angles α_E and α_F, the latter being the feed angle.
- the compaction zone (*C–D* or *C′–D′*, respectively), in which the rolling force becomes effective and the powder mass becomes coherent. Powder particle deformation is of some, but obviously limited, importance in this zone, especially at the strip surface. α_E is the compaction or gripping angle.
- the deformation zone (below *D* or *D′*, respectively), where extended deformation takes place and the green strength of the formed strip reaches its maximum.

α_v is the angle determined by the elastic compression of the rollers, which controls the thickness h_s, of the strip. α_v approaches zero and the strip thickness equals the roller gap h_A, if the elastic roller deformation is zero. The elastic spring-back of the rolled strip, however, generally results in greater thickness than h_A.

The green strip density ranges usually between 80 and 90% TD, the pores being partly open and partly closed (the latter especially near surface). A relationship between roller diameter D and the porosity U of the emerging strip can be formulated as

$$U = 1 - \frac{\rho_{\text{tap}} \{D\,(1 - \cos\alpha_E) + h_A\}}{\rho\, h_A} \tag{6.15}$$

ρ_{tap} being the tap density of the powder and ρ the density of the solid metal. The other symbols are found in Fig. 6.26. Thus, large-diameter rollers provide higher strip densities.

Sintering of the strip is usually carried out in a horizontal furnace, on a moving belt. When the sintering furnace is linked to a green-strip rolling mill, continuous production is possible. This can then be linked to a hot rolling mill in order to achieve full densification or the product is re-rolled to final size. The efficiency of sintering controls the economics of the entire process. The alternative (but not often used) method is to coil the green strip carefully and sinter it in batches in coiled form. Hot rolling of metal powder as a first compaction step has also been investigated, but seems to be less feasible in practice.

A large number of metals and alloys have been processed by roller compaction in the laboratory or pilot plant scale, including Fe, Cu, Co, Mo and several ferrous and non-ferrous alloys. Strip containing a second dispersed phase, not soluble in the solid state, as well as two- or multi-layered strips have also been produced. The mechanical properties of strips from powders compare well with those of conventional products. Especially in cases where the metal powder is produced directly during the extraction a substantial economic advantage can be achieved.

The widespread commercial application of these methods, however, has not yet taken place as envisaged 30 years ago although certain specialist companies are operating the process for a wide range of applications. The most notable example of commercial application is the production of high purity Ni- and Co-strips and sheets, from both hydrometallurgical and atomised powder by Sherrit Gordon Mines of Canada. The main application of nickel strip is for making coin blanks for the mints of Canada and other countries. It is also used for electronic tubes, batteries (where a certain porosity may be important), thermostats and other plated bimetal applications.

6.1.2 Hot Compaction

In hot compaction, high temperature deformation mechanisms of the powder material itself are activated by the simultaneous application of enhanced processing temperature and an external pressure. The main hot compaction techniques are axial and isostatic hot pressing, hot forging and hot extrusion. Most of these techniques include final forming and shaping, as well as consolidation (sintering) of the powder compact. Full density processing is generally the desired goal of hot compaction.

6.1.2.1 Fundamental Aspects

The fundamentals of hot compaction and pressure sintering respectively are closely related to the fundamentals of pressureless sintering, which are outlined in detail in Chapter 7. Pressure sintering can simply be treated as a sintering process, where the internal driving forces of the system (see section 7.1.2) are enhanced by local stresses, which result from external loading of the compact. For a particle–particle contact in a solid state system, the Laplace capillary stress σ, given in equation (7.1) can be considered as a mechanical driving force existing without any external loading on the particles. In pressure sintering, an external pressure acting on the compact is transmitted to the single particle contacts and superimposed on the capillary stress, thus increasing the effective contact stress. The same holds for the later stages of sintering, where isolated pores tend to be eliminated under the capillary pressure given by equation (7.3). In pressure sintering, this tension stress in the curved surface is again enhanced by the external pressure. It may be reduced to some extent, if gas becomes entrapped in the pores.

The way in which external loading contributes to the stresses in particle contacts or pores depends on the type of the external stress, the compact geometry, and the microstructural situation under consideration. Modelling is limited to solid state pressure sintering of random dense packing arrangements of monosized spherical particles under an external hydrostatic pressure. This model has already been explained in section 6.1.1.2.1 and Figs. 6.6 and 6.7. The effective contact pressure P_{eff} induced by the external pressure P is acting normally on the contact planes between the particles. It is plotted in Fig. 6.27 as a function of fractional density. At the onset of densification, P_{eff} tends towards infinity, due to the zero dimensional nature of the contacts (point contacts). With increasing density, i.e. increasing number of contacts and size of contact planes, P_{eff} tends towards the external hydrostatic pressure P. In the density range $0.9 \leq D \leq 1$, the compact can be regarded as a matrix containing equal-sized spherical pores. The pore pressure component resulting from the external pressure can then be assumed simply to be equivalent to P.

Common average capillary pressures are in the range 10^{-2}–1 MPa. The total contact and pore surface stresses are thus largely increased by the external pressures, which range from ~ 5–500 MPa for the various hot compaction techniques. Such high local stresses favour material transport mechanisms, which play only a minor role in pressureless sintering (see Table 7.1). The important hot pressing densification mechanisms in solid state systems are:

a) plastic yielding

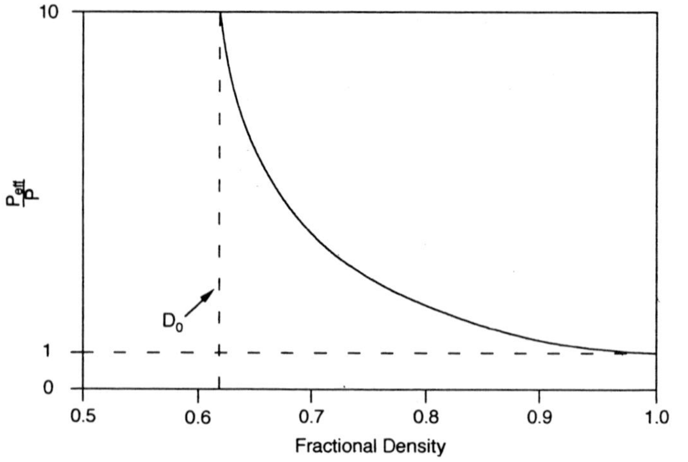

Figure 6.27 Normalised effective pressure in particle contacts during HIP densification (see Fig. 6.6; after Ashby)

b) power law creep
c) diffusional creep
d) grain boundary sliding

Plastic yielding is practically independent of time, i.e., it occurs spontaneously, when the total contact and pore surface stress goes beyond the yield strength of the material. Yielding proceeds until contact flattening and/or work hardening have balanced the contact stress with the deformation resistance of the material. The other mechanisms result in finite rates of local (contact- or pore-) deformation, and thus in time dependent densification. Power law creep deformation can be based on several principles, e.g. climb-plus-glide of dislocations or dynamic recrystallisation. Diffusional creep results from stress-induced vacancy gradients which induce a vacancy flow from the grain boundaries in tension to those in compression. This is called Coble creep, if diffusion is through grain boundaries, and Nabarro–Herring creep in the case of diffusion through the crystal lattice. Grain boundary sliding is a relative movement of adjacent grains, resulting in high deformation rates. It has always to be accompanied by other creep mechanisms for shape accommodation of the grains, or by boundary cavitation.

The contribution of the various mechanisms to local deformation and thus to densification depends on material and microstructural parameters and on the process variables, temperature and pressure. The role of these different parameters can be illustrated, e.g. in HIP diagrams, based on the geometrical model mentioned above and the deformation mechanisms a) to c). For a given material, these diagrams represent the functional correlation between the fractional density *D* reached in a given time under the external pressure *P* and temperature T. They also specify the parameter fields where a certain deformation

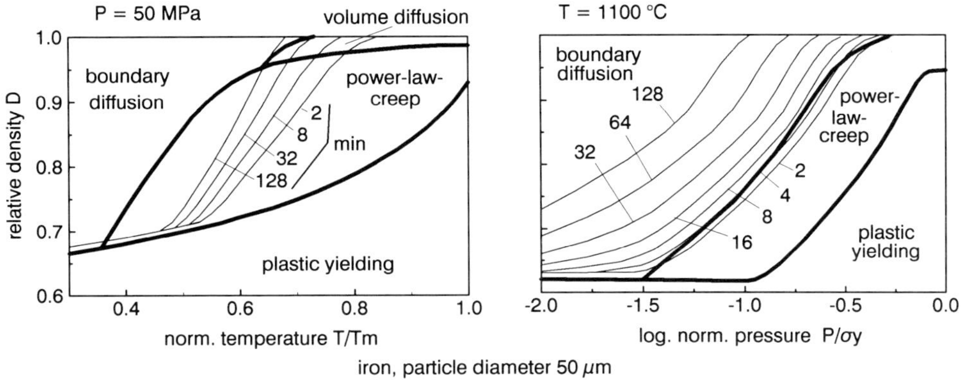

Figure 6.28 Hot compaction diagrams (isostatic pressure conditions, after Ashby)

mechanism dominates, i.e. yields the highest densification rate. P and T are normalised by the material's yield strength σ_y and absolute melting temperature T_m, respectively. Examples are given in Fig. 6.28.

It becomes obvious that hot compaction is relatively ineffective at temperatures below $0.5{\cdot}T_m$. Spontaneous plastic yielding is the mechanism which is most pressure sensitive and thus dominates at high P/σ_y. The temperature influence on yielding is determined by the temperature dependence of the yield strength σ_y. It takes place mainly in initial densification, when the local stress concentrations are high.

Dislocation or power law creep is next sensitive to pressure and thus dominates at medium P/σ_y. Its densification rate increases progressively with pressure, depending on the power law creep exponent of the material. Both plastic yield and power law creep densification are found to be independent of the powder particle and grain size. When these mechanisms dominate up to full density, P and T are the only variables controlling the hot compaction process.

Diffusional creep densification rates only linearly increase with pressure and become dominant at low P/σ_y. The temperature dependence of diffusional and power law creep obeys exponential laws. Thus, they are much more sensitive to variations in temperature than in pressure. Diffusional creep densification is influenced strongly by the grain size of the compact with a densification rate inversely proportional to the square of the grain diameter. In most practical cases of low rate, hot compaction processes (axial and isostatic hot pressing), diffusional creep is the rate-controlling densification mechanism, at least in the final part of the process. Its contribution to densification can be enhanced by small grain sizes. Grain growth during compaction is highly detrimental to densification.

As during pressureless sintering, also during pressure sintering of multicomponent system liquid phases may occur. The present understanding of liquid phase pressure sintering (LPPS) is relatively poor, although LPPS is widely applied in the production of cemented carbides and silicon nitride. Important pressure activated densification mechanisms are:

a) viscous grain–liquid flow
b) viscous flow of the liquid phase alone
c) pressure enhanced dissolution and reprecipitation (contact flattening)

Mechanism b) is undesirable, because it results in grain–liquid separation and thus creates inhomogeneities. It depends on pressure, viscosity of the liquid phase, grain size and pore channel width and can be avoided by limiting the pressure at a given viscosity and microstructure. In capsule-free HIP densification of liquid phase systems, the pressure has to be kept below the capillary pressure to avoid viscous flow of the liquid phase alone. This can become a limiting factor with high liquid contents at large grain sizes.

6.1.2.2 Practical Aspects of Axial Hot Pressing

Axial hot pressing is one of the oldest methods of PM hot consolidation. The powder or a precompacted preform is placed in a rigid die and compacted by single- or double-action punches (Fig. 6.29). Due to the enhanced processing temperatures, the complexity of tooling is limited in comparison with cold pressing. Common hot pressed parts exhibit simple single level geometries.

The tooling material has to be chosen taking into account strength, and compatibility with the compact under hot pressing conditions. Graphite is the most common material, enabling pressing temperatures of up to 2500°C. Heavily loaded dies are sometimes also made from carbon-fibre reinforced carbon (CFC). Other suitable materials include tungsten, TZM (dispersion-hardened molybdenum), Stellites (cobalt alloys), heat resistant alloy steels, and ceramics. Lining the tool faces with replacable sheet or foil materials, or spray coating are measures to reduce wear and ejection problems and to achieve economic tool lifetimes.

Heating of the compact can be carried out by direct resistance heating or by

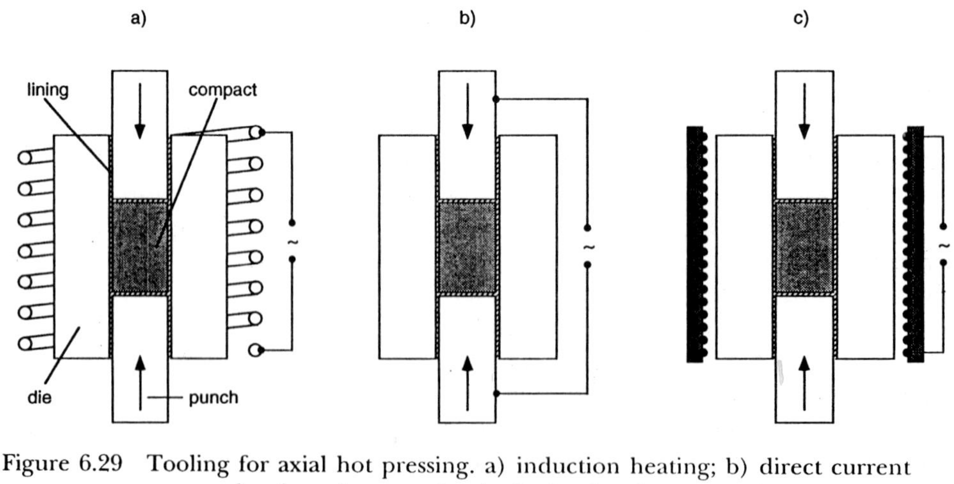

Figure 6.29 Tooling for axial hot pressing. a) induction heating; b) direct current heating; c) convection/radiation heating

heat transfer from the tool, which can itself also be resistance heated by direct or induction currents, or by radiation and convection from an external heat source. The atmospheric environment must be adapted to the compact material (see also Chapter 8, 'Sintering Atmospheres'). For most metal and alloy powders inert or reducing gases or vacuum are suitable atmospheres. If graphite is used for tooling, the carbon monoxide formed by oxidation of the tool enables the tool to operate in air for certain materials with moderate oxygen and carbon affinity.

Typical compaction pressures are up to 100 MPa in steel alloy dies and < 50 MPa in graphite dies. For liquid phase systems, lower pressures are preferred, to avoid squeezing liquid out of the compact. The cycle times vary from several seconds up to about an hour. Axial hot pressing is an important fabrication method for many non-oxide ceramics, e.g. Si_3N_4, SiC, BN, B_4C. It is also generally used for metal matrix composites, such as metal-bonded diamond tools. Other applications include beryllium pieces and blocks and cemented carbide parts. Isostatic hot pressing is increasingly used for carbide parts, as well as for a variety of other metallic materials which require hot compaction. This is because of the lower production rates, lower maximum pressure and non-uniform pressure of the axial hot pressing technique.

6.1.2.3 Practical Aspects of Isostatic Hot Pressing

HIP Process Variants

Hot isostatic pressing (HIP) is a hot compaction process under isostatic pressure conditions. Near-isostatic conditions can be realised in axial hot pressing by embedding the compact into a suitable pressure transducing medium, e.g. a ceramic powder bed. Fully isostatic conditions require a fluid for ideal hydrostatic pressure transmittance. In the temperature range for hot compaction of most materials of interest (> 0.5 T_m) gases are employed for this purpose. The external pressure can only become effective if the pressure medium cannot penetrate the open pore channels of the compact. This makes sealing the pores necessary, which can be achieved by flexible bag tooling as in cold isostatic pressing. At enhanced temperatures, however, this tooling becomes one of the limiting and cost controlling factors. Another method is the elimination of the open porosity (see Fig. 7.21) by pressureless sintering and the application of gas pressure only in the final reduction of the residual closed porosity. This leads to the three principal HIP process variants shown in Fig. 6.30.

Encapsulated HIP, i.e. using a suitable tooling or canning system, is the only processing route which allows pressure-aided densification over the complete range from green compact density to theoretical density. Containerless HIP includes two sub-variants, Post-HIP and Sinter-HIP. Post-HIP is the hot isostatic compaction of preforms which are sintered in a separate cycle to the density level (90–93% T.D.) whereby the pores are closed. Sinter-HIP is a continuous

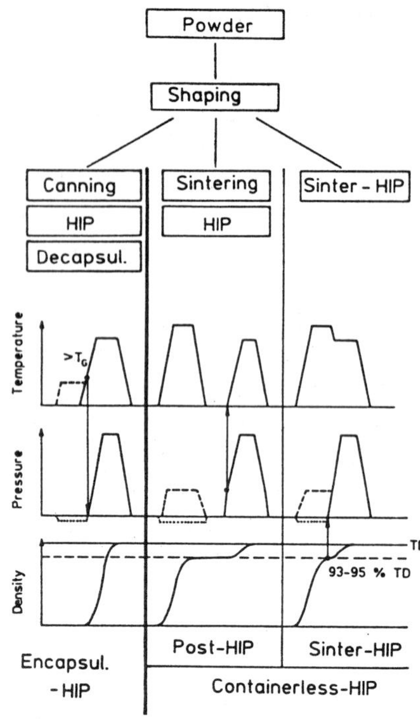

Figure 6.30 Process variants of hot isostatic pressing (HIP)

process combining pressureless sintering to allow pore closure and subsequent HIPing without intermediate cooling of the compact.

HIP Process Equipment

Compared with the liquid pressure media used in cold isostatic pressing gases are much more compressible and thus have a much higher energy content under pressure. Therefore safety aspects are important in vessel design and operation. Figure 6.31 gives a schematic overview of a HIP unit. It consists of a cooled-wall pressure vessel containing a furnace and the workload. Pressurisation of the process gas is performed by a gas compressor system. Depending on the process, the thermal expansion of the gas upon heating can make a significant contribution. Larger HIP installations are equipped with a gas purification and recycling unit. Depending on the pressure range, systems are sometimes differentiated between 'gas pressure furnaces' (5–20 MPA) and HIP vessels with pressure capabilities up to 400 MPa. Under such pressures gases behave like liquids in respect of their thermal conductivity. This results in a highly convectional heat transfer and convectional currents and therefore requires sophisticated insulation to prevent unacceptable heat losses to the cooled vessel walls. Conversely, this effect can also be used for rapid cooling or even quenching of the workload after its consolidation, as realised in rapid cooling heat exchange systems in industrial equipment.

Inert gases like argon and helium are the most common process gases. Nitrogen is used as an inert, as well as a reactive process atmosphere, e.g. in Sinter-HIP consolidation of Si_3N_4, where it suppresses high temperature decomposition of the nitride. Recent developments even include oxygen or oxygen-containing atmospheres for hot compaction of oxide ceramic materials for electrical applications. Other gases can be employed to incorporate an *in situ* coating.

The furnace and insulation material has to be selected according to the desired pressing temperature, the capsule or compact material and the process gas. Graphite or CFC heating elements and graphite insulation parts are widely used and operated up to 2200°C. Molybdenum furnaces are used in the temperature range up to about 1600°C for workloads sensitive to carbon impurities. Oxygen-containing atmospheres require oxidation resistant heaters such as platinum metal or $MoSi_2$, which can be used up to 1400 and 1600°C respectively.

Encapsulation Technology

The flexible tooling, which in cold isostatic pressing can simply be made from polymers, is a key factor in encapsulated HIP. It has to be designed to be plastically deformable at the pressing temperature without reacting with the compact or penetrating into the pores. For most metal powders, containers

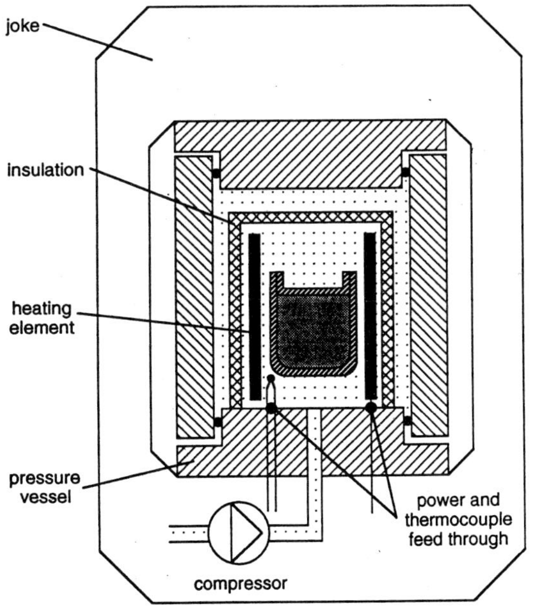

Figure 6.31 HIP components (schematic)

made of thin metal sheet are suitable for near net shape parts production. The container metal should have good weldability for safe sealing of the capsule. Figure 6.32 shows an example of such a container. It is made by welding the single-formed sheet parts. After helium leak testing of the welds, the capsule is filled through the evacuation tube. After degassing, the evacuation tube is sealed by local heating and squeezing. The capsule shape must be designed according to the local shrinkage during HIP. Shrinkage is not ideally isotropic, due to the rigid nature of the container walls. For the same wall thickness, shrinkage is less at smaller compact diameters than at larger ones, due to the higher deformation resistance of the capsule with increasing wall-thickness/diameter ratio. The high deformation resistance of sheet elements along their own plane results in corner effects, constriction, and differences in axial and radial shrinkage for compacts with height/diameter ratios significantly different from unity. A few examples of these problems are presented in Fig. 6.33. The linear axial and radial shrinkage given in this figure is based on calculations assuming cylindrical compacts with an initial powder fill density of 66% of TD, which is a typical value for vibrated metal powder beds of not too fine particle sizes.

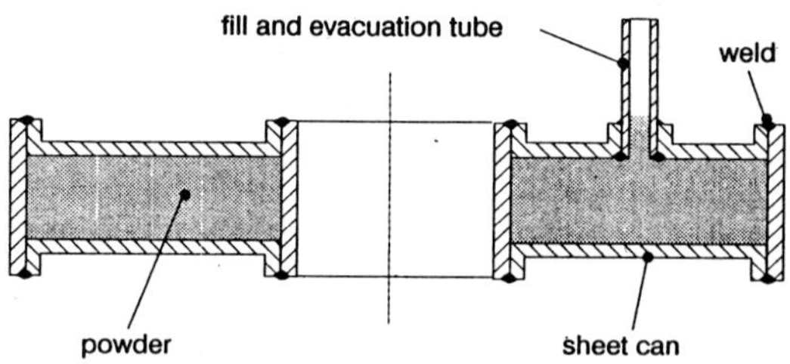

Figure 6.32 Metal sheet container for HIP application

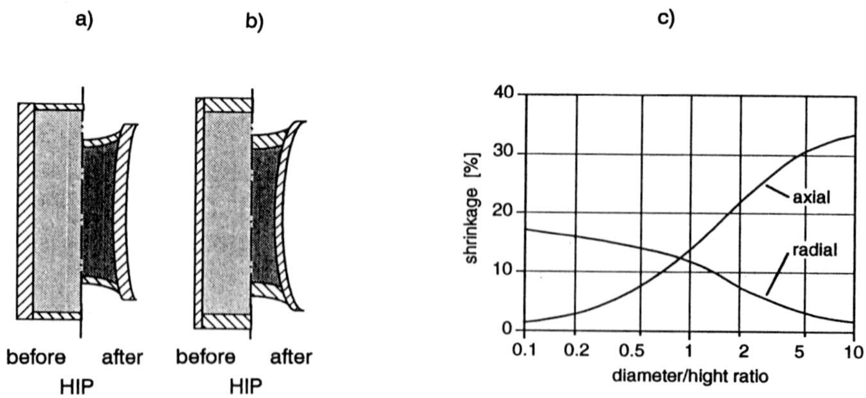

Figure 6.33 Problems related to HIP capsule design. a), b) effect of sheet thickness relations; c) constriction effects in cylindrical capsules (after Grewe *et al.*)

Low carbon steel, austenitic stainless steel and titanium are typical sheet canning materials. Steel capsules can be used up to 1400°C, titanium is suitable in the temperature range from 1400–1650°C, e.g. for hot compaction of refractory metals.

Other encapsulation methods of practical relevance include ceramic investment moulds and glass encapsulation (Fig. 6.34). Ceramic investment moulds are made in the same way as in the investment casting technique. A wax pattern is coated with a ceramic slurry, followed by drying, dewaxing and firing. The ceramic shell is filled with the powder to be compacted, and packed into a pressure transmitting ceramic granule bed inside a steel can which is evacuated and sealed. This method is industrially applied for making complex titanium alloy parts for aerospace applications.

Glass encapsulation is mainly used in hot consolidation of ceramics and refractory compounds, where the high processing temperatures do not permit the use of metallic capsule materials. In glass encapsulation, shaping of the compact is carried out by a cold compaction method such as cold pressing or injection moulding. A layer of glass-frit is applied to the surface of the compact. After degassing of the compact, the temperature is raised to melt the glass powder, which forms a dense surface layer. The glass has to be selected with regard to the desired pressing temperature. At this temperature its viscosity should range from ~ 10^6–10^3 Pa·s. Very pure SiO_2 can thus be used up to ~ 1800°C. For higher temperatures, the molten glass has to be prevented from

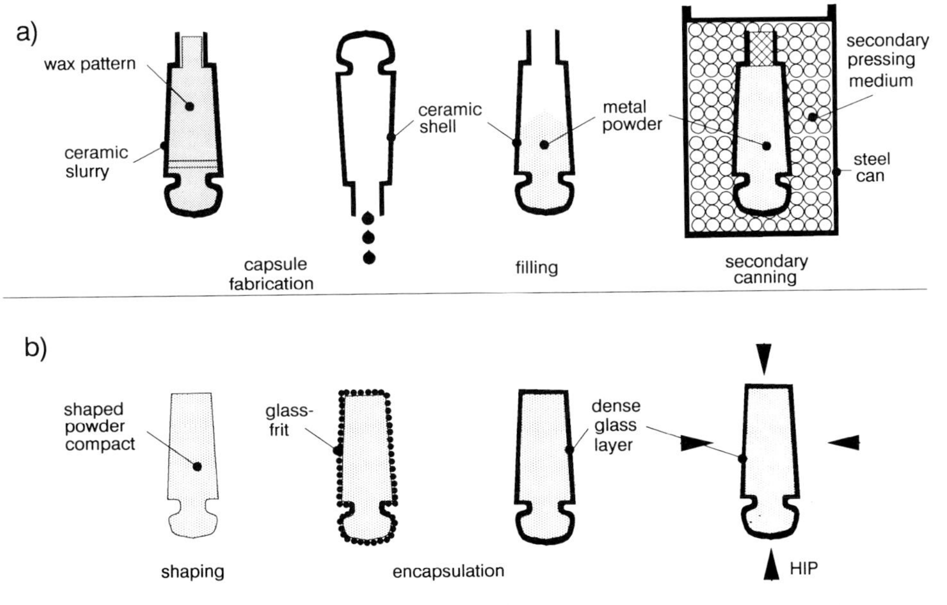

Figure 6.34 Special HIP encapsulation methods: a) ceramic investment moulding; b) glass-frit encapsulation

running off the part by an additional layer of glass powder or foam. In this way, processing temperatures of about 2000°C become feasible. Glass capsules are somewhat difficult to remove from the consolidated part, however.

Containerless HIP

Like glass encapsulation, containerless HIP starts with preforms shaped in a separate process stage requiring completely sealed surface layers. In principle, this can be achieved by a graded structure of the compact with enhanced sintering of near surface layers. This method, known as 'sinter canning' is limited in application to non-net-shape compacts, where the surface layer can be removed, or to systems where a gradient structure of the final part is appropriate. In general practice, the complete volume of the compact is sintered to the state of closed porosity. Depending on the microstructure, this occurs at fractional density levels of 85–95% to which sintering must take place without pressure assistance. Containerless HIP is therefore of interest for materials where this density level is achievable by pressureless sintering without unacceptable grain coarsening.

The decision whether to consolidate a material by separate sintering and subsequent Post-HIP, or by a continuous Sinter-HIP treatment is mainly dictated by the overall process economy, which favours Sinter-HIP mainly for moderate HIP pressures. However, there are also materials aspects from which a general advantage can be expected from Sinter-HIP. These relate to irreversible microstructural changes during intermediate cooling and reheating in the Post-HIP process.

Processing Sequence and Parameters

A flow chart of the main HIP process cycles is shown in Fig. 6.30. The main difference in the process cycle for the different process variants is in the pressure–temperature relations. In encapsulated and Post-HIP treatment, the process gas can be charged into the vessel at low temperatures, making it possible to use gas thermal expansion for further pressurisation. In Sinter-HIP, pressurisation has to be carried out at the hot pressing temperature, which requires a high compressor capacity.

The HIP temperature and pressure are usually chosen in order to reach full density in a soaking time of the order of one to a few hours. Higher pressures usually allow for lower temperatures and thus result in finer microstructures. Lower pressures are mainly applied in liquid phase pressure sintering, where the desired viscous grain–liquid flow is already strongly activated with low external pressures. A few examples of typical HIP parameters are summarised in Table 6.1.

Application of Hot Isostatic Pressing

Encapsulated HIP is mainly used for full density processing of materials where the powder route either is indispensible or results in superior properties, or

Table 6.1 Typical HIP temperatures and pressures for metals and ceramics.

Material	Temp. (°C)	P (MPa)	Process
Al alloys	450–530	100	encapsulated
Ti alloys	850–950	100	,,
tool steel stainless steel	1000–1100	100–150	,,
Ni alloys	1100–1200	100–150	,,
Mo, Nb, W	1250–1650	100–150	,,
Al_2O_3	1300–1500	100	,,
Si_3N_4	1750	100–200	,,
WC–Co	1350	6–10	containerless
Si_3N_4	1750–2000	10–150	,,

where net shape production leads to economic advantages. An important example in semifinished products is tool steels from gas atomized powders. The microstructure of the densified steels exhibits a very fine, homogeneous and isotropic carbide distribution. This quality is not achievable in wrought products (due to the very coarse primary carbides) nor in pressureless sintered tool steels, where higher process temperatures result in a more pronounced carbide grain growth.

Titanium- and nickel-based alloys for aerospace applications produced by HIP as near net shape components are comparable in properties with their ingot based counterparts while utilising only 10–30% of the material in the final product. The near net shape aspect is also the reason for the application of encapsulated HIP with refractory metals like molybdenum, tungsten and niobium. In oxide-dispersion strengthened (ODS) superalloys, which are dependent on the powder route, capsule HIP is a consolidation method competing with hot powder extrusion. In advanced ceramics, capsule HIP has been developed to a high standard for net shape production of complicated silicon nitride parts, such as gas turbine rotors.

Containerless HIP is less costly and therefore in wide use. A significant proportion of cemented carbides is now produced by Sinter- or Post-HIP with finer microstructures and greater reliability than pressureless sintered counterparts. Silicon nitride for high temperature applications with minimal content of sintering additives is typical of the Sinter-HIP materials in the ceramics field. The major application of HIP outside powder metallurgy is the healing of defects in big and expensive castings.

6.1.2.4 *Powder Forging*

Powder forging is used for both net shape parts production and semifinished products from special powder materials. It's most important application, however, is in the production of structural steel parts, where the net shape aspect dominates. It is used to obtain higher densities and thus higher strength, ductility and fatigue lifetime than are possible with cold compaction and subsequent sintering.

Ferrous Structural Parts

A flow chart of powder forging is given in Fig. 6.35. Preforms are made by axial cold compaction of prealloyed powders. The preforms are sintered at 1100–1250°C in a hydrogen containing atmosphere to give sufficient strength and plasticity for forging and to ensure complete reduction of the surface oxides of the particles. The parts are cooled down for intermediate storage, or directly transferred in the hot state to the forging press. Before forging, the parts are coated with a graphite film for lubrication and to prevent decarburisation. The cooled parts are reheated by induction heating. Forging takes place in a closed die. After forging controlled cooling of the parts is carried out in a protective atmosphere.

Forging results in macroscopic deformation and densification at high strain rates. Of the densification mechanisms discussed in section 6.1.2.1, plastic flow is by far the most important. Plastic flow is favoured by a high deviatoric (shear) stress component. The shear stress and strain can be varied by the preform design, e.g. by the relation of the preform and the forging die diameter d_0/d_f,

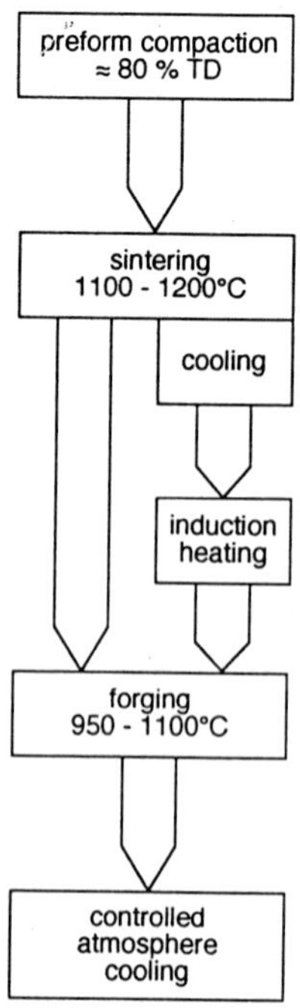

Figure 6.35 The powder forging process

which controls the lateral material flow in the case of the simple tooling shown in Fig. 6.36. The re-pressing mode with $d_0/d_f=1$ is much closer to hydrostatic conditions than upset forging which occurs with $d_0/d_f<1$ as long as the compact diameter remains below d_f. Microscopic shear deformation is claimed to be essential in powder forging for welding the plastically collapsed pore surfaces which are often covered by thin residual oxide layers. However, practical powder forging is frequently very close to the re-pressing mode dictated by shape complexity of the parts.

The preform density is usually in the range of 80% of TD. Forging takes place above the dynamic recrystallisation temperature in the range 950–1100°C. Tooling is kept at ~ 250–300°C. Powder forging tools are completely closed, in contrast to conventional forging dies (see Fig. 6.36).

Powder forging results in mechanical properties comparable or sometimes even superior to those of conventionally forged parts. Advantages derive mainly from the nearly isotropic behaviour of powder forged parts. Powder forging increases the material use and process economics by requiring less machining operations and is used commercially for several highly stressed parts such as gears, stator cams, roller bearing races, etc. A breakthrough for wider application is expected from the recently commenced mass production of powder forged connecting rods.

Forging, Rolling and Extrusion of Metal Powder Billets

Forging, rolling and extrusion are directional hot working operations where the material undergoes shear deformation. This type of deformation is suitable for breaking up particle boundary layers, e.g. oxide films, and for forming a deformation substructure, which allows for a directional recrystallisation treatment. These hot working operations are carried out starting either from powders, or from powder preforms which are consolidated already to near theoretical density. Examples for the latter are components from Ti- and Ni- alloys for

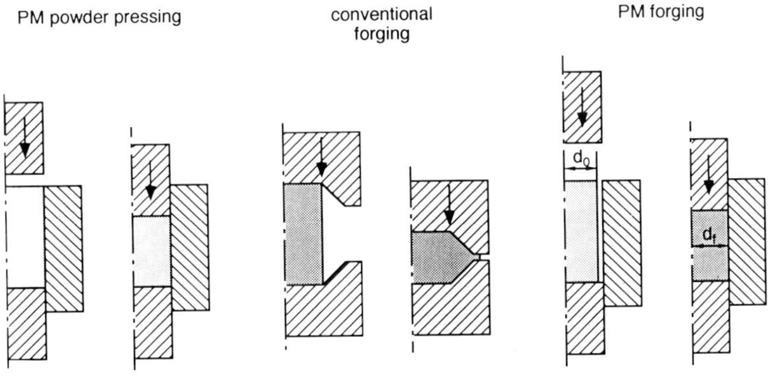

Figure 6.36 Powder forging in comparison to powder pressing and conventional forging (after Weber)

aerospace applications, where the tolerable failure probability is extremely limited. In the case of the superalloys, a subsequent directional recrystallisation treatment promotes the desired coarse, elongated grain microstructure.

Hot powder extrusion is usually carried out with vibrated or cold precompacted powders enclosed in a metal capsule. The capsule is heated and hot-extruded as shown in Fig. 6.37. This method is applied mainly for the production of dispersion-strengthened materials, such as ODS superalloys and dispersion strengthened aluminium, but also for composite materials, high speed tool steels and seamless stainless steel tubing.

6.1.3 HIGH COMPACTION RATE PROCESSES

In most compaction methods, generally low rates of pressure increase are applied. With the possibility of high energy rate metal forming, several methods of high rate powder compaction have been developed. This may be advantageous for achieving high densities (above 97% of T.D.) for compacting large powder masses as well as for powders which are difficult to compact. One major possibility is to accelerate the upper punch in a conventional die, so that it acts as an impactor to the powder mass (or as a projectile in a gun-like mechanism), or to rapidly increase the pressure in an isostatic unit. For both methods the detonation of an explosive has been used, while for die impacting the discharge of a capacitor system is also applicable.

6.1.3.1 Explosive Compaction

Figure 6.38 shows equipment for direct compaction of powders by explosion, in which the high pressure after detonation acts directly on the powder mass. The powder is enclosed in an easily deformable sheet. Indirect explosion methods with an intermediate liquid have been developed, but they are less significant.

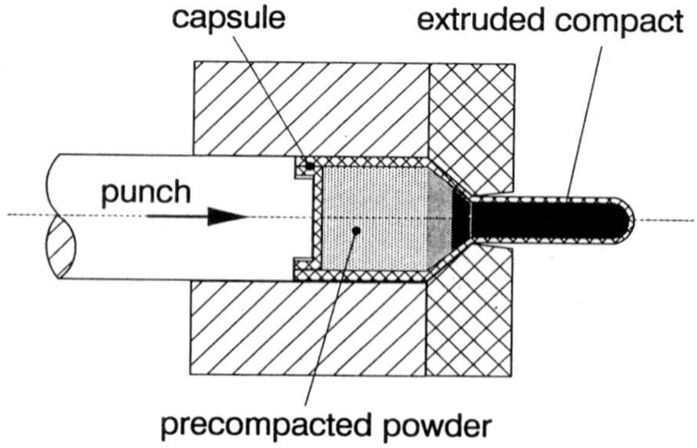

Figure 6.37 Hot powder extrusion of canned precompacted powder

The detonation pressure P can be calculated from the detonation rate v_D according to

$$P = \frac{1}{4} v_D^2 \rho_0 \tag{6.16}$$

where ρ_0 is the density of the explosive. This pressure is effective within the shockwave, acting on the powder mass during a time of ca. 2 μs. Under these conditions a strong compression occurs resulting in the generation of new dislocations, point defects and deformation twins. When, at very high shock wave pressures considerable heating takes place, defect healing and local melting may occur. A short-period liquid phase may occur which sometimes makes subsequent sintering unnecessary.

Although big differences have been observed in the densification mechanisms as between conventional and high rate explosive compaction, the densification curves of both often look similar, or one curve is a continuation of the other. This is illustrated in Fig. 6.39 for powders of W, Mo and stainless steel. The highest density values are obtained (up to 100%) when there is a transient liquid phase. It is not easy, however, to avoid density differences in larger explosive-compacted parts without destroying the part from strong decompressing effects caused by the shock wave (overcompaction).

Ceramic powders have also been densified in this way, but the densification has been limited (to e.g. 95% of T.D. in Al_2O_3) due to the occurrence of shear bands.

Many materials undergo phase transformation under high pressure, and this has also been observed during explosive compaction. Most interesting are the

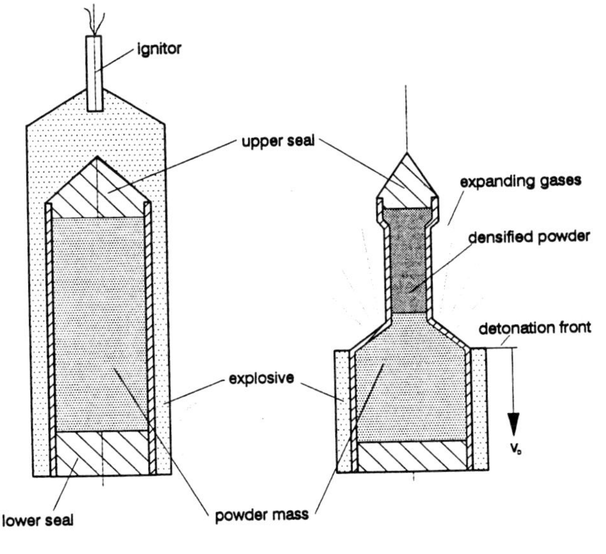

Figure 6.38 Device for direct explosive compaction of powders: before ignition (left); after ignition (right), (Prümmer)

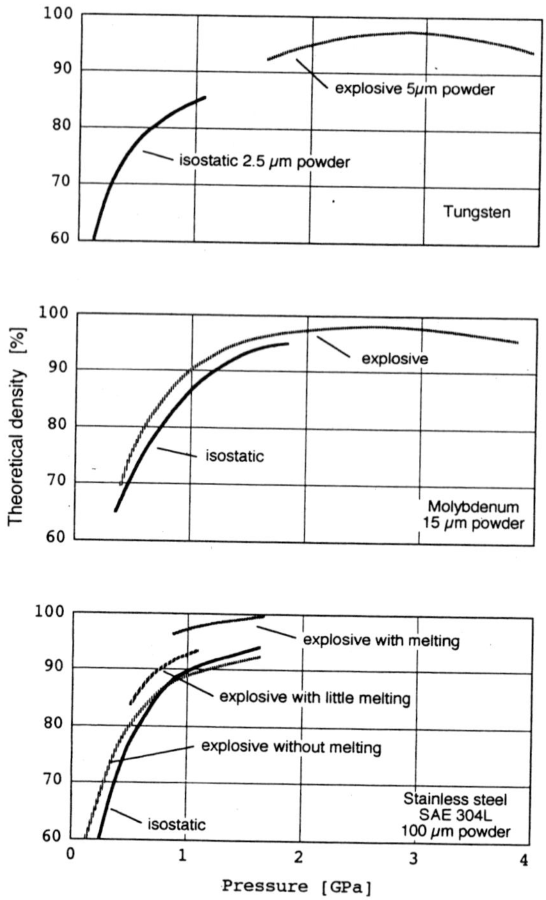

Figure 6.39 Density-pressure relation for isostatic and explosive compaction of powders from W, Mo and stainless steel (Prümmer)

graphite–diamond and hexagonal BN–cubic BN transformations, which have been investigated on a production scale at Du Pont. The resulting diamonds are fine grained (10–100 μm) with a high concentration of lattice defects.

Irrespective of some continuing industrial interest the explosive powder-compaction methods are of very limited practical importance. They seem to have, however, a potential for refractory metals and superalloys.

6.1.3.2 Impact and Discharge Compaction

Other than by explosives, the energy for rapid densification of powder masses can be generated by mechanical and by electromagnetic means. High-velocity pneumatic presses have been developed, in which compressed air is used to accelerate a ram piston into a die in which the powder mass is densified at one blow. The potential energy of the compressed air is converted into the kinetic energy of the piston, which can be varied by the pressure (e.g. 7.5 MPa).

The densification of powders by pulsed magnetic fields has been developed for different geometries, e.g. for densification of thin-walled tubes (up to 1 m in length), for parts with length-to-diameter ratio up to 5 with a profiled inner surface, and for billet densification. The discharge of a high-voltage electric capacitor by a high conductivity coil provides a strong transient magnetic field which collapses the powder mass. Equipment of 20 to over 140 kW with a voltage of up to 20 kV have been used and powders of W, Mo, Al_2O_3 and others have been densified up to 85–97% of TD. The time of discharge is in the range of 10–100 μs and the maximum magnetic pressure up to 400 MPa. The densification data of several powders by electrical discharge are given in Fig. 6.40. The sudden discharge of electrical energy through metal powders leads to the breakdown of oxide layers and provides good bonding between the powder particles. A magnetic field may be also used to accelerate an iron punch which will densify a powder mechanically in a thick walled die. The discharge energy has to be carefully adjusted.

The densification of massive rods by this method requires a high conductivity metallic cladding, which has to be removed before sintering. This limits the process to simple geometries. When separate coils are used they have to sustain strong mechanical forces and have a limited life time. These methods therefore seem hardly applicable to mass production.

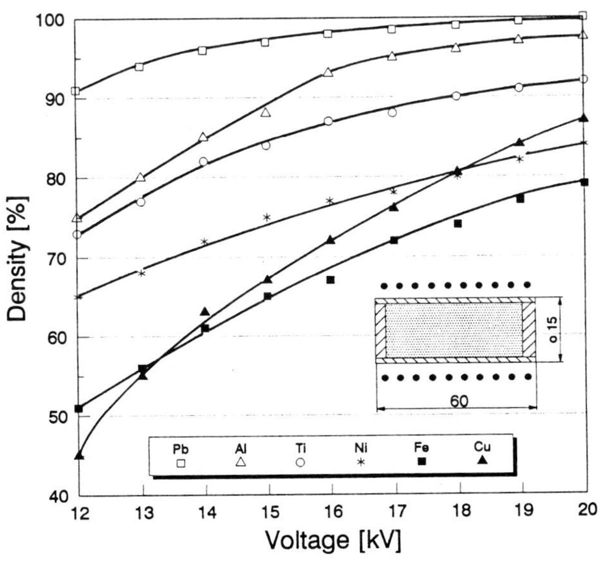

Figure 6.40 Green densities of electromagnetically compacted metal powders dependent on voltage and discharge energies. Capacitor 45μF coil with 30 turns, 0.2mm gap clad: Cu 15mm diametre, 0.75mm wall thickness (Mironov)

6.1.3.3 Rotary Swaging

Thin walled metal tubes filled with powder, preferably by vibratory compaction, can be reduced in diameter by rotary swaging. Commercial swaging machines with two or four hammers can be used while the tube passes through the swaging dies once or several times. Strong densification of the powder is achieved. This process was developed in the context of UO_2-fuel rod development as an alternative to tube filling with sintered pellets. Because of the danger of tube damage during swaging this idea was not realised for mass production. For manufacturing metallic (e.g. iron) bars from powders the swaged rod has to be sintered and the metal sheet removed. Although this process has been tried experimentally in combination with densification by high voltage capacitor electrical discharge (see above), practical applications have not been reported.

6.2 PRESSURELESS SHAPING

Although powder compaction methods by the use of pressure are by far the most important, pressureless forming is of practical significance and is used for mass production of metallic and ceramic parts. Pressureless shaping can be done by simple pouring of powders, by vibration compaction and by slip casting.

Common features of these methods are:

- No expensive or complicated tooling is necessary,
- Very large parts can be fabricated, as long as handling and sintering procedures are controllable,
- No substantial density gradients occur
- No particle deformation occurs. Consequently, the specific particle contact areas are much smaller than in the pressed condition and
- no fracturing or penetration of surface oxide films on the particles takes place.

The first three features are advantageous, the last two more disadvantageous. High green as well as sintered densities can nevertheless be achieved, when favourable particle size distributions are used and optimized particle packing (i.e. high filling or vibration densification) is provided. But pressureless shaping is favoured for manufacturing of highly porous parts, such as bronze filters and diaphragms, from spherical powders.

6.2.1 Powder Filling and Vibration

When powders are to be sintered in a 'loose' state, the mould should be filled simply by pouring the powder or by vibration. The powder has to be sintered in the mould, which is usually made of graphite or stainless steel. It was shown as early as 1931, that spherical carbonyl iron powder can be sintered at 890°C under hydrogen to a density of 7.12 g cm^3. This is not possible, however, with irregular powders. The production of bronze filters (see section 11.8) and porous nickel membranes are the most important examples of pressureless shaping. Spherical powders of closely controlled particle size range are used for bronze filters.

Vibration compaction was extensively investigated in the context of nuclear fuel element development from coarse spherical UO_2 powders. The aim was to achieve high vibration densities in stainless steels tubes without subsequent sintering, as an alternative to pellet production. By using mixtures of three distinct particle size ranges (44–74 μm, 149–222 μm and 840–1400 μm) with low surface roughness, vibration densities of ~ 83% of theoretical density have been achieved. This can be compared with the maximum packing density of monosize spherical powder, which is theoretically 73.9% in a regular close arrangement and ~ 63% in a random dense arrangement. Despite of these theoretical possibilities, however, the vibration compaction of size-optimized powder fraction mixtures is limited in practice, because the ideal size range including the morphology is difficult and expensive to achieve. However, vibration compaction is an important step in encapsulated HIP or hot extrusion of powder materials. It is used to achieve a homogeneous density distribution at a high fill density level within the capsule. The fractional fill density should be over 60% to avoid buckling of the capsule, which excludes the possibility of near net shape production.

6.2.2 Slip Casting

Slip casting (or slurry casting) is a conventional method for shaping of preferably large or complicated clay-ceramic parts (e.g. high voltage insulators) and has been used commercially for a long time. Oxide and non-oxide, as well as metal powders, can also be slip cast by a similar technique. The mould is usually made of plaster of Paris with high porosity and adequate strength, from partly dehydrated calcium sulphate ($CaSO_4.0.5\ H_2O$) and water, which react to $CaSO_4.2\ H_2O$. The slip is an aqueous or non-aqueous (alcohol) suspension containing the ceramic or metal powder in the highest possible concentration – 70–80wt%. The slip is poured into the mould, where it loses part of its water by capillary forces and forms a semi-hard layer on the mould surface (Fig. 6.41). The casting (still containing ~ 15% water) is further dried, with additional shrinkage, and then sintered. As shown in Fig. 6.41, not only drain but solid castings can also be made. While solid castings require the slip over the full cross section of the mould, for drain castings part of the slip is poured out of the cavity after some time, depending on the wall thickness required.

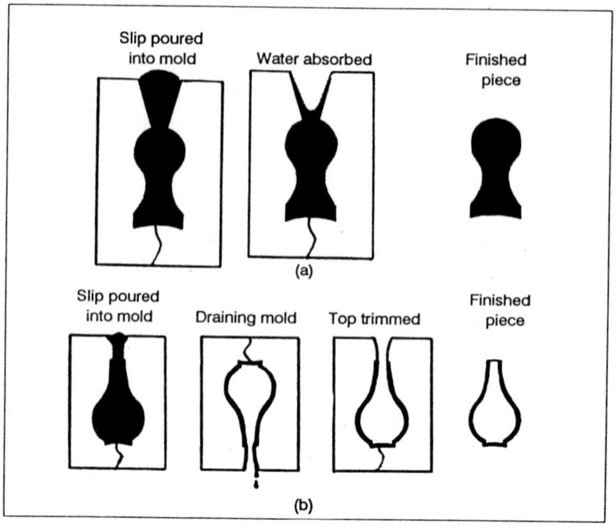

Figure 6.41 Principles of slip casting: a) solid b) drain casting

For porcelain and sanitary ware pressurized slip casting (20–40 bar) has been developed since 1970, in order to increase the rate of water removal and to decrease production times. Removal from the mould, which in this process is made of porous plastic (or sometimes porous ceramic) is facilitated by applying pressure (underpressure for removal). This process can be regarded as pressurized filtration.

In suspensions or slips of fine particles in a liquid, attractive forces between the particles occur in most cases, which may lead to agglomeration and enhances sedimentation. The most important properties of the slip are its stability against sedimentation during processing, its castability and the resulting green density after drying. These properties are controlled mainly by the type of powder and its concentration, the stabilizing additives, the viscosity and the pH value. An adequate stability against sedimentation is achieved by the following means:

- Electrostatically, by using liquids with dissolved surface active ions or polarized molecules, which are adsorbed to the particle surfaces. Repulsion between the equally charged particle surfaces prevents agglomeration and sedimentation, at least for the time interval necessary for handling and pouring the slip.
- Sterically, by absorbing surface active neutral organic compounds. This provides a barrier between the particles and the smallest possible distance between the particles is increased, which diminishes the attractive forces between them. This method is also applicable in tandem with electrostatic stabilisation.

The slip stability depends on the relation between the attractive (van der Waals) and the repulsive (electrostatic or steric) forces between the particles on the one, and their kinetic energy from particle motion on the other hand. The

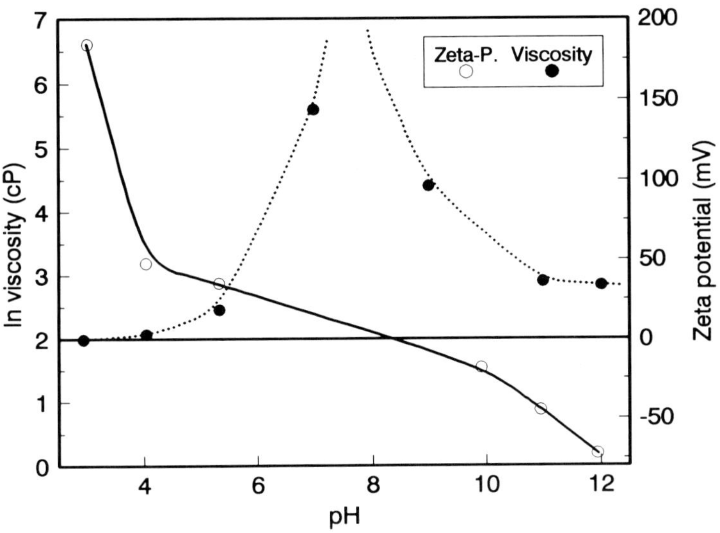

Figure 6.42 Zeta potential and viscosity of alumina slips in the pH range near the isoelectric point (after data from Karaulov)

kinetic energy relates to the shear rates in slip handling operations. With increasing shear rates, the intensity of particle collisions is enhanced and higher repulsive potentials are necessary for preventing particle agglomeration.

For oxide powders, electrostatic stabilisation is the usual way of stabilising slips. Metal powders can often be treated similarly, as thin oxide layers cover the metal surface. In aqueous suspensions these oxides form surface OH groups, which undergo acid and base reactions, forming OH_2^+ by accepting or O^- by releasing a hydrogen ion. The concentrations of these species and thus the resulting surface charge depends on the hydrogen ion (H^+) concentration in the suspension, which is characterised by its pH. Depending on pH, the so called Zeta-potential, which is an indicator for the surface charge, changes from positive to negative values as given in Fig. 6.42. The slip viscosity correlates to the Zeta-potential. It decreases with the pH-difference to the isoelectric point, where electrostatic repulsion has its minimum and particles tend to agglomerate. Thus, slip properties can be adjusted by controlling the pH of the suspension.

The surface charge of the particles can be dramatically changed by addition of small amounts of polyelectrolytes, which form multivalent ions from their ionizable side groups. Their adsorption to the particle surface with one or a few side groups reverses and multiplies the particle charge. Examples for such deflocculants are sodium silicate and -pyrophosphate (inorganic), or sodium- or ammonium polyacrylate (organic compounds).

Optimized slips are castable to green densities (e.g. 70%) much higher than those of poured or vibrated powders. After liberation of the parts from the deflocculant, conventional sintering of the dried parts follows. Only bodies completely free of casting defects yield good mechanical properties in the final product. The main disadvantage of slip casting, however, is its unsuitability for mass production of small parts. Consequently, it has so far a very limited importance for powder metallurgy. On the other hand slip casting is an important forming method in the ceramics industry and has also been applied for short fibre and whisker-containing composites and for coatings.

An interesting speciality is *continuous slip casting* of thin sheets, strips or even foils with thicknesses between 0.2 to 1.5 mm (*tape casting*). This process has been developed for the manufacture of electronic parts such as ceramic substrates, housings, capacitors etc. and is now even used for heat exchangers and catalyst carriers. The slip (slurry) contains a plasticiser and is cast on a hot, stainless steel conveyor belt where it loses most of its moisture and remains as a plastic, rubber-like strip, containing ~ 50 vol% powder, 35 vol% organics, and 15 vol% porosity. The tape can be cut or punched as well as coiled and stacked and becomes a rigid ceramic foil after dewaxing and sintering. This process is applied in mass production, e.g. for Al_2O_3 substrates for electronic devices. A new modified slip casting process is called '*wet pouring*', in which a solvent-powder mixture with a few percent of binder is poured into a mould. After evaporating the volatile (mostly organic) solvent, the green body is mechanically stabilised by the binder. The low volume fraction of the binder (e.g. 2%) allows fast thermal debinding, followed by conventional sintering.

6.2.3 Electrophoretic Forming

The well-known effect of electrophoresis is the transport of charged (preferably colloidal) particles in an electric field. This phenomenon can be used for controlled deposition of fine particles from dispersions at an electrode. This is not practicable in aqueous slips because electrophoretic transport and electrolytic processes are difficult to separate. The use of nonaqueous slips with only a few percent of water, however, offers the possibility of controlling the charges at the particle surface by additives and avoiding electrolysis, so that the electric transport is almost exclusively performed by the charged particles.

A range of 50–1200 V has been used to deposit 5–10 mm thick layers of Al_2O_3 or SiC, with green densities of ~ 60% of TD and with very homogeneous particle packing and narrow pore size distributions. At the higher voltage range the deposition time for the above thickness is only a few minutes. Extremely fine powders can be compacted (which is probably of particular interest for nanosized powders), but only simple geometries, depending on the shape of the electrodes (e.g. plates, tubes) can be achieved. The sinterability of these compacts is very good.

6.3 SPRAY FORMING

The spray deposition of metals, alloys, composites and compounds (oxides) has been practised for several decades, especially in order to obtain surface coatings with different functions. These coatings are mainly applied in order to increase wear, corrosion and oxidation resistance, for thermal or electrical insulation, or to provide special electrical, chemical or optical properties. Among the deposition methods the thermal spray technologies — flame and (since 1957) plasma spraying — are the most important.

Over the last decade spray technologies have also been developed for the manufacture of free standing near-net-shape parts, semi-finished products, especially tubes and rods of large diameter and length, and clad billets. The principle routes are the direct atomizing of molten metal onto solid substrates by pressurised gas, or the plasma spraying of powders or powder mixtures, where the particulate material is partially or fully melted within the plasma. The particulate material is highly accelerated in the gas stream in the form of small droplets (30–40 μm), when strong particle deformation and densification of the deposit takes place during impacting. Thus, spray deposition by either method is an alternative route for conventional casting or for powder shaping in dies. However the plasma route is much slower and extremely costly for large components.

The best known spray forming process, patented and commercialised now for several years, is the *Osprey Process*, described first in 1981. This is the direct conversation of molten metal into a fine grained, semi-finished product by means of a combined gas-atomising/deposition process. The principle is illustrated in Fig. 6.43, in which the five stages of the process are shown on the left and the process parameters which determine the integrity of the preforms produced on the right. The density of the spray deposit can be > 97% or even > 99% of theoretical density depending on alloy and atomisation gas. The solidification conditions, together with the continuous destruction of dendrites in the semi-solid surface generate abundant nuclei resulting in microstructures with fine grain sizes similar to those found in other powder metallurgy products.

According to the mathematical analyses of the thermal behaviour of the droplets a continuous decrease of the proportion of liquid occurs within the flight distance. The exact amount of liquid present on impact depends upon the alloy (Fig. 6.44). While solidification during flight is rapid, the fraction of liquid which remains in the deposit undergoes relatively slow cooling at the high deposition rates used in practice (e.g. several tens of kg per minute).

The Osprey Process is applied mainly for the production of stainless steel tubes up to 8m in length (Sandvik), for high-speed steel roll preforms (up to 250 kg), for rod and wire mills (Sumitomo), for superalloy billets (up to 220 mm Ø), copper alloys and by several companies for the production of special aluminium alloys and composites. While the high-speed steel preforms are subsequently forged using standard equipment, the aluminium billets are further processed by extrusion. The microstructure of spray-deposited and forged

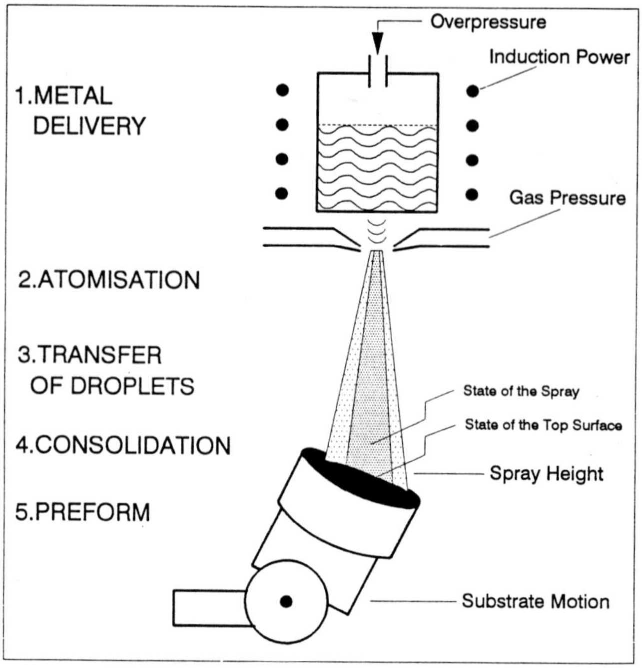

Figure 6.43 Schematic of Osprey spray deposition showing the five stages, five independent process parameters and two critical dependent parameters (Mathur *et al.*)

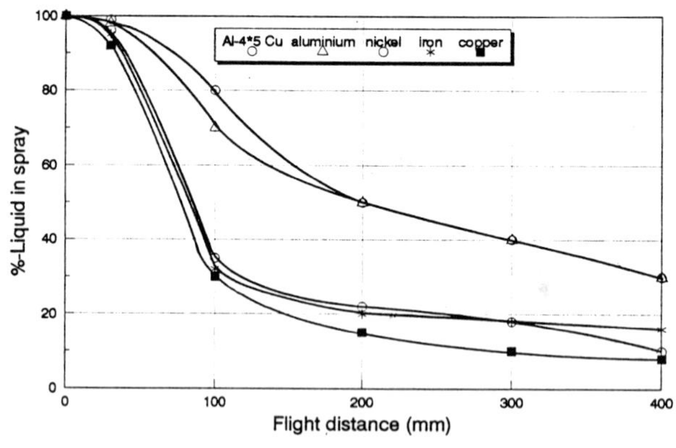

Figure 6.44 Predicted variations of percentage liquid in the spray with flight distance computed from the temperature history of individual droplets in flight (Mathur *et al.*)

high-speed steel shows fine and uniform carbide particles, equivalent to or only slightly coarser than in sintered products, but significantly finer than that of cast and wrought alloys. Consequently, the hot workability of such products is excellent. After heat treatment properties similar to conventional powder metallurgy products are achieved.

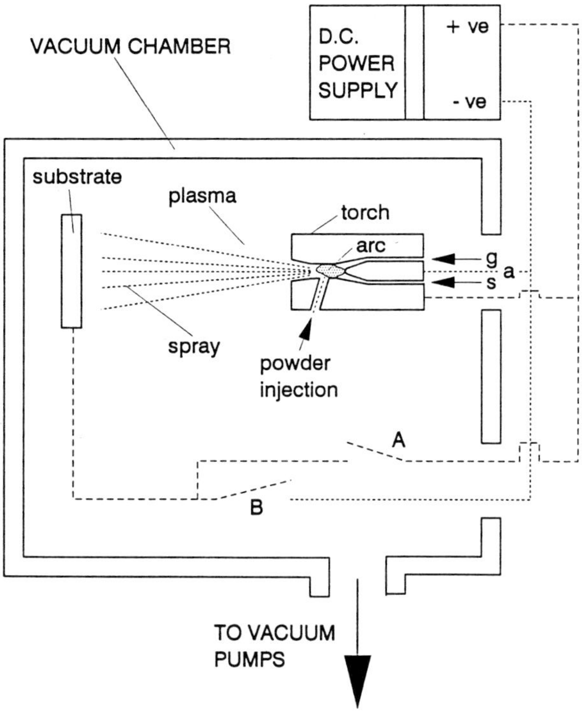

Figure 6.45 Schematic of vacuum plasma spraying

The process allows for a considerable reduction in the number of processing steps with improved economy (especially for large billets) compared with normal powder routes. Powder handling and consolidation are not involved as separate procedures. Some powder overspray (not hitting the billet) is difficult to avoid, however, and ranges from 10–30%. In many instances this can be recycled by powder injection, a technique which has been developed to create in situ composites. Thus a wide variety of particle-strengthened metal matrix composites are available. The solidification conditions minimise segregation or agglomeration of particulates. For example, dispersion-strengthened Al–Al_3C_4 and Al–20% SiC mixtures can be sprayed successfully.

For vacuum plasma spraying, powders have generally to be used as raw materials. This method is advantageous when materials are involved which are difficult and costly to process by other means. It is claimed that a plasma spray process can compete with conventional powder route fabrication methods when high deposition rates (say 100 kg/h) are achieved. The powders used should not be too fine (> 100 μm), otherwise an undesirably high oxide content would be introduced into the deposit. A schematic of the process is given in Fig. 6.45.

FURTHER READING

Metals Handbook, 9th edn, Vol. 7: Powder Metallurgy, ASM, Metals Park, Ohio, 1990.
D.J. Cumberland and R.J. Crawford, 'The Packing of Particles', J.C. Williams and T. Allen, eds, *Handbook of Powder Technology,* Vol. 6, Elsevier, New York, 1987.
R.M. German, *Particle Packing Characteristics,* Princeton NJ. MPIF, 1989.
A.D. Rosato, T. Vreeland, Jr. and F.B. Prinz, 'Manufacture of Powder Compacts' , *International Materials Review* (1991), **36**, (2), 45–61.
J.M. Capus, R.M. German, eds, *Compaction and Other Consolidation Processes,* Vol. 2 of *Advances in Powder Metallurgy and Particulate Materials,* Princeton NJ, MPIF, 1992.
R.M. German, *Powder Metallurgy Science,* MPIF, Princeton, N.J., 1984.
W. Schatt, *Pulvermetallurgie Sinter- und Verbundwerkstoffe,* Hüthig Verlag, Heidelberg, 1986.
G. Zapf, 'Pulvermetallurgie', G. Spur, ed., *Handbuch der Fertigungstechnik,* Vol. 1, Carl Hanser Verlag, München, Wien, 1981, 823–977.
E. Ernst, *Axiale Preßvorgänge in der Pulvermetallurgie,* VDI-Fortschrittsberichte 259, VDI-Verlag, Düsseldorf, 1992.
DIN-Taschenbuch 247, Pulvermetallurgie, Beuth, Berlin, Köln 1991.
P.J. James, *Isostatic Pressing,* Elsevier Applied Science, London and New York, 1991.
M. Koizumi and M. Nishihara, eds, *Isostatic Pressing,* Elsevier Applied Science, London and New York, 1991.
E. Arzt, and H. Fischmeister, 'Fundamental Aspects of the Compaction of Metal Powders,' *Mémoires scientifiques revue métallurgie,* Oct. 1979, pp. 573–579.
A.S. Helle, et al., 'Hot Isostatic Pressing Diagrams: New Developments,' *Acta Metall,* 1985, **33**, (12), pp. 2163–2174.
M.F. Ashby, 'The Modelling of Hot Isostatic Pressing,' *Proceedings Hot Isostatic Pressing – Theories and Applications,* T. Garvare, ed., Centek Publishers, Lulea, 1988, 29–40.
H. Kolaska, ed., Fortschritte bei der Formgebung in Pulvermetallurgie und Keramik,' *Pulvermetallurgie in Wissenschaft und Praxis, 7. VDI-Verlag, Düsseldorf, 1991.*
R.M. German, *Powder Injection Moulding,* MPIF, Princeton, N.J., 1990.
G.Y. Onoda and L.L. Hench, eds., *Ceramic Processing before Firing,* John Wiley & Sons, New York, 1978.
M. Takahashi and S. Suzuki, 'Compaction Behaviours and Mechanical Characteristics of Ceramic Powders,' *Handbook of ceramics and composites: Syntheses and properties,* N.P. Cheremisinoff, ed., Marcel Dekker, Inc., New York, 1990, Vol. 1, pp. 65–98.
J.S. Reed, *Introduction to the Principles of Ceramic Processing,* John Wiley & Sons, New York, 1987.
R. Prümmer, *Explosivverdichtung pulvriger Substanzen,* Springer Verlag, Heidelberg, 1987.
P.R. Roberts and B.L. Ferguson, 'Extrusion of Metal Powders', *International Materials Review* (1991), **36**, (2), 62–79.
A.G. Letham and A. Lawley, 'The Osprey Process: Principles and Applications', *The International Journal of Powder Metallurgy,* (1993) **29**, (4), 321–329, and the following papers in this volume.
G. Naeser and F. Zirm, 'Das Walzen von Bändern aus Eisenpulver,' *Stahl und Eisen,* 70 (1950), S. 995–1004.
R.K. Dube, 'Metal Strip via Roll Compaction and Related Powder Metallurgy Routes,' *International Materials Reviews,* (1990), **35**, (5), 253–291.
H.A. Kuhn and B.L. Ferguson, *Powder Forging,* Princeton NJ, MPIF, 1990.
V. Mironov, 'Pulververdichten mit Magnetimpulsen; Plansee-Berichte für Pulvermetallurgie,' (1976), **24**, 175–190.
Spray Forming, Science, Technology and Applications, Princeton NJ, MPIF, 1992.
P. Mathur, D. Apelian and A. Lawley, 'Fundamentals of Spray Deposition via Osprey Processing,' *Powder Metallurgy,* (1991) **34**, (2), 109–111.

compact the pores (assumed to be spherical, Fig. 7.1b) are subjected to compressive stresses according to:

$$\sigma = \frac{2\gamma}{r_{pore}} \quad (7.3)$$

Furthermore, an excess vacancy concentration (ΔC) over the equilibrium concentration (C_0) exists beneath the concave surface according to the Thomson-Kelvin-equation:

$$\frac{\Delta C}{C_0} = \frac{\gamma V_0}{RT\rho} \quad (7.4)$$

as well as a reduced vapour pressure ($-\Delta P$):

$$-\frac{\Delta P}{P_0} = \frac{\gamma V_0}{RT\rho} \quad (7.5)$$

where V_0 = molar volume, R = gas constante, T = temperature in K, P_0 = vapour pressure over the flat surface.

This leads to vacancy diffusion into regions with lower vacancy concentrations (flat or convex surfaces) or to vacancy sinks like grain boundaries. This may also cause an isothermal transport via the gas phase (Fig. 7.2.). Driving forces for surface and grain boundary diffusion into the neck curvature can also be defined.

The aim of the sintering must lie in the elimination of these stresses, the disappearance of the zones with excess vacancies and the equalising of the vapour pressures by transport of atoms. These *thermodynamic* considerations do not include preferences for the *kinetics* of the material transport, however.

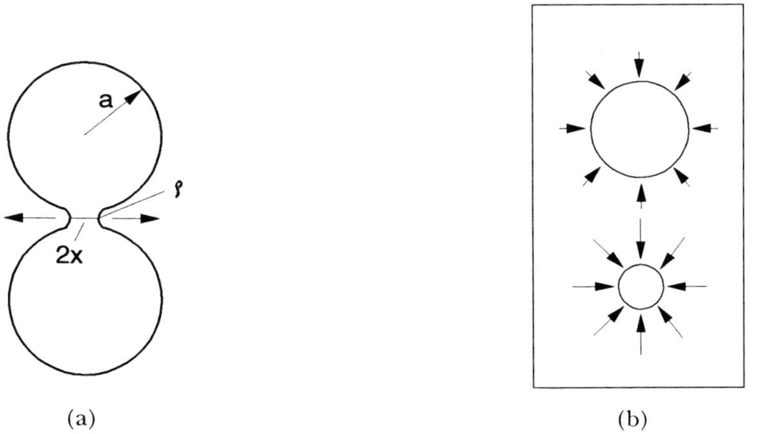

Figure 7.1 Laplace stresses (a) in a two-particle neck (a = particle radius, x = sintered neck radius, ρ = neck curvature radius) (b) around pores

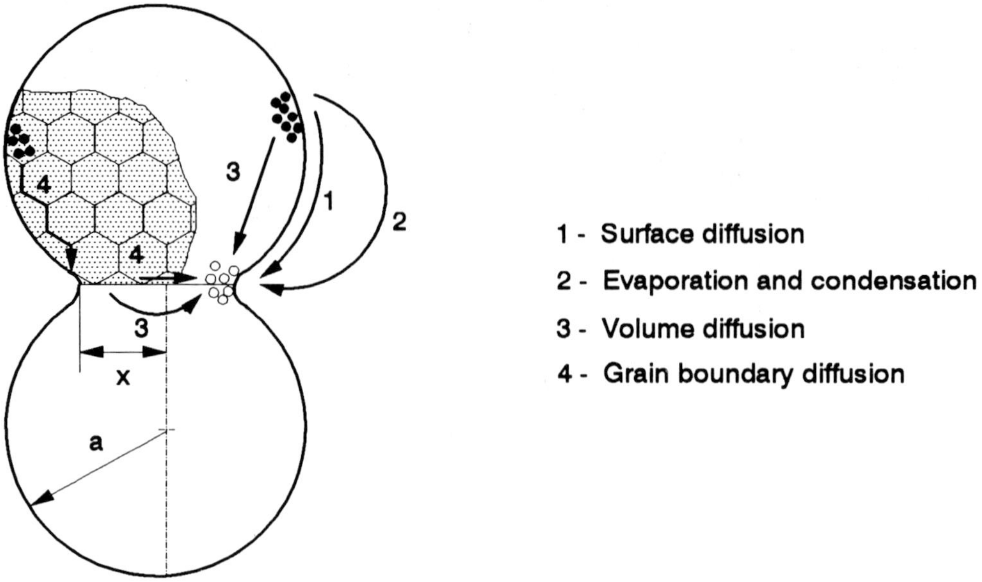

Figure 7.2 Possible sintering mechanisms in a neck area (except plastic and viscous flows)

7.1.3 Stages of Sintering

The sintering process can be subdivided phenomenologically into three stages, according to Fig. 7.3 which holds mainly for solid state sintering.

- *First stage*: The particle contacts are transformed to sintered bridges, called 'necks'. Even before sintering, the contacts exist in form of microplanes instead of point contacts, their extension dependent upon compacting pressure. According to model experiments (see section 7.2.2), the neck growth follows an exponential time law. During this stage the powder particles remain discrete. Grain boundaries are often formed between two adjacent particles in the plane of contact. The centre of the particles approach only slightly (equating to very little shrinkage).
- *Intermediate stage*: When the x:a ratio exceeds a certain value after strong neck growth, the single particles begin to loose their identity. A coherent network of pores is formed and grain growth occurs, which results in a new microstructure. The grain boundaries usually run from pore to pore. Porosity is greatest open, i.e. connected with the surface. Most of the shrinkage takes place at this stage.
- *Final stage*: Between 90 and 95% of the theoretical density, the relative proportion of the closed pore space increases rapidly. The isolated pores become increasingly spheroidised. In cases where gases cannot diffuse out, they

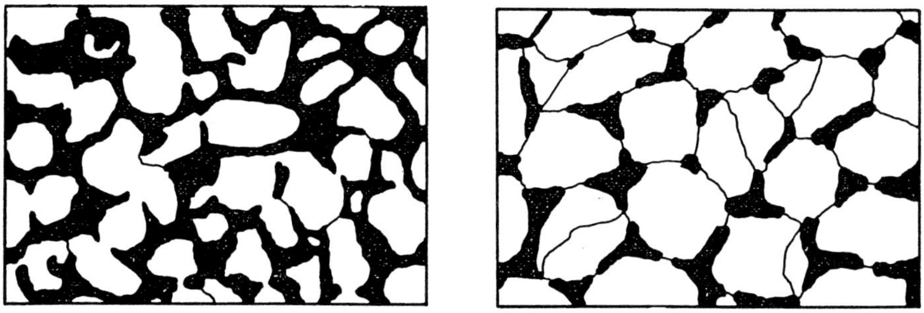
(1) (2)

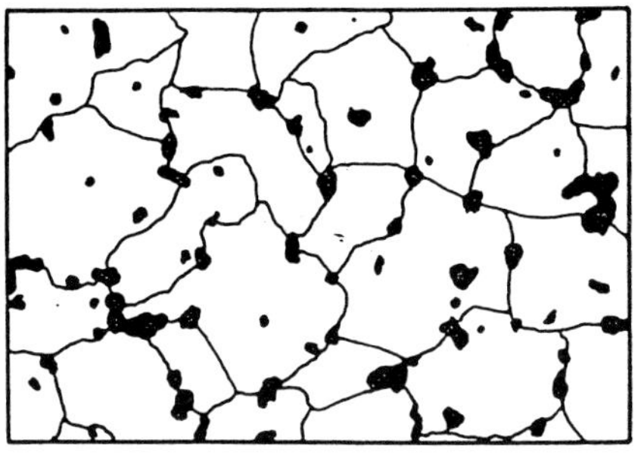
(3)

Figure 7.3 The three stages of sintering 1. First stage 2. Intermediate stage 3. Final stage

are enclosed in the porosity and further densification becomes impossible as soon as the gas pressure reaches equilibrium with the pressure due to surface tension. When the pores are essentially empty (sintering in vacuum) or the entrapped gases are easily diffusible in the solid matrix, further slow densification may occur, especially in fine grained microstructures.

Most of the sintered products, like steels, magnets, cemented carbides, which are specified to high densities and high strength, have reached (but mostly not completed) the final stage or at least the end of the intermediate stage. In products with high porosity and high permeability for gases or liquids, like filters and diaphragms, which are sintered from coarse, often spherical powders, the first stage is essentially the only one reached (see section 11.8). The finer a powder is, on the other hand, the more difficult the first stage is to identify: shrinkage and grain growth commence at low temperatures and sintering proceeds to a large extent in the heating-up phase (see section 7.2.2.5). It must be remembered that in highly compressed powders very extended particle contact areas exist even before sintering, and during sintering very little shrinkage occurs (e.g. in sintered steel parts). In these cases, the early and the intermediate stages cannot be distinguished in that way.

The sintering models devised by many authors are related only to specific stages: The neck growth model for the early stage, the pore shrinkage models for the final stage and powder shrinkage investigations for the intermediate and final stages.

7.2 SINGLE COMPONENT SINTERING

7.2.1 Mechanisms of Material Transport

Table 7.1 summarises the conceivable mechanisms that can occur during solid state sintering. Adhesion processes, which occur in every type of particle contact, may constitute the dominating sintering principle only in impact sintering. When the effective impact period is very short at low temperature, it is unlikely that particle transport will take part in the true sense. Agglomerations in very finely divided powders and the cold welding of powders or compact bodies can also be attributed to the capture of free surface valencies and to van-der-Waals forces, being responsible for adhesion. During 'conventional' sintering processes, one or more of the material transport mechanisms are effective.

Surface diffusion probably occurs during all sintering processes, especially at low temperatures, in fine powders and in the first sintering stages, when the specific surface is still high. Surface diffusion requires the smallest activation energy (slope of the curve) of all types of diffusion (Fig. 7.4). It is generally accepted, however, that it cannot cause pore shrinkage and hence densification, but it can cause growth of necks, smoothing of surface roughness and rounding-off of the pores. Obviously, surface atoms have their greatest mobility on convex and their smallest on concave surfaces. Consequently they can be caught in necks with sharp curvatures (see Fig. 7.2). During intermediate and final stages, surface diffusion may control pore spheroidisation. Also grain boundary diffusion is a low-temperature sinter-

Table 7.1 Possible elementary processes (transport mechanisms) during sintering

Without material transport	Adhesion	
With material transport	Surface diffusion	movement of individual atoms or ions
	Volume diffusion via vacancies	
	Volume diffusion via interstitials	
	Grain boundary diffusion	
	Vaporisation and re-condensation	
	Plastic flow	collective movement
	Viscous flow	
	Grain boundary sliding	
	Particle rotation	

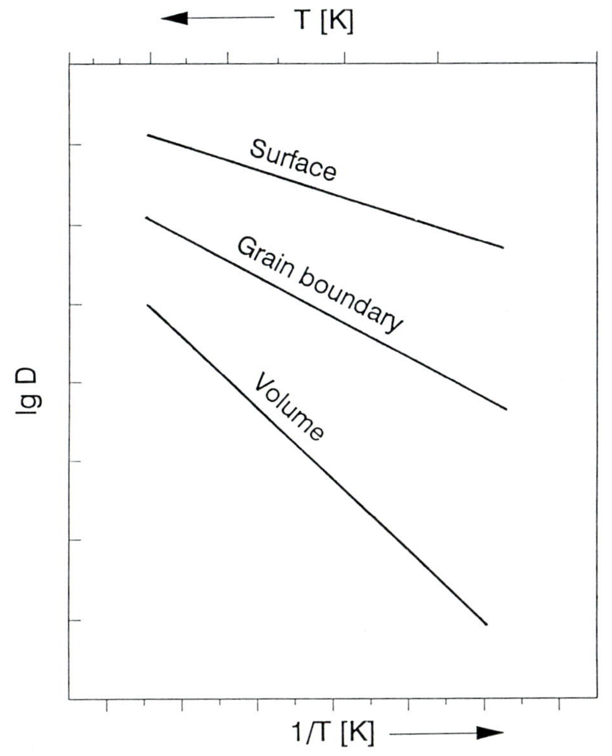

Figure 7.4 Self diffusion coefficients and their temperature dependence in a polycrystalline solid (schematic)

ing mechanism, its activation energy ranging between surface and volume diffusion. In any material, however, grain boundaries are not uniform, their atomistic structure and atom diffusivity depend on the angle between the adjacent grains. So, the statements of activation energy and diffusion coefficients in Fig. 7.4 are mean estimates only. Besides acting as diffusion paths, grain boundaries are sinks for the directed vacancy flow during volume diffusion from pores, which is most important for pore shrinkage in the intermediate and later stages of sintering.

Volume diffusion by vacancies seems to be dominant in many practical cases. Its high activation energy (Fig. 7.4) rules out more or less surface and grain boundary mechanisms at higher temperatures and at higher sintered densities, when the internal surfaces become small. Vacancy gradients exist between the undistorted lattice and both curved surfaces and pore edges, between surfaces having different curvatures (Fig. 7.2) and between distorted and undistorted parts of the lattice. The diffusional flow during sintering is controlled by the type and geometrical arrangement of the vacancy sources and sinks, which can be defined as follows:

Vacancy sources	*Vacancy sinks*
concave surface	plane or convex surfaces
small pores	large pores
dislocations	grain boundaries
	dislocations

The participation of dislocations, by climbing, can be effective as well as sources and sinks, a fact which has been disputed for a long time. The appearance of high concentrations of dislocation, formed in the neck area during the earlier stages of sintering (cf. 7.2.2.3), obviously underlines their efficiency in sintering. They provide a particle contact area of increased diffusivity as well as of low viscosity which may cause movement or rotation of entire particles.

7.2.2 Attempts to Define the Sintering Mechanisms

The theoretical as well as the experimental approach for determination the transport mechanisms utilised models with simple geometry. This facilitates to obtaining definite data characterising the time and temperature dependence of sintering. The particle models (Fig. 7.5) used for the earlier stages of sintering consist either of two spherical particles, a single particle lying on a flat plate (b) (both with 'point' contacts) or of a bar – wire combination, consisting of a wire spiral wound on a cylindrical bar (c) or a wire-wire-model, consisting of several superimposed close-packed wire layers (d) or of three twisted wires (e) (all with 'line' contacts). The growth of the contact width is measured microscopically after metallographic preparation. While a two particle model never can simulate a pore, three- or multiparticle models may represent pores. For investigation of the later stages models with cylindrical pores in the form of drilled holes (f) have been used. Close packed wires (d) are also suitable for this purpose, when the behaviour of the 'pores' instead of the necks is observed.

Although all these earlier model investigations have a very limited relevance for practical sintering processes, they helped a lot to achieve a deeper insight into the sintering process.

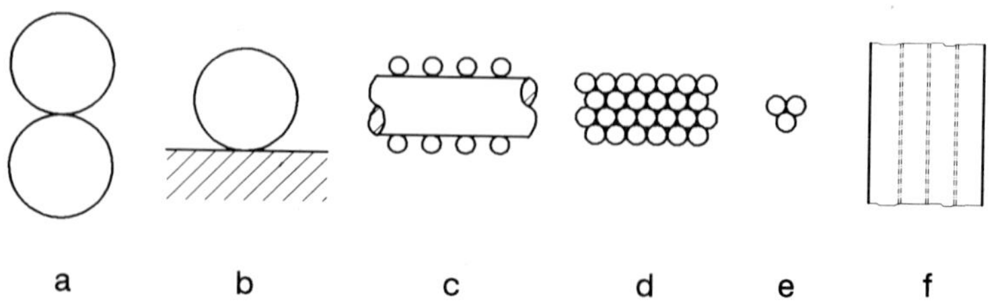

Figure 7.5 Models utilised for investigation of sintering

Table 7.2 Exponents controlling neck growth in the two-particle models

Mechanism	n	m	Particle centre approach
Viscous or plastic flow	2	1	yes
Evaporation and re-condensation	3	1	no
Volume diffusion	4–5	2	yes
Grain boundary diffusion	6	2	yes
Surface diffusion	6–7	3	no

7.2.2.1 *The Two-Particle Model*

Frenkel, Kuczynski and others have shown, that different transport mechanisms are related to different kinetics of neck growth at a given temperature. The basic equation is

$$\frac{x^n}{a^m} = F(T) \cdot t, \tag{7.6}$$

simplified:
$$\left(\frac{x}{a}\right)^n = F(T) \cdot t_i, \; x^n \sim t \tag{7.7}$$

(x = neck radius; a = particle radius; t = sintering time; $F(T)$ = numerical function, depending on model geometry) with the exponents n and m for the different mechanisms according to Table 7.2.

Especially flow, volume and grain boundary diffusion processes may provide particle centre approach, yielding shrinkage in a powder mass.

These correlations have been investigated in many metallic and nonmetallic materials. During sintering of amorphous substances (glass spheres) an exponent n=2 is generally observed, as shown in Fig. 7.6. The material is transported by viscous flow according to the following equation:

$$x^2 = \frac{3\,\gamma\, a}{2\,\eta} \cdot t \tag{7.8}$$

(γ = surface tension
η = viscosity)

η is related in amorphous materials to the self diffusion coefficient D by:

$$\frac{1}{\eta} = \frac{D \cdot \delta}{K \cdot T} \tag{7.9}$$

(δ = atomic distance, K = Boltzmann-constant, T = Temperature in K)
The capillary stresses in the contact area according to the Laplace-equation are acting as the driving force, as well for neck-growth as for particle centre approach.

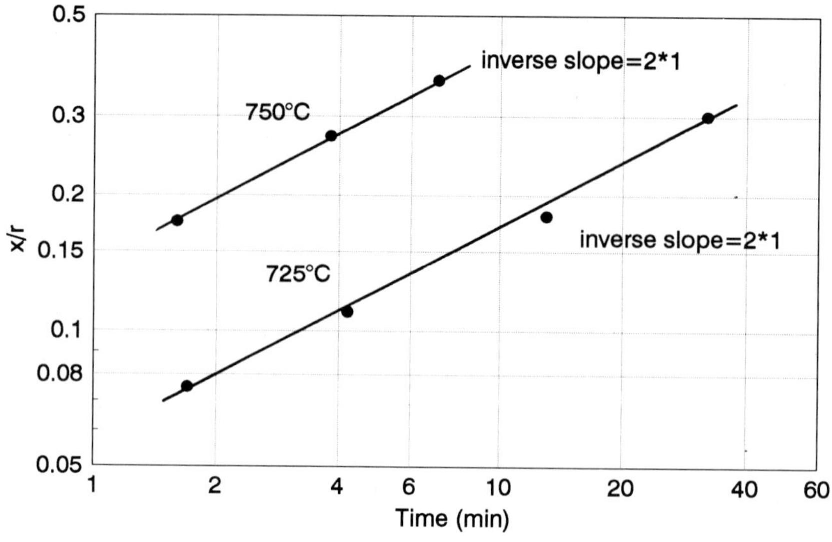

Figure 7.6 Neck growth between glass spheres in accordance with $x^2 \sim t$; inverse slope 2.1 (Kingery and Berg)

Evaporation and condensation as the dominating principle (n=3) are seldom observed, because the vapour pressure of most materials at their sintering temperature is very low. There are a few exceptions, however, namely during sintering of ZnO (above 1050°C) and NaCl in which $n \approx 3$ is experimentally observed. The results are supported by the observation of no particle centre approach.

In most metallic and oxidic materials, volume diffusion has been defined as the dominating process, e.g. in the sintering of copper wire-bar models (Fig. 7.7) which exhibit clearly the relation $x^5 \sim t$ under most conditions investigated. Although some uncertainty in the deviation of this $x{:}t$ relationship as well as in the experimental procedure exist and criticisms have been published, the comprehensive factual material confirms the growth law $x^5 \sim t$ for volume diffusion.

$$\frac{x^5}{a^2} = k \frac{\gamma\, \delta^3\, D_v\, t}{K\, T} \qquad (7.10)$$

(k = constant, depending on model geometry, D_v = volume diffusion coefficient)

It was originally claimed that this type of model experiments allows, when the constant k and γ in the solid state are known, in the determination of volume diffusion coefficients D_v without using radioactive tracers, as far as the process is essentially controlled by volume diffusion. In several cases, D_v-values obtained from tracer and from model sintering experiments have been determined as being within the same order of magnitude, whilst other experiments gave different results, however. In summary, there are several uncertainties: the flux equations for the atom transport from surface and grain boundaries to the

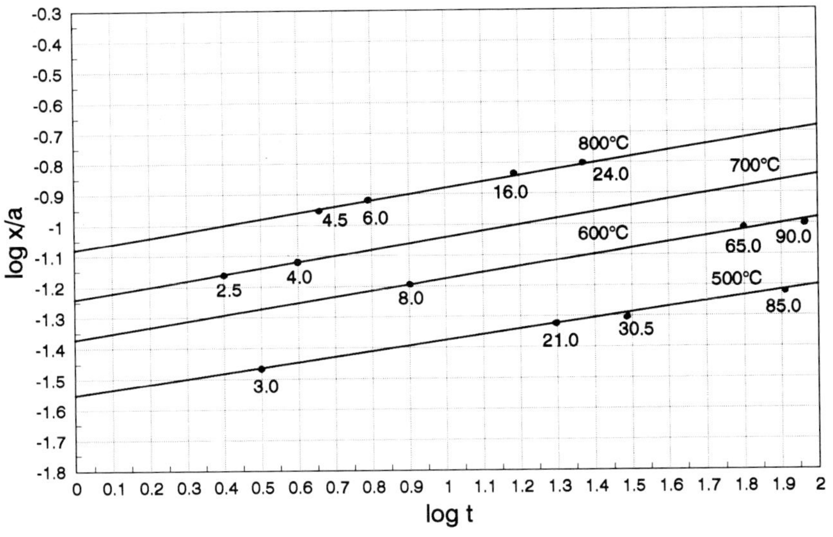

Figure 7.7 Neck growth in a copper wire-rod model in accordance $x^5 \sim t$ and $x^7 \sim t$ (500°C). Inverse slope 5 and 7, resp. (Kuczynski)

neck, the disregard of defect-generation in the neck area (see section 7.2.2.3) and the sometimes badly-defined neck geometry (see section 7.7) does not allow the general application of a special equation. In addition it can rarely be assumed that atom transport takes place only through the lattice, when a grain boundary is present in the neck area.

Figure 7.7 shows at the lowest temperature an inverse slope of 7 which defines surface diffusion as the controlling mechanism ($x^7 \sim t$). Because the activation energy of surface diffusion is lower than for volume diffusion, its contribution increases with decreasing temperature.

7.2.2.2 The Three- and Multi-Particle Model

The step from the two-particle – to the three- or multi-particle-model leads to the recognition of more important details on sintering, especially the simulation of pores. During sintering of three spherical particles, forming an angle (Fig. 7.8a), or a straight row of particles, the angle between the particles can change with time. Consequently, pores can shrink as well as grow (Fig. 7.8b) depending on the geometrical variations within the particle mass. This phenomenon of pore growth is sometimes observed during sintering in practice and was formerly difficult to explain. Particles as such are able to move or to rotate to some extent since the surface areas around the sintering necks are of much lower viscosity than neighbouring areas and the interior (see section 7.2.2.3). As a consequence the neck growth may occur non-symmetrically and necks already formed may be even destroyed as well as new contacts generated

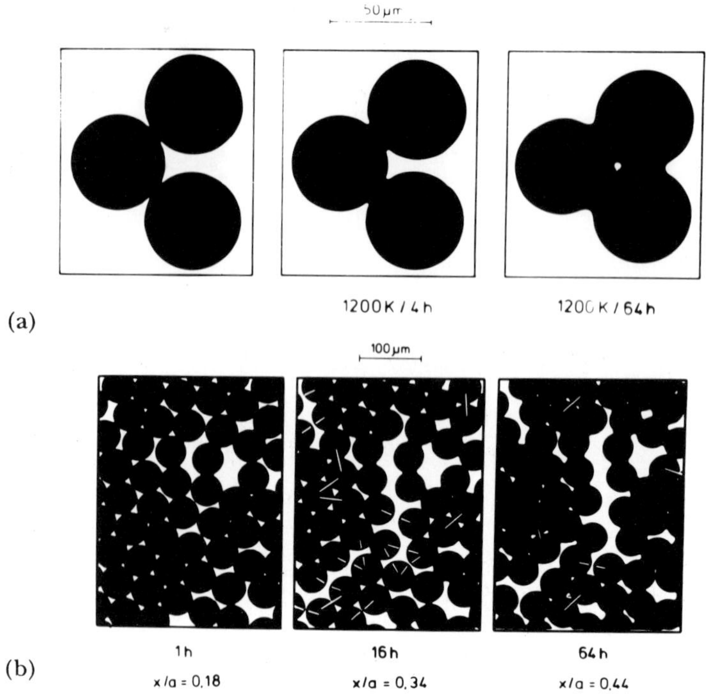

Figure 7.8 Behaviour of a multiparticle Cu-sphere array during sintering (Exner) a) change of angle between particles b) neck destruction and new pore formation

during the course of sintering. The driving force for the rotation is the grain boundary energy in the neck area, being different for each neck due to different orientation of adjacent grains. This energy is minimised by the rotation process. Thus, only multiparticle models can describe the realistic 'disturbed' neck growth between the particles while the two-particle models represent only an idealised, undisturbed growth process. However, even the simplest model provides difficulties for a rigorous mathematical treatment.

7.2.2.3 The Role of Dislocations and Decreased Neck Viscosity

As early as 1963 Morgan observed the spontaneous formation of dislocations in the neck area during sintering. This was confirmed later by several authors in metals as well as in ionic crystals and was investigated in detail by Schatt and co-workers. Around the grain boundary formed in the neck between Cu-spheres high dislocation concentrations ($> 10^9$ cm^{-2}) are observed, their geometrical arrangement being time dependent and still observed after long sintering times of 96 h. Their concentration does not exceed, however, values necessary for recrystallisation, being the reason for their stability. The dislocations reach a depth of > 20 μm and extensions larger than the neck width (Fig. 7.9). They are obviously formed due to the Laplace stresses in the neck area and demonstrate the occurrence of microplastic processes.

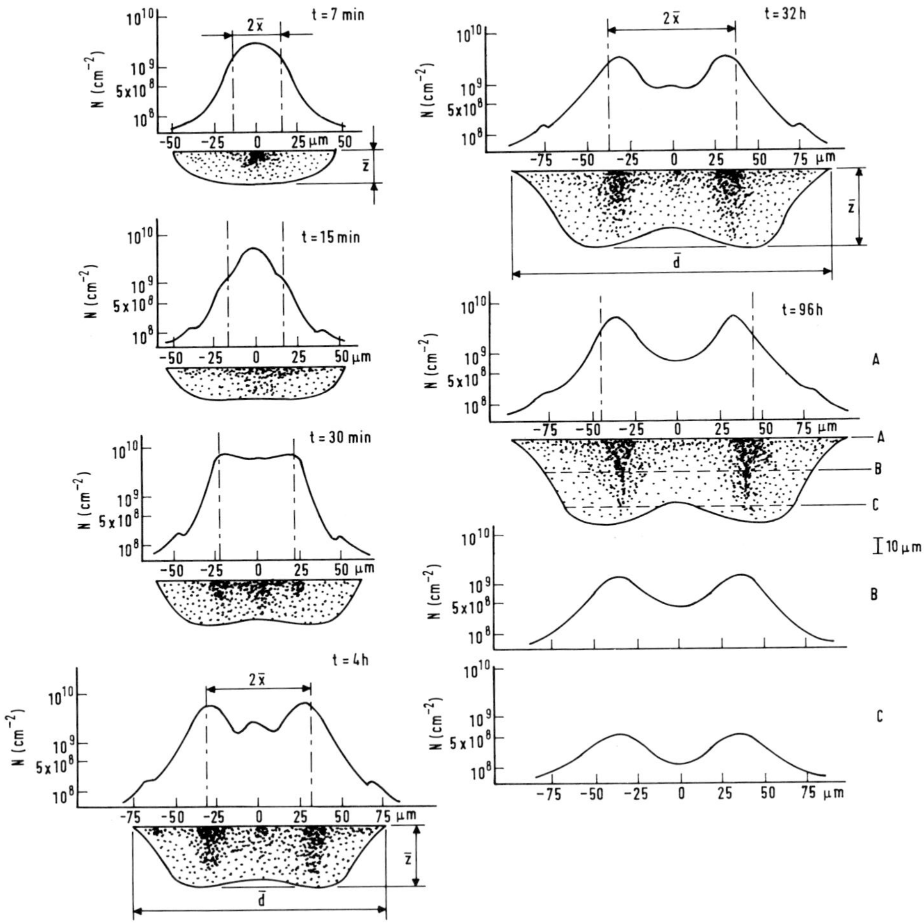

Figure 7.9 Generation of dislocation during sintering of a Cu-sphere-plate model (Schatt) (N = dislocation density [cm^{-2}], $2x$ = neck width, t = sintering time, A, B, C = dislocation density distributions at the Cu-plate surface (A) and different depths (B, C), z = maximum depth for dislocation formation, d = dislocation area diameter

The findings are in contrast to former statements, that the Laplace stresses are unable to generate plastic processes in the necks, apart from the very early stages. On the other hand plastic flow as a defined gliding process seems not to be the controlling mechanism, since the generation of dislocations by microplastic processes is not identical with plastic flow as a controlling mechanism for the entire sintering process. It is believed that dislocation-controlled creep takes part in this sintering stage: The driving force for diffusion as a neck growing mechanism is not only the vacancy gradient (according to the Thomson-Kelvin equation 7.4) but also the healing of dislocations by climbing, which is also diffusion controlled. There is experimental evidence that the diffusion coefficients acting during sintering D_{eff} can be higher than the equilibrium values D_v

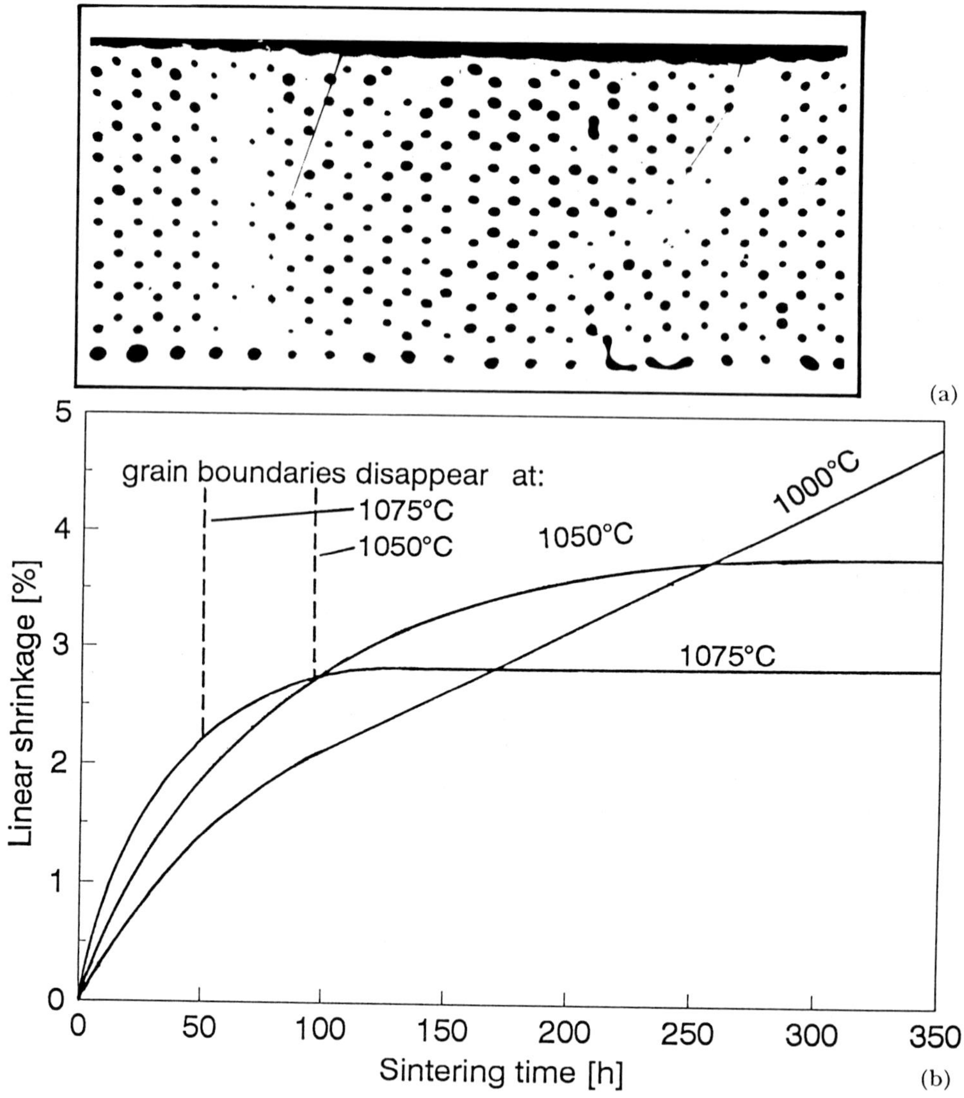

Figure 7.10 Preferred pore shrinkage near grain boundaries in a Cu close-packed wire model (Alexander and Balluffi) a) sintered wire model b) shrinkage vs. time

measured in conventional diffusion experiments, values of D_{eff}/D_v being claimed between 40 and 90. This is in contrast to earlier measurements of D_v – values for sintering models, as explained in 7.2.2.2.

The consequence of the increased defect concentration in the neck area is a strongly decreased viscosity in the neck regions. These 'softened' contact regions may promote the movement of entire particles ('particle rotation') during sintering. This result is transferable to the contact areas of powder compacts (see section 7.2.2.5) and provides even similarities to liquid phase sintering (see section 7.3.3).

7.2.2.4 *Experiments with Pore Models*

Strong experimental evidence has been obtained for the role of grain boundaries as vacancy sinks in diffusion controlled pore shrinkage, as demonstrated in Fig. 7.10a on a close-packed wire model. The decrease of pore size is enhanced near grain boundaries, while isolated pores being far away from grain boundaries are nearly stable over long periods. The shrinkage completely stops when, due to grain growth, grain boundaries are no longer present in the wire pack (Fig. 7.10b). The role of grain boundaries for enhancement of sintering is one of the most important features in the intermediate and later sintering stages.

7.2.2.5 *Powder Shrinkage Experiments*

Several powder shrinkage experiments demonstrate directly the source-sink-mechanism between pores and grain boundaries, as shown in Fig. 7.11. Not only pores attached directly, but also near grain boundaries, are subjected to preferred shrinkage.

A number of mathematical correlations have been derived to describe *isothermal powder shrinkage*, especially for volume diffusion as the controlling mechanism. They are derived partly from neck-growth models, which can be extrapolated to particle-centre approach, being equivalent to powder shrinkage. Others come from single pore shrinkage models including grain boundaries, whilst others are derived from stereological considerations, some are purely empirical. All these shrinkage equations are related to the isothermal period only.

Figure 7.11 Preferred pore shrinkage near grain boundaries in Al_2O_3 (Burke)

A few examples of these attempts are given as follows: Linear shrinkage ($\Delta l / l_o$) has been derived by Kingery and Berg from Kuczynski's model as:

$$\frac{\Delta_l}{l_o} = \left(\frac{20\gamma \cdot \delta^3 \cdot D_v}{\sqrt{2} \cdot a^3 \cdot k \cdot T}\right)^{\frac{2}{5}} \cdot t^{\frac{2}{5}}; \quad \frac{\Delta_l}{l_o} \sim t^{0.4} \tag{7.11}$$

and by Pines as

$$\frac{\Delta_l}{l_o} \sim t^{0.5}, \text{ when } D_v \sim \frac{1}{t^{0.5}} \tag{7.12}$$

This involves the assumption of a decreasing D_v during sintering, which is sensible to some extent due to excess defect healing, originated from powder preparation or defects generated in the neck area. Grain growth is not considered. Coble derived a shrinkage equation from a stereological pore model considering grain growth during sintering according to

$$d^3 - d_0^3 = k\,(T) \cdot t \tag{7.13}$$

This often takes place in fine powder compacts, and leads to shrinkage kinetics according to

$$\frac{\Delta_l}{l_o} \sim lg\, t \tag{7.14}$$

Several metallic and ceramic fine powders seem to obey this shrinkage relation. Neither equation seems to have a general significance, however, which partly is the result of the complexity of the powder system. The main reason for the limited significance of *isothermal* shrinkage equations, however, is the extended shrinkage in the heating-up period, which is obviously not controlled by conventional diffusion coefficients.

Analysis of the *non-isothermal shrinkage* has shown that this range has a considerably higher significance than estimated earlier. A great part of the shrinkage achieved during the full sintering cycle as well as a maximum of shrinkage rate $\dot{\varepsilon} = d\varepsilon/dt$ can be observed during the heating-up range, as demonstrated in Fig. 7.12. When two maxima are observed a change of the controlling mechanism is said to take place. The distinct $\dot{\varepsilon}$ maxima being as high as 10^{-2} min^{-1}, are connected with a strong decrease in the defect (vacancy and dislocation) density in the compact, as analysed by positron annihilation spectroscopy. The defects are formed during compaction as well as in the sintering neck area during the conversion of the high energy contact boundary into a low energy high angle grain boundary. It is claimed that the material transport in the intensive shrinkage stage is based on particle rotation being possible due to the low viscosity of the contact zones (Schatt). Thus, a realistic analysis of a sintering process with respect to shrinkage has to consider the non-isothermal period. These results obtained with the model (cf. 7.2.2.3) can be transferred to powder compacts in principle, the amount of particle movement is dependent, however, on the internal geometry of the compact and is most significant at fine particles and low

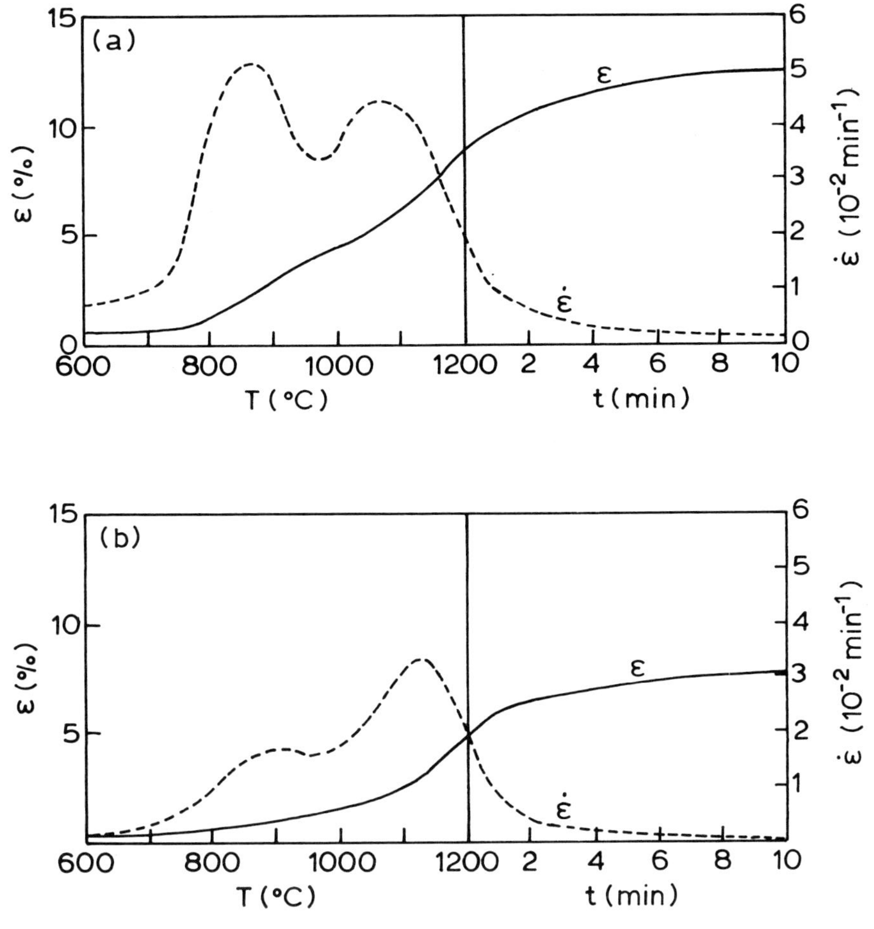

Figure 7.12 Shrinkage (ε) and shrinkage rate ($\dot{\varepsilon}$) of Ni-powder compacts of different green density during non-isothermal and isothermal sintering (Schatt et al.) (a) pressed at 300 MPa (~60% TD) (b) pressed at 700 MPa (~75% TD)

compacting pressures. After high compacting pressure (i.e. at higher green density) the particle movement is reduced or even suppressed.

Detailed microfractographic analysis of pressed and sintered powders demonstrate that a pressed contact is not 'homogeneous' but consists originally of many point- and line shaped microcontacts with micropores inbetween (plane porosity see section 9.2). The contacts change during sintering and transform to more plane contacts, while the micropores are filled, as claimed, by particle movement.

7.2.2.6 *Sintering Diagrams*

In several ranges of temperature and time only one sintering mechanism is dominating while under other conditions two or more mechanisms may control the process. This is represented graphically in sintering diagrams, developed at

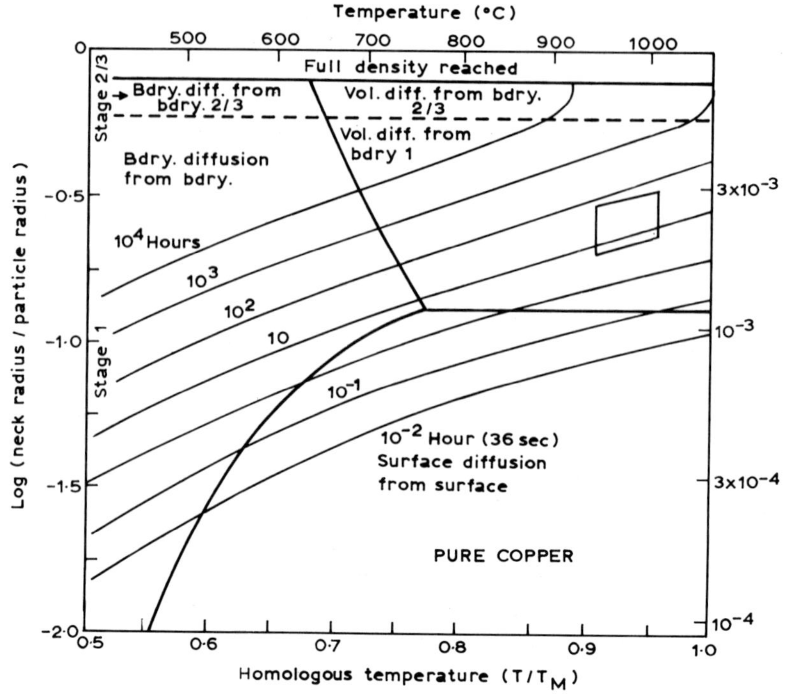

Figure 7.13 Sintering diagram of an 88μm Cu powder compact (Ashby)

first by Ashby. An example is given in Fig. 7.13 representing the sintering behaviour of an 88 μm Cu-powder compact, scheduled as isothermal for the temperature range given. The relative neck radius (x/a) or the sintered density (D/D_{th}) is plotted vs. homologeous temperature while the sintering time is indicated within the diagram. The thick lines define temperature and neck radii values at which two competing mechanisms provide equal contributions, while within these areas one single mechanism dominates. The neck growth or shrinkage values are calculated from the relevant neck growth or powder shrinkage equations, (which are not given here), whose reliability determine the accuracy of the diagrams. Non-isothermal sintering is disregarded.

7.2.3 Sintering Phenomena in Practice

As shown in Chapters 2 and 3 the morphology of powders including their surface properties can be very complicated and depends on particle size, shape, surface roughness, etc. The contact area and contact quality within a pressed or unpressed powder mass and, consequently, the sintering behaviour is very dependent upon external (temperature, time, atmosphere etc.) as well as internal parameters, namely the properties of the powders. The principle influence of important parameters is outlined in this paragraph. Most of them are less regarded in research, unfortunately, than the fundamentals of material transport.

7.2.3.1 Parameter Dependence and Properties Development

The influence of *sintering time and temperature on densification* of low density compacts or loose powders are shown in Fig. 7.14 and 7.15. Most of the shrinkage occurs in the first period of time, especially at high temperatures and in fine powders. Due to the very strong influence of temperature a 'replacement' of a higher sintering temperature by prolonged time is possible only to a very limited extent. The lower the temperature and the coarser the powder, the less densification occurs, but considerable neck growth may be observed (see Fig. 11.17). The easier sintering of fine powders is a consequence of the larger number of particle contacts and the higher driving force for sintering (7.1.2). The shrinkage behaviour of green bodies, cold compacted under different pressures can be seen from Fig. 7.16. After higher precompacting pressure only little shrinkage during sintering occurs. This is the base for the approximate dimensional stability of sintered iron and steel parts. Entrapped gases are assumed to diffuse out of the pores, which is normally fulfilled in green compacts of commercial powders. Very fine powders of high plasticity (like silver) may yield compacts with more closed porosity in which an overpressure of gas during sintering causes swelling (after curve IIIb). In such cases the compacting pressure has to be reduced or the sintering cycle adjusted.

When a high density is reached during sintering the *grain size* generally

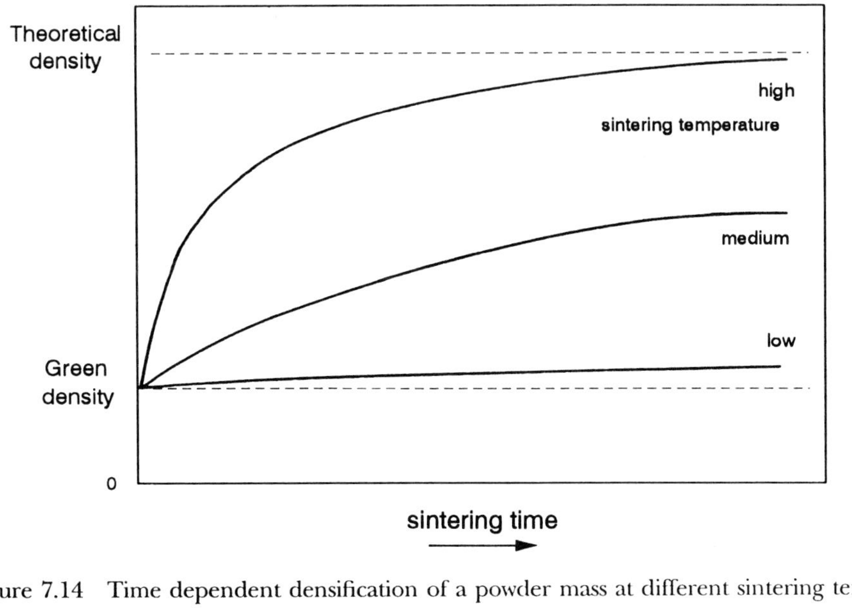

Figure 7.14 Time dependent densification of a powder mass at different sintering temperatures (schematically). Powder properties and compacting pressure are constant

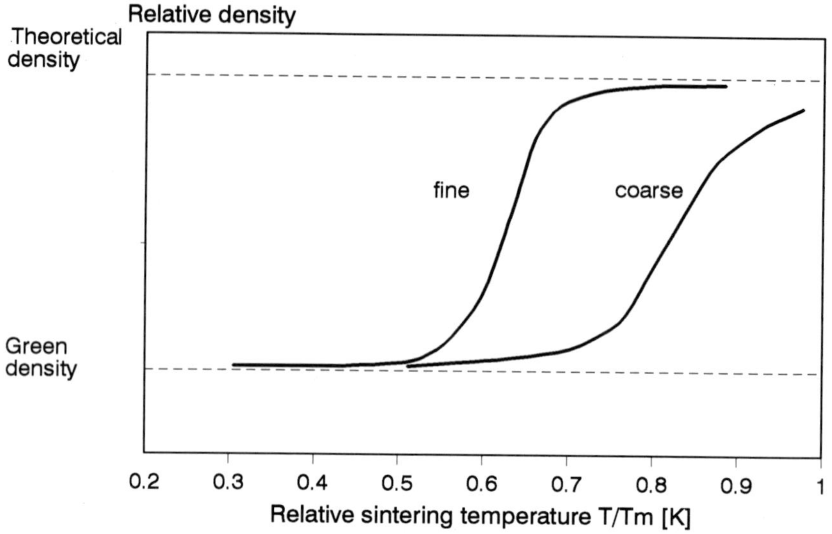

Figure 7.15 Temperature-dependent densification of a fine (1–10μm) and coarse (50–200μm) powder mass (schematically). Sintering time and green density are constant

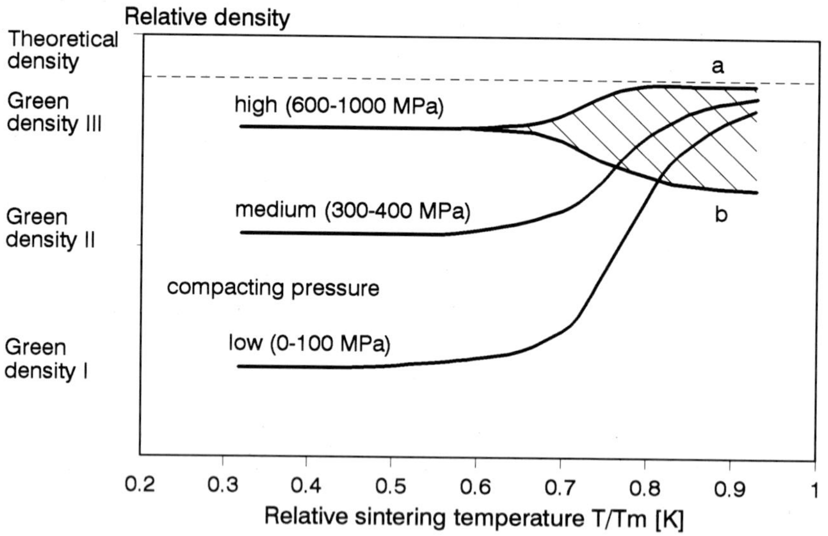

Figure 7.16 Temperature-dependent density change during sintering of powder masses with different green densities (schematically). Powder properties and sintering time are constant

increases. This has to be considered, especially when a sintering cycle has to be optimised to provide high density *and* small grain sizes. This is often difficult to achieve, but always desired for high strength materials, because the strength data are strongly dependent upon porosity as well as upon grain size. This holds

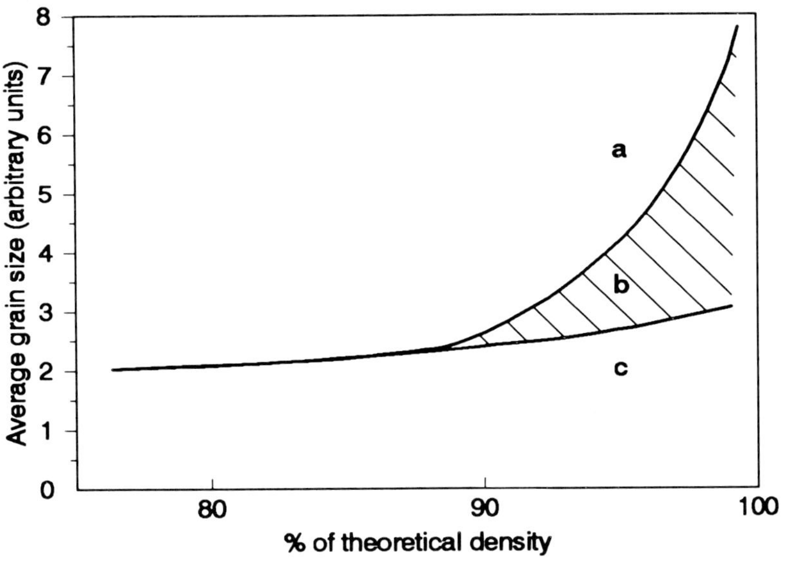

Figure 7.17 Dependence of average grain size on sintered density (schematically). (a) without adding grain growth inhibitor, (b) partial inhibiting, (c) strong inhibiting of grain growth

for both metallic and ceramic products. Fig. 7.17 is a schematic plot of average grain size vs. sintered density for sintering without and with grain growth inhibitors. While at lower densities, the pores themselves retard grain growth effectively, stronger grain growth occurs when the majority of the pores have disappeared and the rest have separated from grain boundaries. Before this stage, sintering additives staying heterogeneous and finely distributed markedly retard grain growth and – at the same time – enhance densification. Pores and second phase inclusions principally act in the same way. A good example is MgO-doped sintered Al_2O_3 which is a commercial product.

It has been shown (Fig. 7.18) that larger pores being linked to grain boundaries strongly inhibit their migration, ('porosity controlled') while small pores move together with grain boundaries ('grain size controlled'). In both cases pore shrinkage occurs by the source-sink mechanism. In a special pore-size–grain-size range, however, the pores easily separate from boundaries and shrink considerably slower, dependent on their distance from the grain boundary. This occurs generally at a critical grain size. During sintering of a porous body (starting from point g_o, ending e.g. in g_f), the grain size–pore size development curve should not touch the separation field, which easily happens when strong grain growth occurs ('coarsening') and the number of pores reduces by densification and pore coalescence. The time-temperature regime is of vital importance to avoid this before high density has been achieved. Figure 7.19 is an

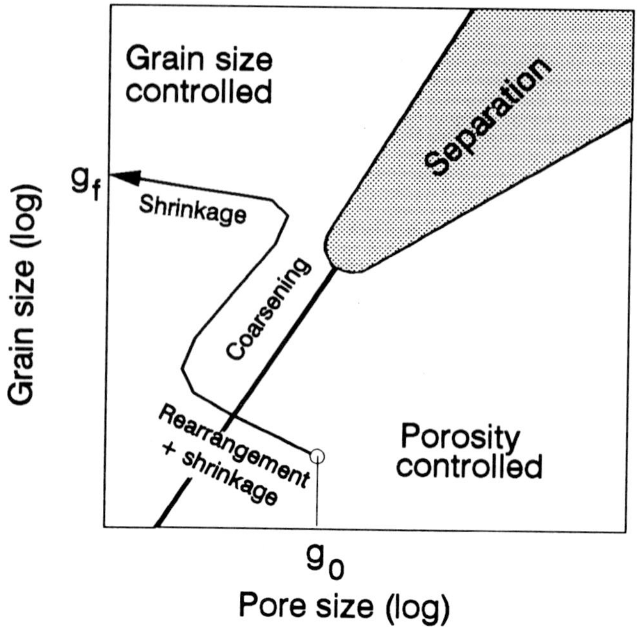

Figure 7.18 Possible grain size–pore size development during sintering. Touching the separation field has to be avoided (partly according to Kaysser) g_0 = grain size before sintering g_f= final grain size

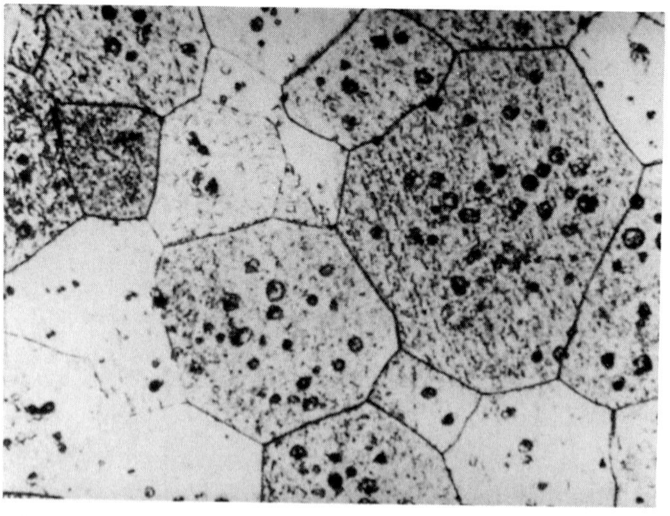

Figure 7.19 Sintered powder compact with gross pore-grain boundary separation

example of a body sintered at a too high temperature with gross pore-grain-boundary separation, see also Fig. 7.24.

A comparison of *different properties developed during sintering*, is shown in Fig. 7.20. The electrical conductivity of metal powder compacts evolves even at very low sintering temperatures, while other properties are essentially still un-

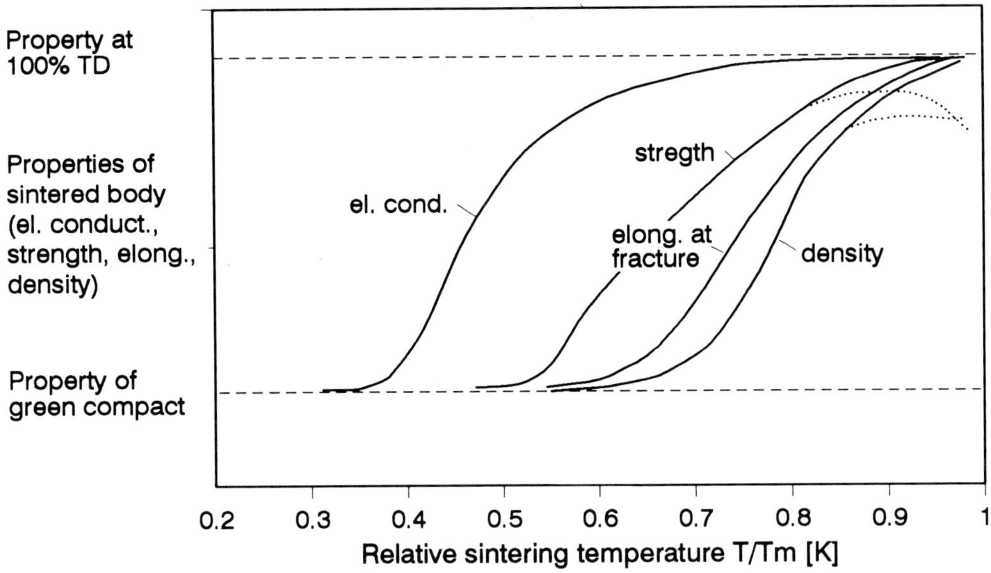

Figure 7.20 Development of different properties of metal powder compacts depending on sintering temperature (schematically). Powder properties, compacting pressure and sintering time are constant

changed. Some neck growth occurs in this stage by surface diffusion together with reduction of oxide films. The strength is developed at higher temperatures, reaching (some times) a plateau earlier than the elongation at fracture. The strength data may even decrease at the highest sintering temperatures due to (mostly undesired) grain growth, while the elongation still increases or remains constant. The sintered density increases only at the higher temperatures, except during sintering of fine powders. These schematic relationships only apply to sieve-range powders and *cum grano salis.*

Sometimes the term '*oversintering*' is used when a too long time or too high temperature can yield undesired properties, such as grain coarsening, density or strength decrease, newly created gas-filled pores, dimensional inaccuracy etc. The reasons may be different, depending on the sintered product and include besides grain growth and gas evolution the occurrence of undesired amount of liquid phase in multicomponent systems (see section 7.3.3). Figure 7.20 includes 'oversintering' as the possible reason for a decrease in properties at the highest sintering temperature.

Another property of general importance is the appearance of *interconnected (open) and closed porosity* in a sintered body, dependent on total porosity. The amount of open pores dominate down to a range of total porosity of 15–12%, (Fig. 7.21) but they disappear completely, when the pore volume ranges below ~7% (5–10%). Between 7 and 12% open and closed pores are present in comparable

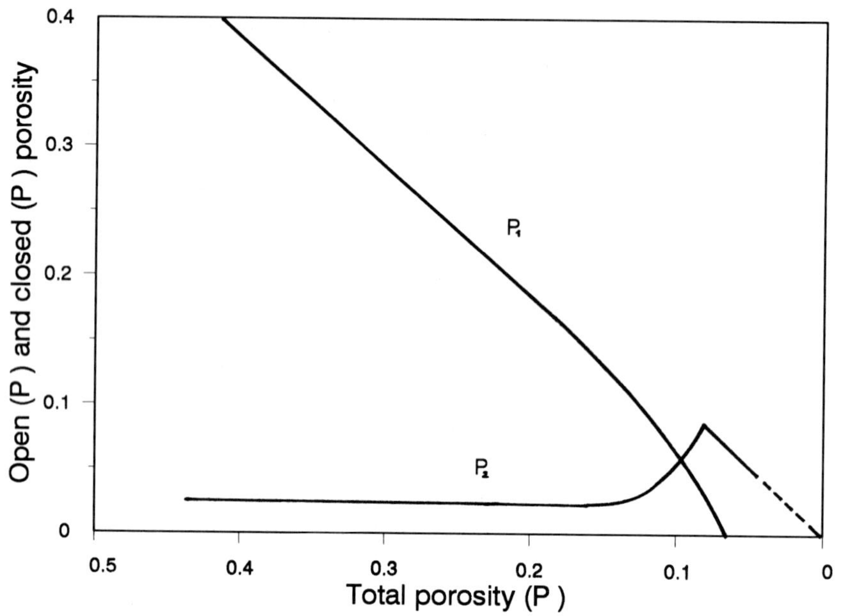

Figure 7.21 Open and closed pores in dependence on total porosity in solid states sintered bodies (schematically)

volumes. These relations were investigated originally in sintered Cu, but have been confirmed in many other materials including ceramics. Nevertheless, Fig. 7.21 has to be accepted as a schematic rather than a quantitative picture.

The open porosity may be subdivided into pores accessible from *one* surface and pores or pore networks being interconnected and accessible to several parts of the surface. This interconnected porosity is of special significance for parts like filters, diaphragms and bearings. For other applications, this porosity is undesired, e.g. for parts of high corrosion and wear resistance, high strength parts or when high surface quality (e.g. hard metal rolls) is needed. In Sinter-HIP-cycles (see section 6.1.2.4) the pore closure has to be achieved completely during the sintering step, otherwise the subsequent HIP is not successful.

7.2.3.2 Sintering Anomalies

Green compacts may be inhomogeneous for different versions: Axially pressed compacts with large height-to-diameter ratio may exhibit density gradients, leading to *differential sintering*: less dense areas undergo a higher sintering rate and larger shrinkage than more dense ones, resulting in poor dimensional accuracy and possible distortions. Also minor constituents in a powder unit (e.g. copper in iron powder) may be inhomogeneously distributed and cause similar effects. The practice has solved such problems by using highly compressible powders and optimising the mixing and compacting process (Chapter 6).

However, inhomogeneities may arise because of different particle packing density over any cross section of a compact, which is observed especially in fine, highly active powders. Such powders are often used in *agglomerated form* for easier handling and die filling. Non-agglomerated powders may agglomerate spontaneously. When the agglomerates partly 'survive' the compaction procedure, then areas with high and with lower densities are neighboured. Sintering stresses (see section 7.3.1) arise at the boundaries between high- and low-rate sintering micro-volumes with following early crack formation, (Fig. 7.22). These cracks, once formed, are enlarged during further sintering and do not heal again. Consequently, the agglomerates have to be destroyed completely during compaction, so that boundaries between agglomerates can no longer be defined. This has to be considered especially in fine ceramic powders.

Monosized, fine, agglomerate-free powders may form areas with near-ideal packing, separated by boundaries, what may lead also to differential sintering and widening of these boundaries to microcracks. Less critical in this respect are powders with a narrow particle size distribution, whose packing is statistically more homogeneous.

Rigid, *non-densifying inclusions* in a mass of fine powder may act similarly to

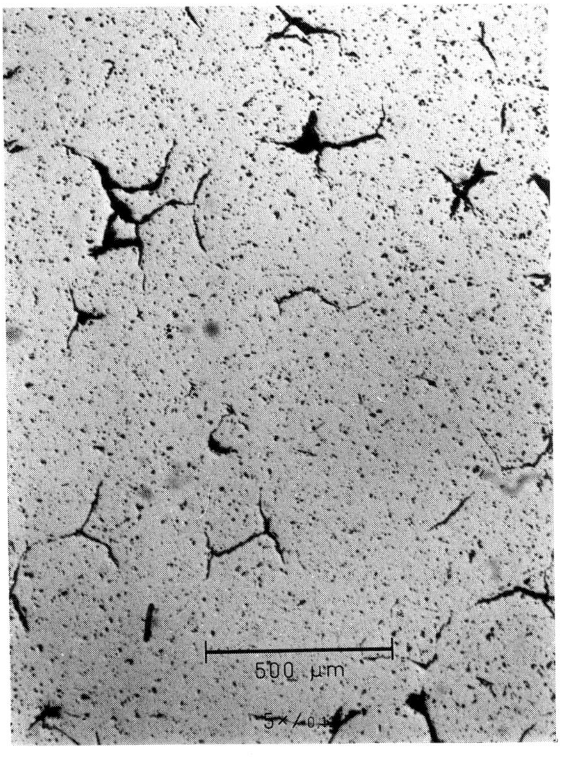

Figure 7.22 Microcracks formed during sintering of an Al_2O_3 powder compact containing hard agglomerates (Agniel)

hard agglomerates: They inhibit sintering of the 'matrix' powder and give danger for nearby microcrack formation. This is of vital importance during the sintering of whisker- or short-fibre-reinforced composites, especially in the presence of clusters ('Desintering' of composites). Thus, sintering without the application of external pressure is seldom successful in such materials.

The *shrinkage* of an axially cold pressed powder compact during sintering is generally *anisotropic*, being often larger in the compacting direction. In practice it is possible to disregard this in most cases, especially when the sintered parts are sized finally (see Chapter 10). Nevertheless, the dimensional accuracy after sintering may be decreased by this anisotropy, and the very high dimensional accuracy of a cold compacted part is reduced after sintering by roughly a factor of 10, depending on sintering conditions. Typically, as-sintered parts may be accurate to a tolerance of ~0.4% parallel to the pressing direction and 0.2% perpendicular to it. Several mechanisms have been proposed for explanation, one of them considering the larger recovery of the deformed particles in the pressing direction. There is no doubt about the larger particle contact areas in this than in the lateral direction, which may cause larger material transport in the direction of compacting pressure.

7.2.3.3 Activating and Inhibiting Influences

Numerous parameters may influence the material transport during sintering and, consequently, the properties of the sintered product. Therefore it is difficult to derive general rules and even more difficult to try quantitative estimate. Most of the work has been conducted with only little theoretical background. 'Activated sintering' (see below) has been investigated in detail and is of some practical significance especially for refractory metals which require very high sintering temperatures. Activating and inhibiting phenomena are observed also during reactions with sintering atmospheres.

Sintering enhancement results from an increased driving force and/or increased atom mobility, induced by physical or chemical means. Special compositions of interfaces (grain or phase boundaries), achieved by additives, or the activation of gas phase transport are examples, in which the activation energy of the overall sintering process is reduced. Inhibition of sintering results from reduced atom mobility and often from inactive films at sintering contacts.

Phase transformations

The bcc-fcc α-γ-transformation of iron at 898/910°C is accompanied by volume contraction of about 0.8% due to the closer packing in the fcc-lattice, as well as by a strong decrease of the self-diffusion coefficient in the lower γ-range compared with the upper α-range. The difference is 2 to 3 orders of magnitude. As a consequence, the shrinkage curve of fine iron powder (measured dilatometrically) is nearly completely stopped when entering the γ-range (Fig. 7.23). Also sintering in the α-range is disturbed when a short intermediate sintering in the γ field occurs. In this case contacts formed during α-sintering

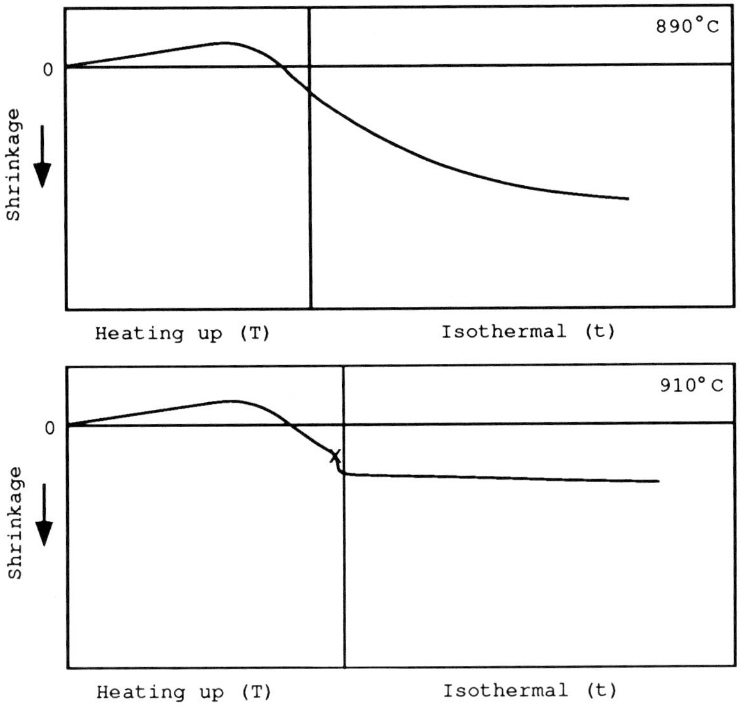

Figure 7.23 Shrinkage of carbonyl-iron powder at 890°C (top) and at 910°C after passing α–γ-transformation (*x*) (bottom)

are partly destroyed during contraction in the γ-range. These effects are not very significant in the production of machine parts, but sintering in the lower part of γ-range (900–1050°C) is impossible due to the low atomic mobility. On the other hand iron alloys in which the γ-range has completely or partly disappeared (e.g. with 0.6% P or a few % Si), are sintered effectively to high densities, depending on the amount of α-phase formed during sintering. Unfortunately, this occurs with relatively high shrinkage and considerable grain growth.

Oxide films

Each metallic surface in contact with ambient air is coated with an oxide film, except that of some noble metals. The higher the specific surface of a powder, the larger is the oxide content, which may have a strong influence on the sintering behaviour.

Oxides with a *high energy of formation* (e.g. Al_2O_3, MgO, Cr_2O_3) are stable under normal vacuum or protective gas sintering conditions. They strongly inhibit sintering since the diffusional exchange between the metallic surfaces is reduced or essentially impossible. When a low melting point metal (like Al or Mg) is covered with its high melting point oxide the actual sintering temperature for the metal, related to its oxide film, is relatively so low that hardly any atom transport takes place in the oxide. It is therefore impossible to produce parts of pure aluminium powder by conventional cold pressing and sintering, even near the melting point of 658°C. The reasonable way is to introduce a liquid phase which destroys the oxide films or to apply a strong deformation process by extrusion, which spalls off the oxide films mechanically and provides fresh metallic surfaces, formed in-situ. Both ways are used in practice, see sections 11.4.1 and 11.9.3.

Metal or alloyed powders with highly reactive constituents have to be manufactured in practice with low oxygen content, e.g. for hard magnets (Alnico, Co-Sm, Fe-Nd-B), superalloys (Ni-Cr-Co-Ti-Al and others) or titanium alloys. The oxygen specification for superalloy powder is in the range of 100 ppm or below. The oxygen content not only inhibits sintering (which may be overcome by modern production technology like HIP) but oxide inclusions degrade physical and mechanical, especially long term properties, including reliability. It is essential to keep the oxygen content low during the entire processing cycle.

Oxide films with *low energy of formation are* not stable during sintering. This refers to powders of Fe, Ni, Cu, W, Mo and others. The oxide is either reduced (like Fe- and Ni-oxides), evaporates (MoO_3), or is partly soluble in the matrix metal (Cu-oxides). The disappeared oxide may leave behind a highly active metallic surface, especially after reduction. Slightly preoxidised (up-to 1% mass gain) iron powder may exhibit improved sintering behaviour and strength properties. Also careful admixing of small amounts of oxide to the metal powder (Cu, W) may improve sintering. Experiments on oxidised copper powder have shown the enhancement of sintering up-to a certain thickness of the oxide film (~ 40 nm). Thicker layers inhibit sintering, because the decrease of contact quality overcompensates the improved sintering.

Nevertheless, these phenomena are seldom used in practice: The highly compressible powders commercially available do not need additional, costly steps to improve sintering.

Sintering additives and impurities

Minor additives (dopants) cause sometimes large effects in enhancement of sintering which is often named '*activated sintering*'. Most spectacular is the influence of a few tenth percent of nickel on the sintering of W- or Mo-powder, which lowers the sintering temperature for W from > 2000°C to ~ 1300°C (Agte-Vacek-Effect) (Fig. 7.24). Other metals like iron and cobalt have similar, but smaller influence. nickel diffuses onto W- or Mo-particles and enters the grain boundaries formed during sintering, resulting in an increase of grain boundary self-diffusion of W or Mo by a factor upto 5000 at 1300°C. Such paths of high diffusivity at interfaces may provide very rapid sintering; it is important, however, that the activating element remains concentrated at the grain boundaries during sintering. Chemical reactions with sintering atmospheres, causing in-situ formation of fresh particle surfaces also result in effects of activated sintering. The addition of HCl gas or a halide to the sintering atmosphere facilitates mass transport by the formation of vapour-phase molecules (e.g. $FeCl_2$ during

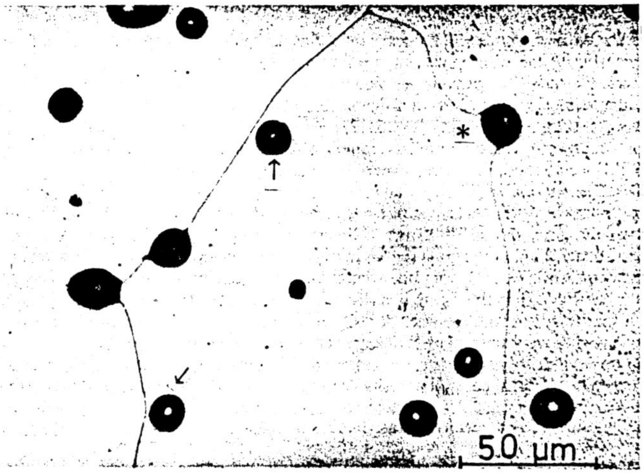

Figure 7.24 Microstructure of *W*, doped with 0.15 wt% Ni, sintered 3 hours at 1400°C to high density, → pore just separated, * pore drag (Kaysser)

sintering of iron). Enhanced pore spheroidisation as well as improved sintering result in improved mechanical properties, mainly in higher ductility.

Additives are often used in oxide ceramics mostly to activate sintering and to suppress grain growth at high sintered densities (see Fig. 7.17), the latter by pinning grain boundaries. The achievement of fine grained microstructures in highly dense ceramics is of vital importance for optimisation of mechanical, dielectric or magnetic properties, resp., of ceramics like Al_2O_3, $BaTiO_3$, ferrites and others. MgO in Al_2O_3 inhibits grain growth and increases pore mobility and shrinkage. The additive has to be insoluble at sintering temperature to inhibit grain growth. Soluble dopants (like CaO in ThO_2 or ZrO_2) may enhance sintering by changing defect chemistry and, consequently, diffusibility in the matrix. Because diffusion controlled sintering in a ceramic material is always a combined cation and anion process (otherwise electrostatic fields would arise), it is essential to increase the diffusivity of the slower (i.e. rate controlling) ion by increasing its defect concentration. The sintering of ceramics with multivalent cations (e.g. UO_2) is very dependent on the oxidation potential of the atmosphere. Strong enhancement can be achieved with UO_2 in slightly oxidizing atmospheres, because a certain oxygen potential form UO_{2+X} and increases both cation and anion mobility. It has been shown, however, that many ceramics earlier thought to sinter purely in the solid state, contain minor amounts of liquid phase (silicates) impurities, forming in fact a multicomponent system. Although improving sintering, they cause variations in properties and may especially degrade high temperature performance, like creep resistance. This requires special attention to the chemical composition of the initial powders. As treated in 7.2.5 covalent materials like Si_3N_4, SiC, B_4C and others generally require additives otherwise sintering is insufficient.

Shock activation and irradiation induced sintering

A very special way to activate sintering is due to mechanical shocks, preferably by explosives. Shock waves may introduce strong plastic deformation into particulate masses causing high defect concentrations (dislocations, vacancies and interstitials). The stored energy, ranging several joules per gramme, leads to low sintering temperatures and high sintered densities, and has been demonstrated with metallic and ceramic powders. Also direct sintering caused by mechanical shocks has been observed.

The activation by particle (e.g. neutron) irradiation has been investigated in the framework of nuclear materials research. Sintering kinetics are enhanced by the production of excess vacancies, generated by collision processes, which reduce the overall activation energy for volume diffusion.

7.2.4 Processing of Ultrafine (Nano-Sized) Powders

When the particle size of powders ranges less than 50 to 100 nm (for definition and preparation see 2.4) their properties become very different from those of coarser powders. These dimensional effects are observed when the particle size becomes comparable with the size or linear dimensions responsible for special properties (e.g. the size of dislocation loops, magnetic domains etc.). This is significant for the processing* as well as for the properties of the sintered products and has been studied theoretically and experimentally mainly during the last decade.

The excessive surface development of, say, 25–100 $m^2\ g^{-1}$ or more, is equivalent to an unusual high surface:volume ratio. This raises the danger of pyrophoricity (see section 5.1), especially for metallic and carbide powders unless they are passivated, (see section 2.4). Under some precautions it is possible to handle such powders as normal.

Nanosized powders below a certain particle size are free of dislocations and cannot generate them by application of stress. Consequently, no essential plastic deformation occurs during compaction, not even in powders of highly plastic metals. The main physical process during pressing is friction between particle surfaces. The pressing characteristics of metallic and ceramic nanosized powders are reported to be similar. The compaction curves of powders well below 100 nm differ markedly from those of powders >100 nm, in which plastic deformation occurs under high pressure. The number of interparticle contacts in a compact reaches $\sim 10^{18}\ cm^{-3}$. The total area of particle interfaces is about $10^4 - 10^5$ times more than in conventional powder compacts. The interfaces represent the main type of defects.

* Processing of ultrafine ceramic powder is often named 'colloidal processing', but this includes also powders larger than 100 nm. This paragraph covers processing steps also before sintering.

Sintering as well as crystal growth and formation of solid solutions or phases occur at temperatures as low as 0.2–0.3 of the melting temperature (Tm) in [K], while similar sintering effects in coarser grained powders are observed only at 0.5–0.7 Tm. The shrinkage rate is $\dot{\varepsilon} = 10^{-3}$–$10^{-4}\ s^{-1}$, being up to 10^2 times higher than in coarse powders. Iron powder has been sintered in H_2 at 700°C to about 98% TD, while during heating up excessive gas desorption and surface oxide reduction (starting at 270°C) take place. Densification started around 400°C. In vacuum, about 900°C are necessary, however, to get a sintered body of high density. Al_2O_3 with a particle size of 160 nm and a narrow size distribution sinters to 99.5%TD at 1150°C. In powder mixtures of Cu–Ni below 50 nm particle size the formation of solid solution progresses to a large extent at 500°C during 5 min, while in coarse powders at 500°C, a nearly complete absence of diffusion exists. The effective interdiffusion coefficient has been calculated as 7 orders of magnitude higher than in coarse powders.

Sintered bodies of very high strength are obtainable at extremely small grain size, which can be achieved exclusively by processing of ultrafine powders. It has been demonstrated with sintered, tetragonal ZrO_2 that with grain sizes of 200–300 nm, a four-point bend strength of 1200-1300 MPa is obtained in mass production, when processing imperfections are avoided. It is difficult, however, to maintain the grain size in the range of the former particle size, when recrystallisation and sintering occur in the same range of temperature, as observed in metallic systems. In nanosized sintered bodies as well as in green compacts the interfaces represent a large part of the entire volume (Fig. 7.25). When the crystallites are too small (< 20–50 nm), however, the volume of the 'open-structured' grain-boundaries becomes very large and may cause weakening of metallic or covalent bond in the entire body. Thus, to get improved mechanical properties is not straightforward.

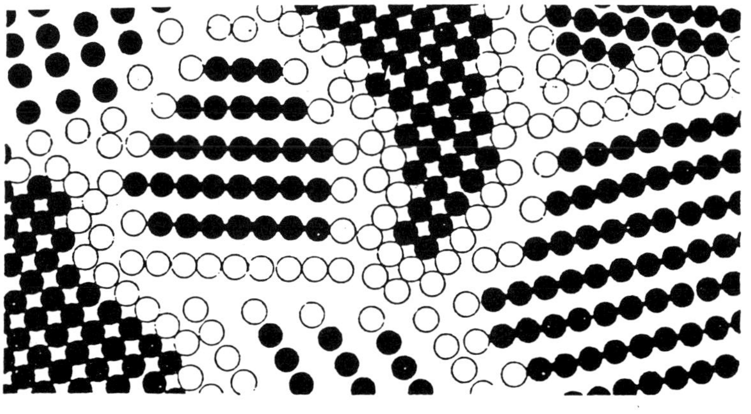

Figure 7.25 Nanocrystalline solid body (Gleiter). The open circles represent atoms belonging to grain boundaries or interfaces, resp.

Another most interesting approach is the processing of *nanoscale composites,* which have also been shown to exhibit superior properties. Their production requires the stabilisation of the particle or grain size of the dispersed phase during sintering. To trap the second phase within the grains of the matrix phase is one possibility to meet this requirement. This approach awaits further development.

The technological potential of processing nanosized powders looks fascinating with respect to mechanical, chemical and magnetic properties, but several difficulties have to be overcome: The powders flow badly, have low filling and tap densities, which hinder automatic charging and compaction, and exhibit large shrinkage with the danger of distortions. Last but not least, the powders are expensive. Nevertheless in hard metals (see section 11.6.3) and ZrO_2-ceramics (see section 11.10.3) ultrafine (or even nanosized) powders are now in use. More application for special products as functional rather than as structural material can be expected in future.

7.2.5 Sintering of Covalent Materials

Most of the materials sintered are metal or ceramic powders, whose bonding is (preferably) metallic or ionic. Commercial sieve-range metal powders like Fe, Ni, Cu and others are sinterable in practice at temperatures between 2/3 and 3/4 of the melting temperature in K, an observation, which can be found already in early Powder Metallurgical textbooks. This is not too different in ionic compounds, especially oxides, although generally subsieve-range powders have to be used. Materials with preferably covalent bonding like elementary Si,

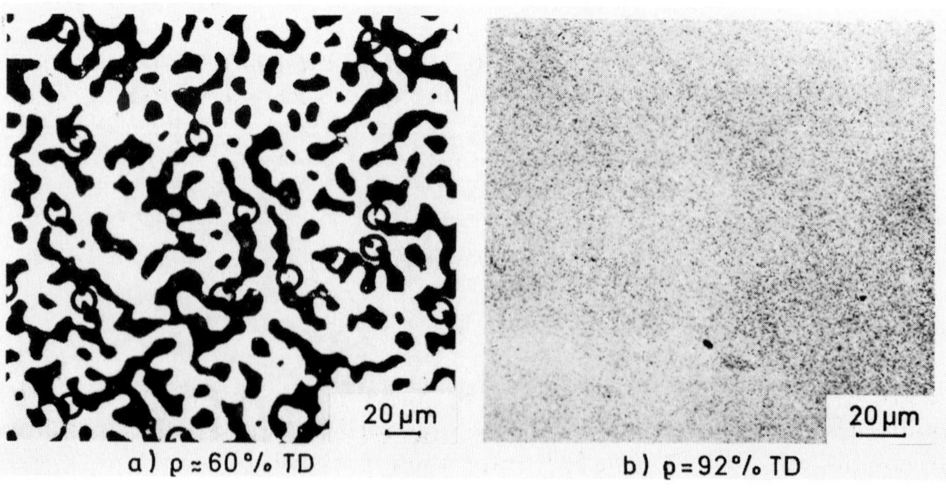

Figure 7.26 Silicon compacts sintered 60 min at 1350°C in Argon. a) 0.23 µm powder b) 0.06 µm powder (Grescovich *et al.*, General Electric)

Si_3N_4, SiC, B or diamond need sintering temperatures close to their melting point (if possible), the use of ultrafine powders and sintering additives. The reason for this poor sintering behaviour is the very strong, directed, covalent bond leading to very low self diffusion coefficients. With respect to possible sintering temperatures, the self diffusion coefficients of Al^{+++} and $O^{=}$ in Al_2O_3 are about four orders of magnitude higher than those of nitrogen in Si_3N_4, even though some ionic bonding has been calculated for Si_3N_4.

It has been demonstrated for solid state sintering of elementary silicon, that 92–98% TD can be obtained by using 60 nm-powder (44 $m^2\ g^{-1}$) *and* temperature near the melting point (0.96–0.98 T_s [K]) at 1350–1380°C (Fig. 7.26). The huge amount of interparticle contacts, and the high defect concentration cause strongly increased atom mobility. The equilibrium volume diffusion coefficient (D_v) is ~ $10^{-12}\ cm^2\ s^{-1}$ at 1350°C, which has been estimated from measurements near this temperature.

This approach cannot be applied, however, to sintering of Si_3N_4. Due to its high N_2 dissociation pressure, reaching 0.1 MPa at 1900°C, the maximum possible sintering temperature is about 1900–2000°C even under N_2-pressure. In this range D_v is about $10^{-16}\ cm^2\ s^{-1}$. Liquid-phase sintering by the use of additives, providing solution-precipitation processes, is the common means to achieve high densities, and with adequate microstructures.

Also for sintering of SiC ultrafine powders (15–20 $m^2\ g^{-1}$) and sintering additives (0.5–1% *C* and 0.2–0.5% *B*) have to be used. The carbon generates fresh, deoxidised (sinteractive) surfaces and forms partly secondary SiC, while *B* diffuses into the SiC-particles and (probably) increases volume diffusion. The sintering temperatures are 1950–2150°C. It is claimed that *B* suppresses sintering mechanisms which do not lead to densification (surface diffusion and evaporation-condensation), so that volume diffusion dominates. Also liquid-phase sintering of SiC by oxide additives at lower temperatures has been developed.

In summary, sintering of covalent materials is more a quantitative than a qualitative problem: When the number of interparticle contacts is drastically enlarged and the atom mobility by what ever means is increased, an acceptable sintering behaviour including strong densification can be achieved.

7.3 MULTICOMPONENT SINTERING

7.3.1 Principal Aspects

Sintering of powder mixtures may occur with or without the presence of a liquid phase and, during solid state sintering, with or without solid solubility of the components. Fig. 7.27 demonstrates these possibilities, including the special procedure of liquid phase infiltration, in which usually a presintered compact is used (see section 7.3.3.3). The common precondition is the decrease of

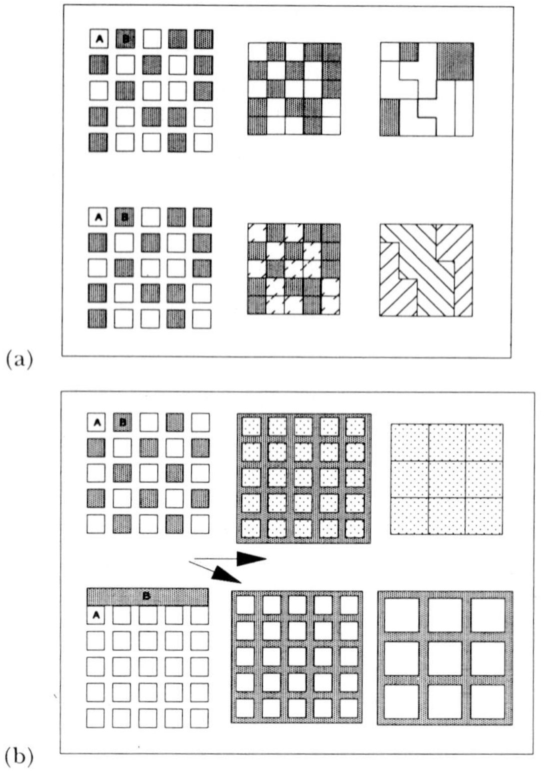

Figure 7.27 Sintering of bi-component system (schematic). (a) Powder mixture, solid state; *top*: no solid solubility *bottom*: complete solid solubility. (b) *top*: Powder mixture, liquid phase, complete solid solubility (→); no solid solubility (↘); *bottom*: Infiltration, no solid solubility

the overall surface energies (γ) in the system. In a binary mixture three types of particle contacts, namely *A–A*, *B–B* and *A–B* are present, with the interfacial energy γ_{AB}. When in a mutually insoluble system

$$\gamma_{AB} > \gamma_A + \gamma_B \tag{7.15}$$

then no sintering of the *A–B* contacts takes place, although the *A–A* and *B–B* contacts may sinter. The condition for sintering of *A–B*-contacts is

$$\gamma_{AB} < \gamma_A + \gamma_B. \tag{7.16}$$

When

$$\gamma_{AB} < |\gamma_A - \gamma_B| \tag{7.17}$$

complete sintering is possible thermodynamically. In insoluble systems the equilibrium state is that the component with the lower γ covers the other component. The interfacial energy controls the process in all cases with and without the formation of liquid phases: the lower the value of γ_{AB}, the greater is the driving force. The effect of γ_{AB} is most clearly seen in the presence of a liquid phase: The characteristic wetting angles allow the easy determination of γ_{AB} (see 7.3.3.1) which is more difficult in systems without a liquid phase. The sinterability of powder mixtures, especially in the solid state, is generally influenced by the generation of *sintering stresses*, resulting from different sources, namely:

1. Thermal expansion anisotropy in non-cubic phases.
2. Inhomogeneous (differential) sintering resulting from long-range or short-range density gradients within the mixture (caused by compaction or agglomerates, resp.)
3. Thermal expansion mismatch between the phases.
4. Differential sintering caused by phase inhomogeneities in the mixture.

Points 1 and 2 relate also to single component sintering. Tensile stresses may range between 0.5 and 2 MPa, being effective during heating as well as during the isothermal shrinkage period and, in case of large thermal mismatch, during cooling. Their magnitude depends on the relative rates of the matrix shrinkage and creep rate under stress, leading to equilibrium between shrinkage and creep deformation on the one hand and stress evolution on the other.

The most important type of damage due to sintering stresses are micro- or even macro-cracks, similar to those near agglomerates (see section 7.2.3.2). Sintering stresses are often disregarded in research and practical work. They may be of less significance in metallic systems with high plasticity of the phases under sintering and cooling conditions, but are probably most important in metal–ceramic composites with an essentially undeformable ceramic component, like SiC-whiskers or fibres. They should be also relevant in ceramic–ceramic composites. Pressure-assisted sintering is a means to avoid this type of damage. Multiphase and ceramic parts are often subjected to *residual stresses*, after cooling.

7.3.2 Solid Phase Sintering

7.3.2.1 Without Solid Solubility

Sintering of powder mixtures without solid solubility (or phase formation) exclude interdiffusion (or reactions, resp.) in the *A–B* contacts. This is the case, e.g., in the metallic systems W-Cu or Mo-Cu below the melting point of Cu or in cermet systems like Fe-Al_2O_3 or Cr-Al_2O_3. Neck formation and shrinkage depend from the number of the *A–A* and *B–B* contacts, related to the number of *A–B*-contacts. When one component (*B*) does not sinter at the temperature considered, *B* inhibits sintering effectively. Consequently, the properties after sintering strongly depend at constant porosity on the amount of the 'inactive' component *B*, as shown in Fig. 7.28. Tungsten is here the inactive component. Although the linear extrapolation of strength to zero porosity may not be generally allowed the experimental result is clear.

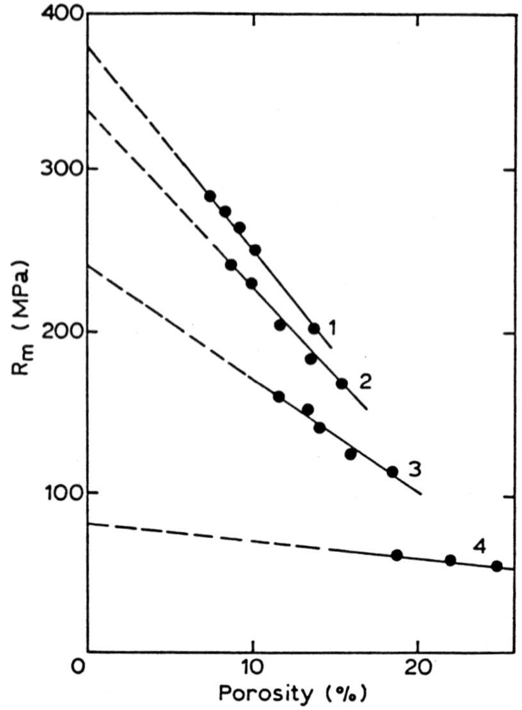

Figure 7.28 Tensile strength as a function of porosity for Cu–W-powder compacts (Pines and Sukhiuin) 1) Pure Cu; 2) Cu+5 wt% W; 3) Cu+20 wt% W; 4) Cu+35 wt% W

When the 'inactive' component *B* is present in the form of much finer particles than *A*, a special type of mixture may result, as shown in Fig. 7.29. Each particle of the coarser component is covered with fine *B*-particles, which prevent direct *A–A* contacts to a large extent. Thus, little or no sintering at all takes place. It is easy to calculate that an admixture of only 1 wt-% of 1 μm Al_2O_3 powder to a 50 μm Fe powder (assumed to be spherical) provides ~ 2.500 times as much Al_2O_3 than Fe particles in the mixture, having a deleterious effect on solid state sintering. Mechanical alloying (see section 2.1.1.1) is a method for implantation of the fine particles into the interior of the matrix particles, which improves the *A–A* contacts.

7.3.2.2 With Mutual Solid Solubility

In mixtures with extended solid solubility we have to distinguish between a system in which each particle represents a homogeneous solid solution before sintering (prealloyed powders) and a heterogeneous mixture. Sintering of *homogeneous solid solutions* (without concentration gradients) was investigated by the bar-wire-model. The sintering is related to hardness and strength properties of the mixed crystal, as shown in Fe-Mo- (Fig. 7.30) and other Fe-based systems

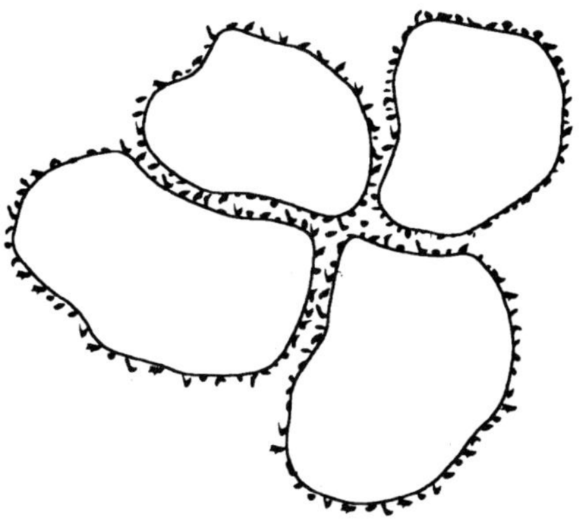

Figure 7.29 Admixture of a fine (1μm) to a coarser (50μm) powder, causing particle separation

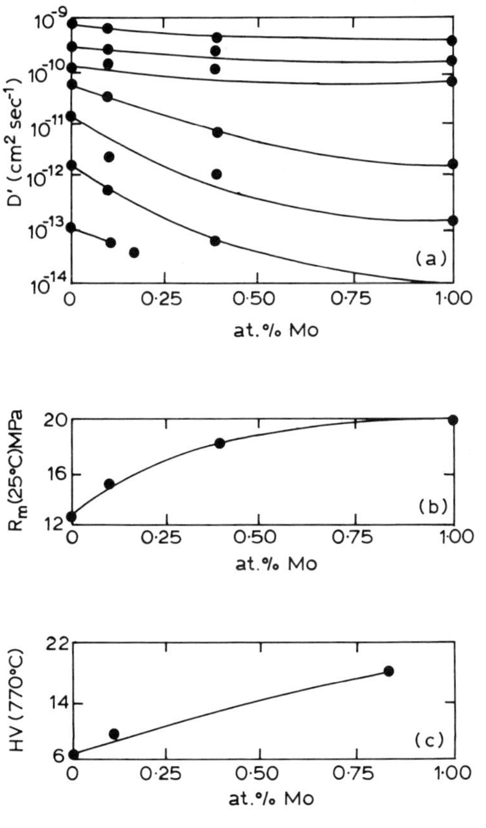

Figure 7.30 a) Mobility coefficients (D') determined in model sintering experiments using Fe–Mo solid solutions b) Tensile strength (R_m) at room temperature c) Hardness (HV) at 770°C

up to 2 at % of the second metal. In cases of higher strength and hardness, which originate from the bonding force in the lattice, a marked inhibition of sintering by the second metal is observed. The effect is related to a higher formation energy for vacancies, and (very probably) to lower self-diffusion coefficients in the Fe-Mo-alloys than in pure iron. Reverse effects should take place, when the atom mobility is increased by the second metal.

Sintering with *simultaneous formation of solid solutions* is much more complicated. Strongly inhibiting (incl. swelling) or promoting effects, resp., due to the second component are observed, depending on the materials used, the stage of sintering and the degree of homogenisation.

The effective interdiffusion coefficients should control the solution as well as the sintering process, in either diffusion path, being dependent on concentration and on the degree of homogenisation. When homogenisation proceeds, self-diffusion coefficients become more dominant. During interdiffusion new vacancies are continuously generated, depending on the difference of the actual partial diffusion coefficients. They can assist sintering by increasing the effective atom mobility, provided they remain 'dissolved' in the lattice. If they coagulate, however, to form new pores ('diffusion porosity') they have an inhibiting action by reducing the sintered density and the cross section for the diffusion flow. This indicates, that non-equilibrium vacancies can mask completely 'normal' diffusion controlled sintering. Only within a thermodynamically ideal binary system (which does not exist in practice) does the sintering behaviour fall linearly between that of the pure components (Fig. 7.31a). In all practical cases the system behaves according to Fig. 7.31b or c, the latter possibly with swelling effects, especially when new phases are formed during sintering (see 7.3.4). Enhancement of sintering may occur at the beginning of the homogenisation and when no steep concentration gradients are active (Fig. 7.31b), otherwise the opposite may take place. The different behaviour is observable in the same binary system at different experimental conditions, as demonstrated in the Co-Ni system.

The equilibrating of concentrations during *diffusion controlled homogenisation* is an asymptotic process, complete homogeneity being reached theoretically after $t=\infty$ (Fig. 7.32). The time necessary to achieve a certain degree of homogenisation depends on the effective interdiffusion coefficient D and the diffusion length $2R_A$ within the compact:

$$t \sim \frac{(2R_A)^2}{D} \tag{7.18}$$

As shown in the model of Fig. 7.33a, $2R_A$ $(= L/2-1/2)$ depends on the distance of the alloying particles *B*. Smaller *B* particles at the same concentration cause smaller R_A values, which is limited, however, by the size of *A* particles. Very fine *B* particles cause a different geometry, namely a shell of *B* particles around *A* (Fig. 7.33b). Any contact disturbance by diffusional porosity is disregarded here.

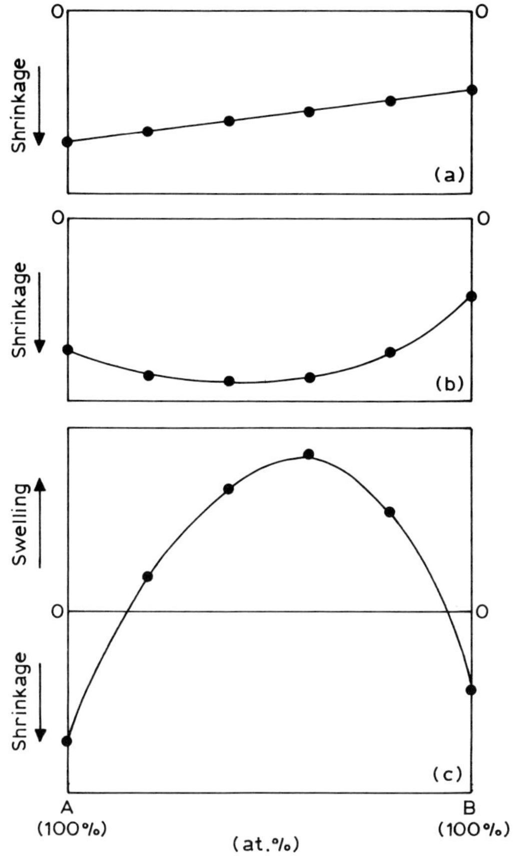

Figure 7.31 Shrinkage of heterogeneous binary powder mixtures *A–B* with complete solid solubility (schematic). Constant processing parameters within each field *a*, *b*, and *c*, respectively. (a) Thermodynamically ideal system; no influence (hypothetical). (b) Mutual enhancement of sintering by non-equilibrium vacancy formation. (c) none-quilibrium vacancy formation, followed by vacancy condensation (diffusion porosity) gross swelling may occur. The volume change dependence on composition may be more irregular as shown in the figure

The base for the calculation of time for a certain degree of homogenisation is Fick's second law adapted to the boundary conditions given by the experiment, which can be assumed as spherical in many cases:

$$\frac{\delta C}{\delta t} = \frac{1}{r^2}\frac{\delta}{\delta r}\left(Dr^2\frac{\delta C}{\delta r}\right) \tag{7.19}$$

The equation can be solved for a defined degree of homogenisation, when:

1) the effective interdiffusion coefficients are known which are in a powder compact often different than in classical diffusion couples. They can be determined separately and may be assumed for simplification as independent of concentration.

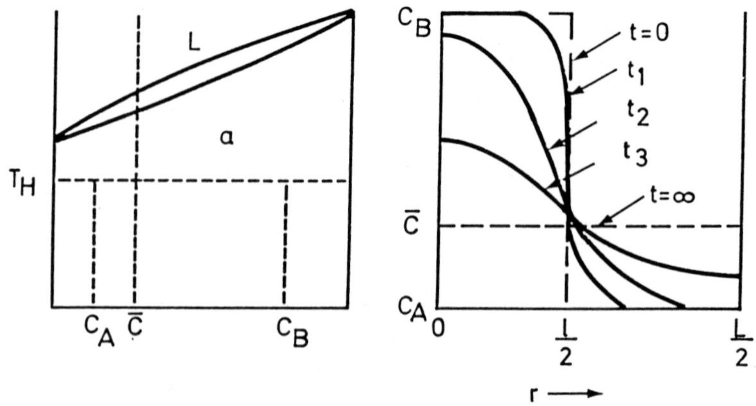

Figure 7.32 Time dependent homogenisation of a binary mixture. C_A, C_B = concentrations of *A* and *B*, respectively, $\bar{C}$ = equilibrated concentration, T_H homogenisation temperature; for *l* and *L* see Fig. 7.33

2) the particle contacts remain essentially undisturbed during the entire process, i.e. no gross diffusional porosity occurs.

The degree of homogenisation can be determined experimentally by microscopical and compositional analysis as well as by X-ray line broadening and peak analysis.

7.3.3 Liquid Phase Sintering

7.3.3.1 Principal Aspects

Liquid phase sintering is a widely used fabrication process in Powder Metallurgy for both metallic and ceramic products. The main advantage is the strong increase of the sintering rate, achieving nearly full density, and allowing short sintering times as long as good wetting of the liquid phase takes place. Even small amounts of liquids may result in considerable enhancement of sintering, as shown in many systems, e.g. WC–Co and ceramics. The densification rate depends, nevertheless, upon the amount of the liquid phase, which ranges up to 30 vol.%, or even more, depending on the system.

Sintering can take place, with a *transient* (or temporary) or with a *permanent liquid phase,* as shown in Fig. 7.34a. When the nominal composition of the powder mixture lies within a solid phase field of the diagram ($\bar{C}_t$) and equilibrium is approached during sintering at temperature T_S, then the liquid phase reacts with the solid and is present only during part of the sintering time. In cases when the liquid is in equilibrium with the solid phase ($\bar{C}_P$) it is present during the entire sintering process.

Transient liquid phase sintering is often used to introduce and homogenize smaller amounts of alloying elements, e.g. in the systems Fe-Cu and Fe-P. The

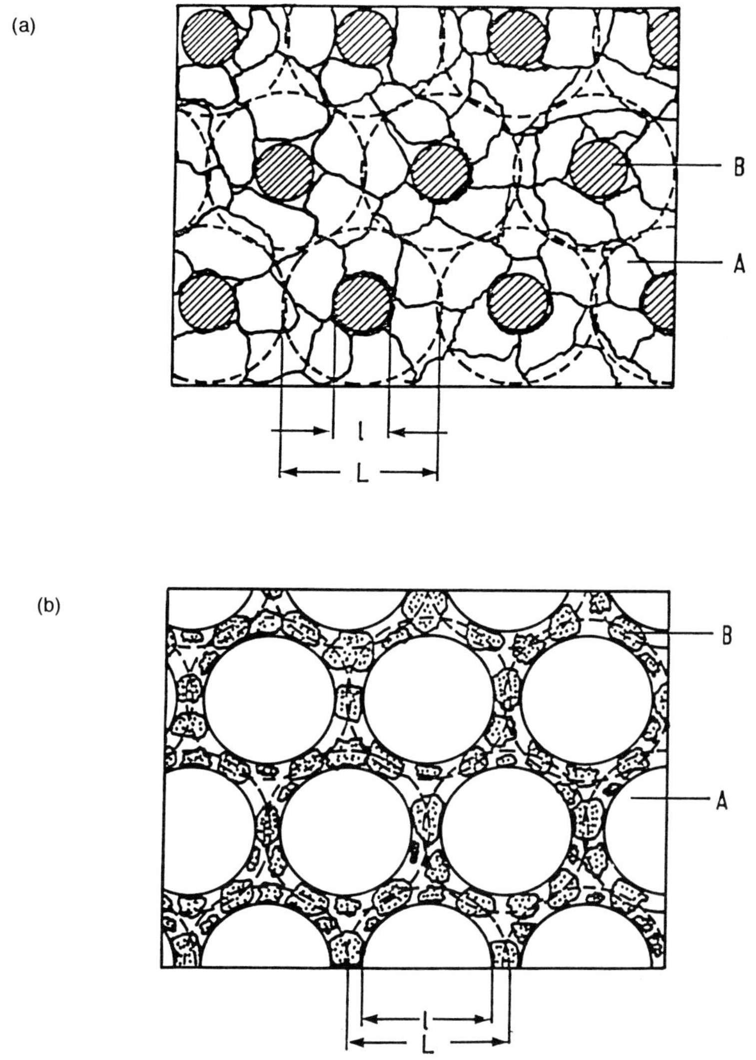

Figure 7.33 Models for homogenisation of binary mixtures. Alloying constituent *B* with minor concentration (a) *B* and *A* particles of similar size (b) *B* particles much smaller than *A*

liquid phase provides large contact areas between the components and facilitates diffusion controlled processes and pore spheroidisation. Very high densification is normally not achieved. Some temporary swelling may occur during the early sintering stages which is used in sintering of iron with 1-2% copper as *shrinkage compensation.* The molten copper particles spread into the iron particle boundaries, (Fig. 7.35) leaving behind additional pores and providing volume expansion by formation of Fe-Cu-solid-solutions. This shrinkage compensation is accompanied by some strength increase by the dissolved copper. Transient

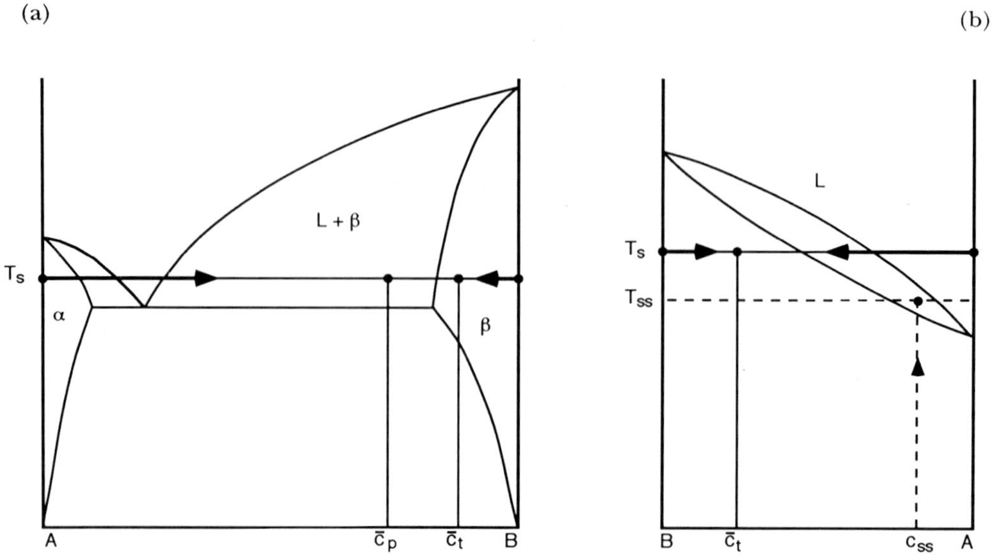

Figure 7.34 Permanent ($\bar{C}_p$) and transient ($\bar{C}_t$) liquid-phase sintering in a binary powder mixture, and supersolids (C_{ss}) sintering in a pre-alloyed powder; $\bar{C}$ are equilibrium concentrations, T_s and T_{ss} are sintering temperatures. Note: The phase diagrams of the relevant multi-component systems are much more complicated

liquid phase sintering is applied also in the production of porous bronze bearings (in which at least a part of the tin is used in elemental in the powder mixture), other copper alloys, hard magnetic materials and others.

Permanent liquid phases occur during sintering of hard metals, high-speed-steels and many ceramic systems, like porcelain and Si_3N_4 with different additives. The aim is to achieve the highest possible densification, combined with high strength and high toughness. The equilibria including homogenisation are reached relatively fast, at least in metallic systems with low-viscosity liquid phases. A special case is the so-called '*super-solidus-sintering*' (Fig. 7.34b) in systems with extended solid–liquid phase fields, using high alloy prealloyed powders. At temperatures above the liquidus, the particles from a liquid phase, preferably at their grain boundaries, which causes particle disintegration and enhanced sintering (see 7.3.3.2). Especially when dT/dx in the phase diagram is small, the sintering temperature (T_{SS}) has to be adjusted carefully because the amount of liquid phase is strongly temperature dependent. The 'sintering window' ranges ±2–±10 K, depending on composition. When the temperature is too low, not enough liquid phase is present, at a too high temperature grain growth and coarsening take place. Super solidus sintering is applied in high speed steels and super alloys, it provides the possibility to utilize coarser powders than in 'conventional' liquid phase sintering.

The decrease of the specific surface energies Δ_γ of the system as the driving force results from the energies involved

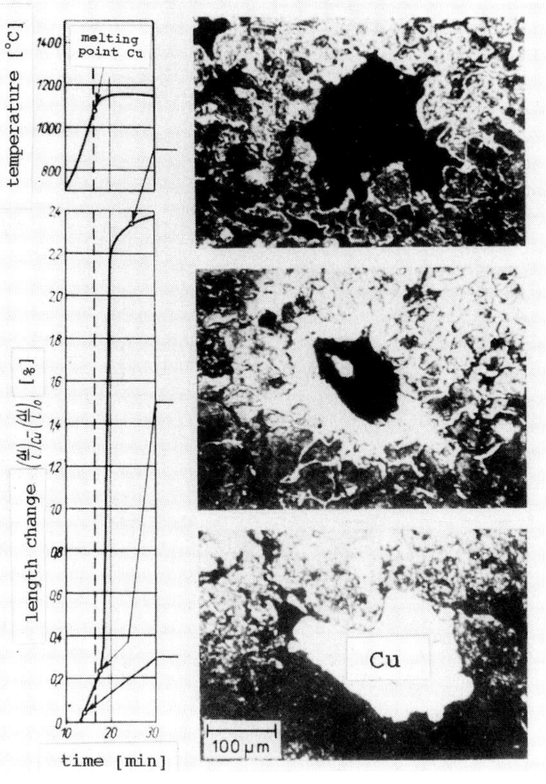

Figure 7.35 Mechanism of shrinkage compensation in Fe–Cu (Bockstiegel). *Bottom*: Cu-particle before melting; *middle*: molten and spread Cu-particle; *top*: pore after complete spreading

$$\Delta_\gamma = \gamma_S + \gamma_L - \gamma_{SL} \tag{7.20}$$

γ_S = specific surface energy solid/gas
γ_L = specific surface energy liquid/gas
γ_{SL} = specific phase boundary energy solid/liquid

An energy decrease only takes place when

$$\gamma_{SL} < \gamma_S + \gamma_L \tag{7.21}$$

Thus, liquid phase sintering is successful only in good wetting systems, in which γ_{SL} is low.

These energies are correlated in the Young-Dupré- quation with the *wetting angle* ω (Fig. 7.36a):

$$\cos \omega = \frac{\gamma_S - \gamma_{SL}}{\gamma_L} \tag{7.22}$$

and combining these equations:

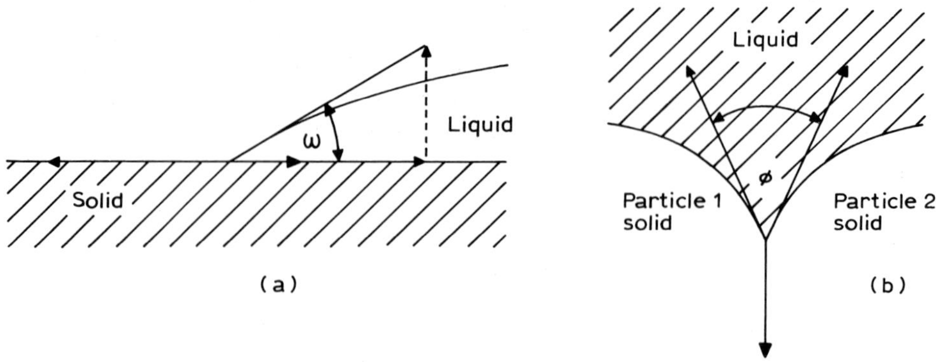

Figure 7.36 Wetting (ω) and dihedral angle (ϕ)

$$\Delta\gamma = (1 + \cos \omega) \cdot \gamma_L \tag{7.23}$$

The higher cos ω (i.e. the smaller ω), the higher is the driving force Δγ. The angle for complete wetting is 0°, occurring at very low γ_{SL} values, as observed in the hard metal systems WC with Co, Ni or Fe. In practice, however, it must be borne in mind that γ_{SL} is very sensitive to impurities, especially thin oxide films which may change the wetting angle markedly. Liquid phases with high wetting angles ($\omega \geq 90°$) are unfavourable and may even inhibit sintering. Parts of the liquid may be squeezed out as droplets on the surface even against gravity.

The liquid phase may also penetrate the solid grain (or phase) boundaries, when the boundary energy

$$\gamma_{SS} > 2\,\gamma_{SL} \tag{7.24}$$

The resulting angles between the liquid and two solid particles are called *dihedral angles* (Φ) (Fig. 7.36b).

$$\cos \Phi/2 = \frac{\gamma_{SS}}{2\,\gamma_{SL}} \tag{7.25}$$

Small dihedral angles result at very low γ_{SL} and high grain boundary energies γ_{SS}, leading to a disintegration of solid (multigrained) particles during liquid phase sintering. This supports the densification by solid phase rearrangement (see 7.3.3.2).

7.3.3.2 Stages and Kinetics

During liquid phase sintering three stages* can be often defined, as shown originally by Lenel, Kingery and modified later by other researchers:

* It has been shown recently, that prior to the occurrance of liquid, considerable solid phase sintering may take place (see below).

1. Particle rearrangement by melt penetration of the solid skeleton due to capillary forces (primary rearrangement). Particle sliding with (mostly) rapid densification occurs, depending on the amount of liquid phase. This process is essentially rate-controlled by mechanical movement in the complex capillary system. Depending on the packed density of the solid particles full densification is theoretically possible above a certain percentage of the liquid (~35vol-%).
2. Solution and reprecipitation, controlled by diffusion processes within the liquid phase, accompanied by shape accommodation and grain growth (secondary rearrangement). Further densification and neck formation between solid particles occurs.
3. Dominance of solid state sintering, leading to particle and pore coalescence, microstructural coarsening with little further densification. Swelling due to gas-filled closed porosity and degradation of properties is possible.

These stages are derived mainly for sintering with permanent liquid phases. Experiments indicate these stages possess different kinetics (Fig. 7.37). In the general isothermal shrinkage equation

$$\frac{\Delta L}{L_0} = \text{const} \cdot t^n \tag{7.26}$$

n amounts ~ 1 as long as viscous flow during rearrangement is rate controlling. n reaches 1/2, when phase boundary reactions dominate and $n = 1/3$ or less for the late stages, like contact flattening and solid state sintering. In reality, however, n may amount to higher than unity (up to 3) in the first stage and the stages may overlap to a considerable extent.

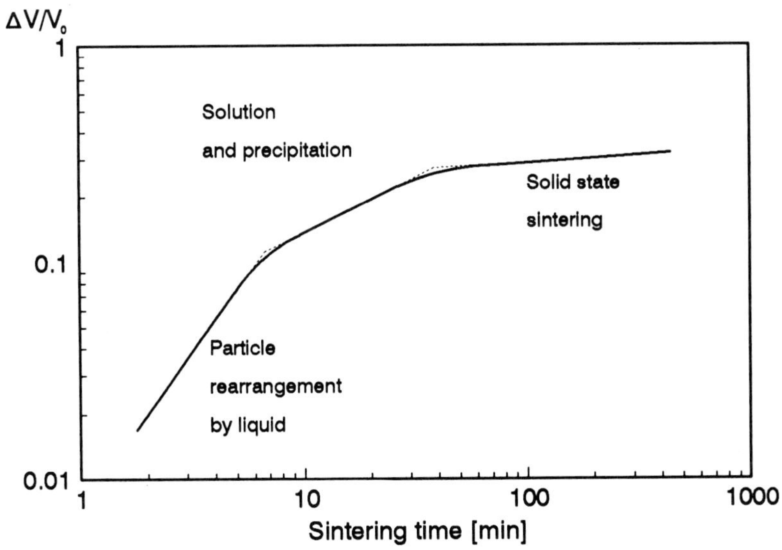

Figure 7.37 The overlapping stages of liquid phase sintering (schematically)

Table 7.3 Phenomena during solution-precipitation in liquid phase sintering (Kaysser)

Mechanism	Relevance on sec. rearrangement	Relevance on shape accommodation
Ostwald ripening	+	++
Contact flattening	−	++
Directed grain growth	+	+
Particle disintegration	++	−
Coalescence	+	−
Liquid film migration	+	++

Difficulties in equal densification may occur with irregular shaped particles due to steric hindrance of the rearrangement process, i.e. surface smoothing by solution and reprecipitation facilitates rearrangement. Although a low viscosity of the liquid phase is favourable for fast penetration, a tendency to particle cluster formation has been observed. This causes inhomogeneities in densification and composition as well as formation of large pores.

The solution-precipitation phenomena controlling the second stage are caused by several mechanisms, as shown in Table 7.3.

Greater local solubilities due to higher free energy may occur in different parts of the compact, namely at crystal planes with high Miller indices, in particle contacts under pressure with liquid films in between, at surface roughness and fine particles. The latter is a consequence of high curvature, leading to higher solubility according to the Thomson-Kelvin equation (see eq. 7.4). The growth of coarse particles or pores on account of fines is known as Ostwald ripening. Shape accommodation of the solid phase is an important phenomenon in this stage which is shown schematically and in a practical example in Fig. 7.38. The processes of Ostwald-ripening and contact flattening are both responsible for the accommodated shape. Also the migration of liquid-phase films (Fig. 7.39) contribute to the final microstructure: Due to concentration

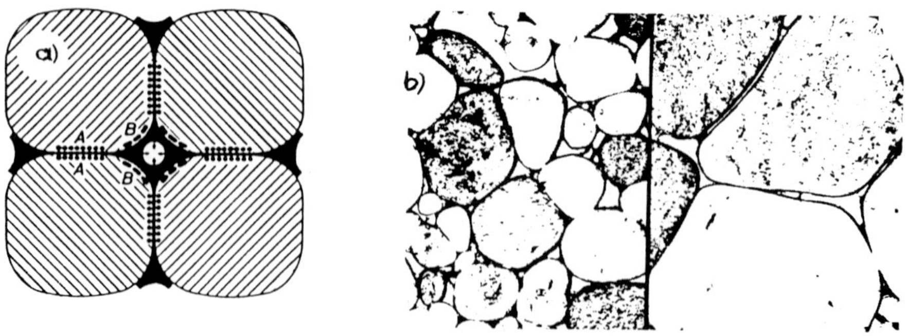

Figure 7.38 Shape accommodation by solution-precipitation a) schematic, A = solution, B = Precipitation zone b) as observed in a heavy alloy, 6 h at 1400°C, accompanied by grain growth (Price, Smithells, Williams)

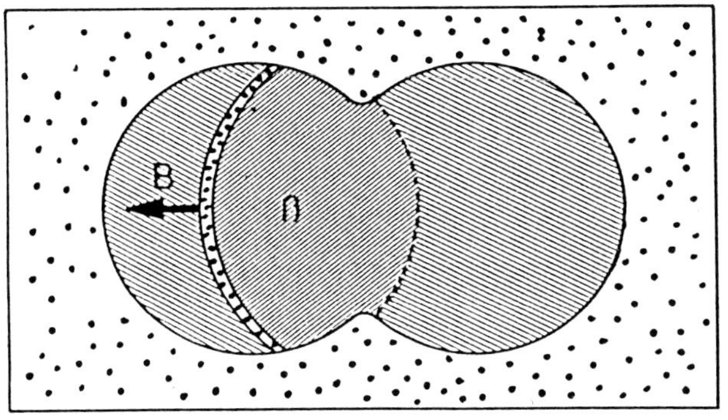

Figure 7.39 Liquid film migration in the solid phase during sintering (B = migration direction) (schematically)

differences between two grains, the liquid film moves with the grain boundary, resulting in directed (anisotropic) grain growth.

During the third stage the action of the liquid phase becomes less important and solid state processes dominate. The rigid solid skeleton allows only little further densification which can be easily overcompensated by secondary effects. The danger of degradation by prolonged final stage sintering ('oversintering') has to be considered and moderate sintering times are therefore typical in practice.

Irrespective of the important role of the liquid phase, sintering of solid state contacts has been underestimated during the past. Considerable microstructural evolution including particle rotation by solid state sintering prior to the liquid formation may take place. The pre-isothermal period should not be disregarded, and is claimed to have a similar significance to that for solid state sintering (see section 7.2.2.5). The combined non-isothermal–isothermal kinetics (see Fig. 7.12) seem similar for liquid-phase and solid-state sintering, since in both a low viscosity phase (i.e. low viscosity contact area in solid-state sintering, see 7.2.2.3) controls the process.

7.3.3.3 Infiltration

Infiltration is a two step liquid phase process, in which the porosity of green or presintered compacts, is filled by a molten metal or alloy. The latter is often deposited in the solid state at the surface of the compact (Fig. 7.40), melts during heating and fills the pores by capillary forces. With interconnected porosity and in good wetting systems the resulting microstructure is essentially free of pores. Metallic infiltrants generally are of low viscosity and require only short infiltration times. The process is easy and reproducible only when the surfaces are free of contamination. Although in presintered compacts only little sintering occurs during infiltration, the process is considered as a special type of liquid phase sintering.

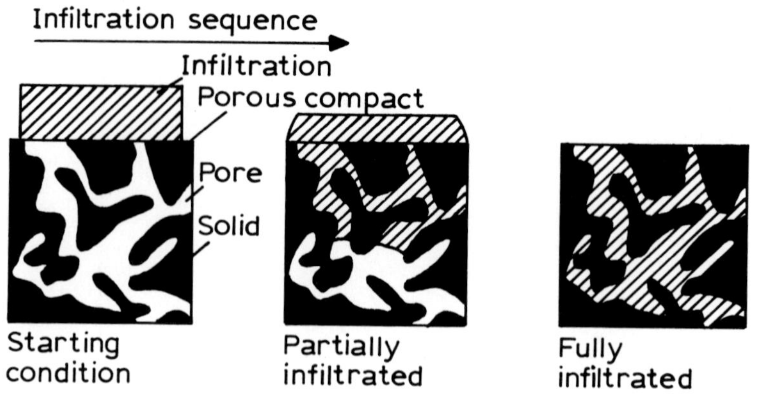

Figure 7.40 The infiltration process (German)

Some difficulties may arise at high mutual solubility. If the liquid is soluble in the solid, the equilibrium is reached only via diffusion in the solid phase and the liquid may even disappear during infiltration. If the liquid has a high solubility for the solid, surface erosion and bad dimensional control may occur. An infiltrant which is saturated with the solid component should be used in this case.

Infiltration has found a limited practical application in powder metallurgy, e.g. in the systems Fe–Cu, W–Cu, W–Ag, Mo–Ag, WC–Cu and WC–Ag. Large, high-voltage electrical contacts are manufactured in this way. The process is performed in vacuum or high purity controlled atmosphere furnaces. Final machining of the original infiltration surface is usually necessary. Siliconized SiC (SiSiC) is an infiltrated and reaction sintered product as well and belongs to engineering ceramics (see section 11.10.3).

7.3.4 Reaction (Reactive) Sintering

This term describes sintering processes of powder mixtures occurring simultaneously with exothermic chemical reactions, i.e. the phase composition is different after sintering than before. The reactions may be solid or liquid state, but also include gas phase reactions. Classical examples are the formation of intermetallic phases e.g. in the systems Ni–Al, Ti–Al etc.:

$$\mathrm{x}A + \mathrm{y}B \rightarrow A_{\mathrm{x}}B_{\mathrm{y}} \tag{7.27}$$

The driving force of the chemical reactions involved is generally orders of magnitude higher than that of pure solid phase sintering which involves a decrease in surface energy only. A commonly observed side-effect (which sometimes can be the main effect!) is an extensive swelling instead of shrinkage (see Fig. 7.31c) during phase formation. The reasons are: the change of the specific volume during reaction, and especially the different diffusion flow of A and B in

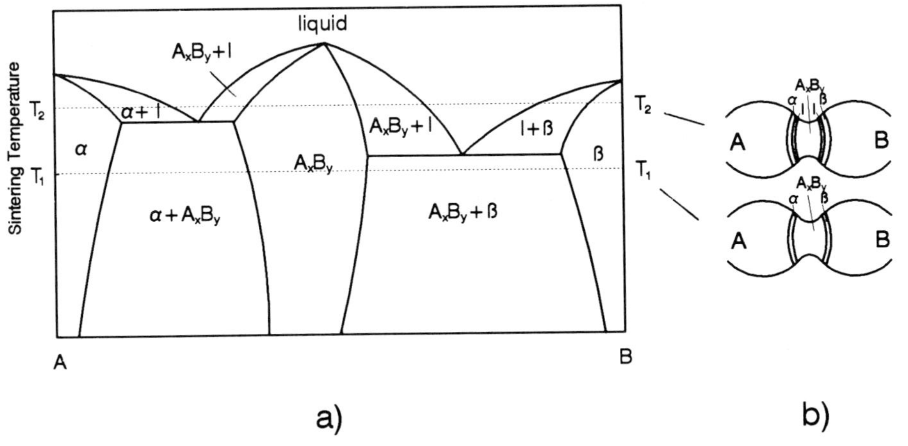

Figure 7.41 Sold state (T_1) and liquid phase (T_2) reaction sintering of an *A–B* powder mixture. (W. Schatt) (a) Phase diagram (b) Phases occurring in an intermediate sintering stage. Note: the process is mostly not iso-thermal as indicated in this figure

A_xB_y. The atomic mobility in A_xB_y is often strongly reduced compared with that in *A* and *B*, as a consequence of stronger bonding forces (partly covalent) in the phase. Elementary powder mixtures corresponding to the nominal composition of the phase formed during sintering exhibit the highest swelling.

To overcome these swelling effects the simultaneous application of pressure is advisable and highly dense materials may be obtained by 'reaction hot pressing'. The occurrence of a liquid phase with good wetting behaviour ('reactive liquid phase sintering') is another possibility to counteract swelling as well as to complete the reaction. Phase formation and sintering can be performed also in separate steps; (1) phase formation under adequate temperature-time conditions with subsequent milling to powder; (2) pressing and sintering of the synthesised phase, usually under conditions different to those of phase formation.

The principle course of solid (T_1) as well as liquid phase reaction sintering (T_2) can be seen in Fig. 7.41; the phases occurring in the material can be derived from the phase diagram at the applied temperature.

A technical ceramic product of some significance is the reaction sintered (bonded) silicon nitride (RBSN) obtained by the exothermal reaction

$$3\ Si + 2\ N_2 \rightarrow Si_3N_4 \tag{7.28}$$

below the melting point of Si (1410°C). This solid–gas reaction takes place under near dimensional stability: The mass gain with the accompanying volume increase proceeds nearly completely within the porosity of the compact (Fig. 7.42).

The production of metal–ceramic or ceramic–ceramic composites is another field of considerable interest for the application of reaction sintering, especially by *self propagating (metallo-thermal) reactions.* Examples are TiC–Al_2O_3–, TiC–Al_2O_3–Al and other composites, as well as intermetallics mentioned above. The reactions

$$3\ TiO_2 + 4\ Al + 3\ C + x\ Al_2O_3 = 3\ TiC + (2 + x)\ Al_2O_3 \qquad (7.28a)$$

$$\text{or: } 3\ TiO_2 + (4 + y)\ Al + 3\ C + x\ Al_2O_3 = 3\ TiC + (2 + x)\ Al_2O_3 + y\ Al \qquad (7.28b)$$

lead to composits with variable amounts of Al_2O_3 and Al. Heating of the mixture is applied mainly to provide the activation energy for the chemical reaction. A wide variety of reaction sintering processes have been developed, but many of them are in an experimental state only at the present time. The future technical production probably includes titanium intermetallics (see section 11.11).

It should be mentioned that, during powder metallurgy and ceramic processing, chemical reactions often take place to some extent without defining the process as reaction sintering. The reduction of residual oxygen of powders, the formation of carbides (from carbon black) during steel-sintering, reactions in SiO_2-based liquid phases in ceramics are well known. The chemical reactions do not represent the major part of the entire process in these cases.

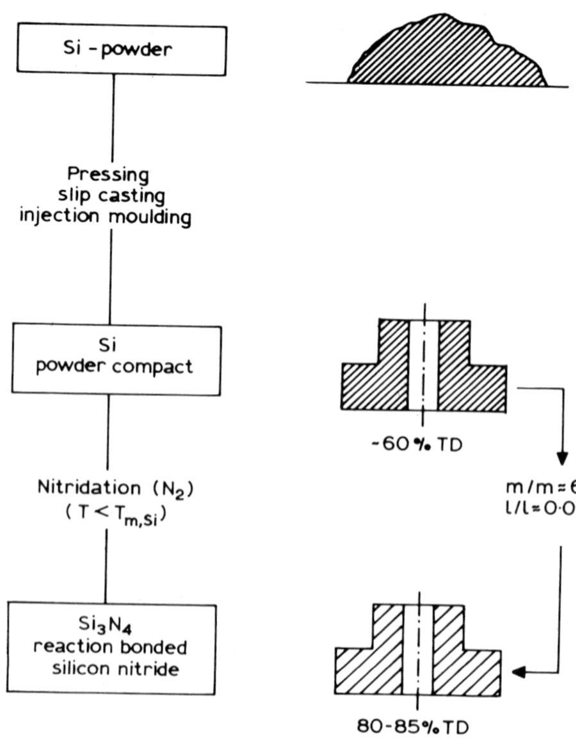

Figure 7.42 Production scheme for reaction bonded silicon nitride (RBSN)

7.4 PRESSURE SINTERING (HOT COMPACTION)

Pressure sintering, hot pressing and hot compaction are synonyms for the same process, namely simultaneous application of final shaping and sintering in one process. Although this field may be treated separately from the viewpoints of compaction and sintering, it seems to be worthwhile not to do so. It is therefore outlined in section 5.1.2.

7.5 SPECIAL SINTERING PROCEDURES

7.5.1 Rate Controlled Sintering (RCS)

It has been shown, mainly by Palmour III, that for a wide range of ceramic (and perhaps some metallic) materials the conventional sintering cycle (with heating, isothermal soaking and cooling period) does not allow an optimised schedule with respect to final density, microstructure and properties. An RCS-cycle, i.e. time-temperature schedule with optimised shrinkage rates over a wide range of densification may be advantageous. Especially refinements of the microstructures as well as better thermal efficiencies during firing can result from their proper application. Since the approach taken for RCS is mainly experimental, sophisticated research methodologies, namely precision digital dilatometry and computer aided design for optimal paths for sintering have to be applied. Likewise solid-state and liquid-phase sintering processes have been found responsive to RCS, e.g.: Al_2O_3, MgO, ZrO_2, clay-based products, porcelain and $YBa_2Cu_3O_7$ – superconductors. One of the important issues is probably the more reliable removal of pores in the vicinity of grain or phase boundaries.

A typical 4-stage RCS profile is shown in Fig. 7.43. It is important to approach (but not to exceed!) the 'maximum safe rate' of densification and later the lower rates by temperature–time adjustment. The strength distribution data obtained after RCS of ZrO_2–(Y)–TZP powder compacts, compared with conventionally sintered samples are shown in Fig. 7.44. Mainly in the lower and medium strength levels the RCS-products are found to be superiour. An overall refinement of Al_2O_3, ZrO_2 composites and of artware porcelain have also been observed after RCS. RCS seems to apply when high atomic defect concentrations in the powder compact can be transferred to the temperatures of the 'intensive shrinkage' stage (see section 7.2.2.5), where they provide the maximum contribution to sintering.

Each RCS optimization is valid only for one material with its specific preprocessing green condition. This obviously hinders to introduce the process in mass production.

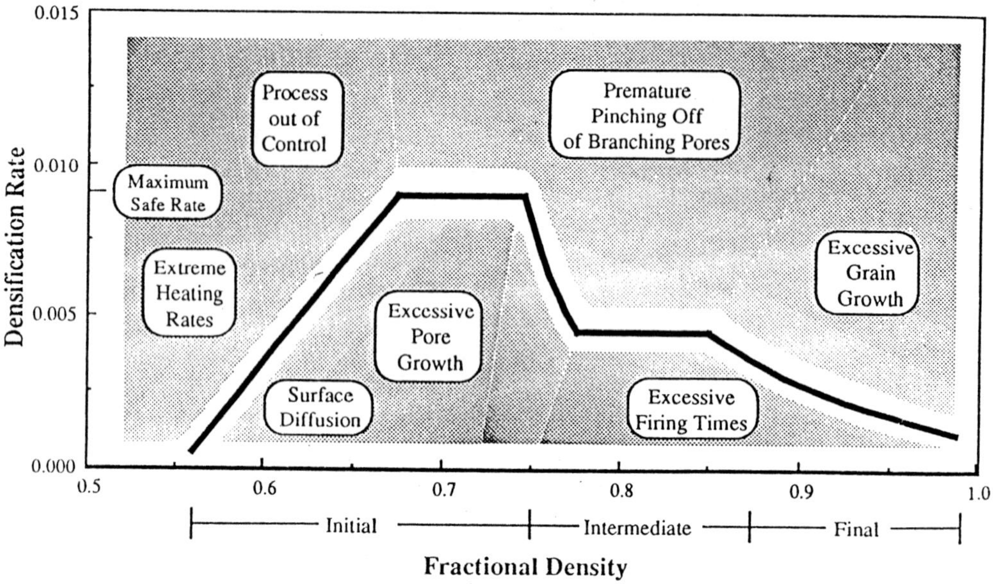

Figure 7.43 Typical RCS profile, with indications of practical constraints as well as issues that can influence path selection (Palmour III)

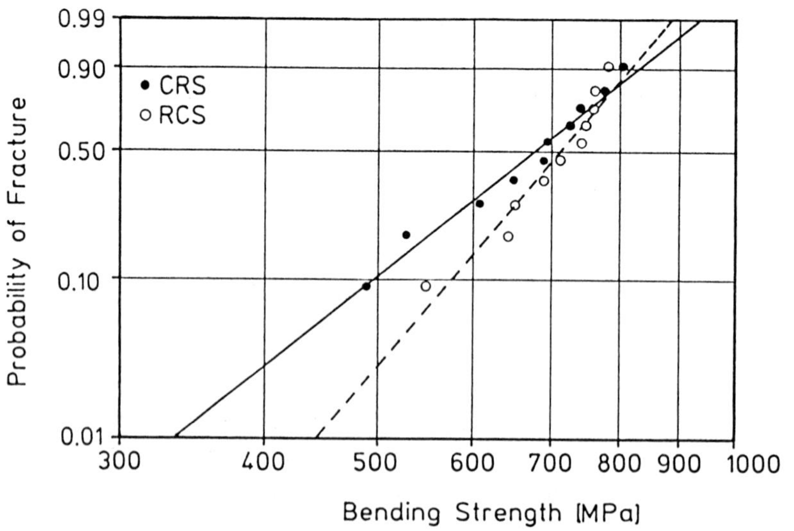

Figure 7.44 Strength distribution of Y–TZP (ZrO_2) after three stage RCS and constant rate sintering (CRS)

7.5.2 Induction Sintering

Sintering by direct coupling of pressed compacts in a high frequency induction field (e.g. 4 KHz) has been investigated several times in the past. Its principal

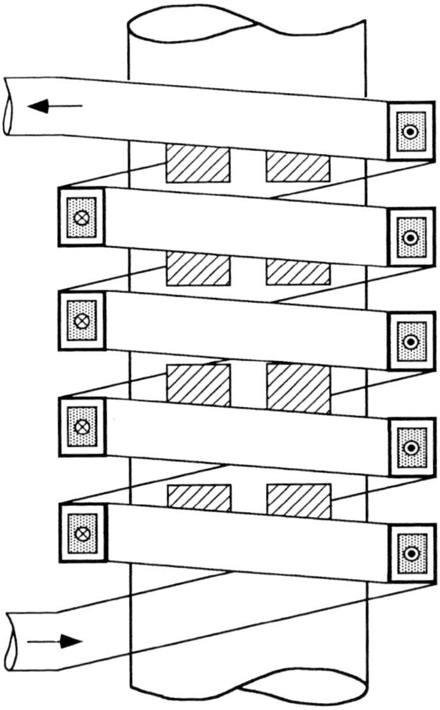

Figure 7.45 Scheme of induction sintering equipment, using cylindrical tube and coil (→ water flow)

simplicity with low investment costs, low energy consumption due to high heating rate, short sintering times (3–10 min) and the absence of large amounts of insulation materials to be heated, looks interesting. Fig. 7.45 shows the principle of induction sintering. The basic mathematical expressions are:

$$i_X = i_0 \exp\left(-\frac{x}{\sigma}\right) \quad (7.29)$$

where i_X = current density at the depth x, i_0 = current density at the surface (x = 0) and σ = penetration depth, defined as $1/e = 0.368 \cdot i_0$. δ characterises the skin effect, which depends on frequency (f), electrical resistivity ρ and magnetic permeability (μ_r) of the material:

$$\delta = 503 \cdot \sqrt{\frac{\rho}{\mu_r \cdot f}} \quad (7.30)$$

This leads to one principle disadvantage, namely, the strong temperature gradients in the parts, mainly during heating-up. Due to the high resistivity of the particle contacts (oxide films) very high local temperatures can arise at the beginning at these points. A critical heating-up rate must not be exceeded, otherwise local melting or cracking may occur.

Induction sintering has been successfully developed on a laboratory scale for simple-shaped parts of unalloyed and some alloyed steels as well as hard metals. The sintered properties show some improvement compared with material conventionally sintered. Nevertheless, the method seems not to be applicable for mass production with high-reliability standard. It is applied, however, as a preheating method for compacts subjected to powder forging.

7.5.3 Plasma and Microwave Sintering

Sintering in plasmas, which may be performed as zone sintering in a hollow cathode discharge has been developed to produce thin walled tubes and small diameter rods of MgO-doped Al_2O_3. Very high heating rates in excess of 100 K s^{-1} provide low sintering times in the range of a few minutes to achieve densities of 99% *TD* at final grain sizes of 2–5μm. The success of this ultra-rapid sintering to get homogeneously sintered, crack-free bodies is critically dependent upon the characteristics of the starting powder, which has to be of fine (below 1 μm) and narrow particle size distribution. The high temperature gradients lead to high linear shrinkage rates of 1–4% s^{-1}. Practical application seems to be possible under very special precautions.

Microwave sintering of ceramics is proposed on the basis of several anticipated advantages, namely: enhanced diffusion processes, reduced energy consumption due to short sintering times and reduced overall sintering temperatures, and, at least in principle, reduced risk of thermal cracking as the parts are heated within their volume and not by heat transfer from the surface. Experiments indicate an increase of diffusivity with increasing frequency, being probably dependent on the material investigated. Good results have been obtained at 28 GHz. Plasma is often used for specimen heating to bridge the temperature region with low microwave losses. Obstacles, however, are thermal runaway effects and unequal energy distribution which should be overcome by optimised green compact preparation, improved experimental devices and control units as well. Theoretical as well as practical work is in progress in order to establish this method.

7.5.4 Cold Sintering

Irregular shaped powders with high plasticity, like Fe, Cu and Al can be densified by high pressure compaction to 97 to 99% *TD* at about 3 GPa compacting pressure. With a proper mechanical design dies and punches made from tool steel or cemented carbide may be used at this pressure, the latter even up to 4 GPa. Under these conditions extensive plastic deformation within the powder particles and shear deformation occur, leading to fresh in-situ formed surfaces which are free of oxide and other contamination. This facilitates the approach of surfaces up to atomic distances, resulting in strong bonding similar to sintered contacts achieved at higher temperatures. The term 'cold sintering' can be understood in view of the transverse rupture strength properties obtained in these compacts, being reported as 100 to 150 MPa by using irregular shaped Fe powders. This is not too far from the strength of a comparable sintered product. The 'physical bonding' leading to high strength improves the higher the compacting pressure and the more irregular shaped the powder particles used. Samples prepared from equiaxed or spherical powders show much lower strength.

The microstructure of cold sintered samples exhibit narrow flat pores and fragments of former oxide layers on the interfaces as islands. Soaking at relatively low temperatures (e.g. 500–600°C for Fe powder compacts) may result in pores and oxides of a more spherical shape (Fig. 7.46). This probably occurs by surface diffusion or by grain boundary mechanisms. Such tempering further improves bonding between the particles.

Prospective application of cold sintering is claimed for the consolidation of rapidly solidified powders (e.g. Al- or Fe-based), keeping the homogeneous distribution of impurities and possible metastable structures unchanged. Advantages are also claimed for composite materials when one of the phases would otherwise be damaged during high temperature processing as in metal-bonded diamond or carbide composites.

7.6 EXPERIMENTAL INVESTIGATION OF SINTERING PROCESSES

A thorough investigation of the sintering process is necessary not only for understanding and defining the mechanisms involved, but also in a more practical sense for optimising the process. The different stages as well as the chemical reactions, phase changes, mass gain or loss, etc. which may occur during

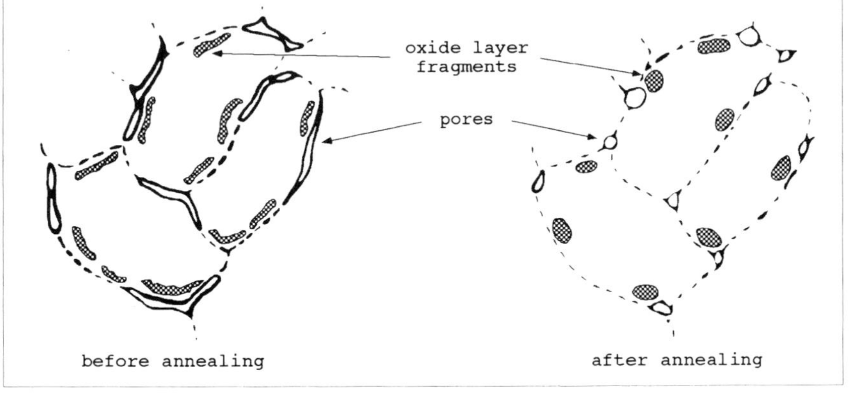

Figure 7.46 Microstructure of a cold sintered plastic metal powder (schematically, Gutmanas)

sintering, require different methods of investigation. Since they are well known to materials scientists, only a few methods of special significance for the sintering process will be discussed.

7.6.1 Dimensional Changes

Dimensional changes occur in all stages of sintering, in the early stage predominantly as *neck forming* and *growth*. The kinetics of this process has often been analysed microscopically mostly in model experiments from ground and polished sections. In order to get concise results a large number of single necks have to be evaluated. Early stages of sintering, especially in metallic powders, can be analysed sensitively by measuring the electrical conductivity of loose or compacted powders. The results often depend, however, apart from neck growth also on oxide film reduction at the neck-forming surfaces.

The most important changes during sintering of powders are the isothermal and non-isothermal *densification kinetics* (shrinkage) and the relationship between time and temperature and density. These measurements include phase transformations with volumetric effect. Such information can be obtained by a considerable number of interrupted tests which, however, require much time and material. Continuous measurement usually with dilatometer gives more and better information within a shorter time and with less experimental effort. Assuming constant sample mass m and specific weight, one linear dimension, generally the sample length, is continuously measured during sintering, instead of the material volume. If the shrinkage behaviour of the sample investigated is known to be isotropic, the actual volume can be calculated from the sample length:

$$D_1 = D_0/(l_1/l_0)^3 \tag{7.31}$$

with D_1 = actual density, D_0 = initial density, l_1 = actual length l_0 = initial length

Dilatometers for continuous measurement are generally designed as separate systems comprising a furnace, length-change as well as temperature measurement and a closed loop control unit. Depending on the heating element dilatometers may work up to 2000°C at atmospheric pressure or in some cases at slight overpressure. The sample is positioned within the furnace and linked to a rod transmitting the length changes to a displacement pickup located outside the hot region (Fig. 7.47). For high accuracy the intrinsic behaviour of the dilatometer is determined and used for correction of the measured signal. Shrinkage curves of powder compacts in the non-isothermal and isothermal range, can be seen in Figs 7.12 and 7.23.

For monitoring high-pressure-assisted sintering HIP-dilatometers have been developed. In Fig. 7.48 the shrinkage behaviour of a zirconia toughened alumina (ZTA) during a sinter/HIP cycle as measured with a HIP-dilatometer is shown. Besides the temperature and pressure the length change ΔL and

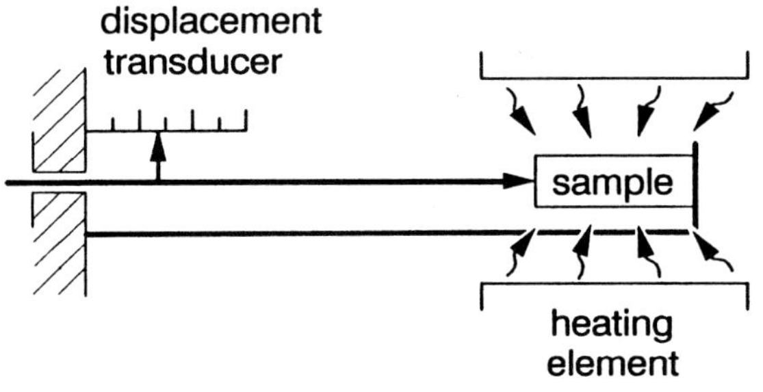

Figure 7.47 Interior of a dilatometer (schematically): temperature dependent length change of the sample is transduced to a measured signal

shrinkage velocity $\Delta L/\mathrm{d}t$ are given. Different stages of length change can be identified. After thermal expansion during heating and the ensuing sintering shrinkage and pressure-assisted densification, a phase transformation during cooling is clearly visible. The stage of closed porosity (for this material about

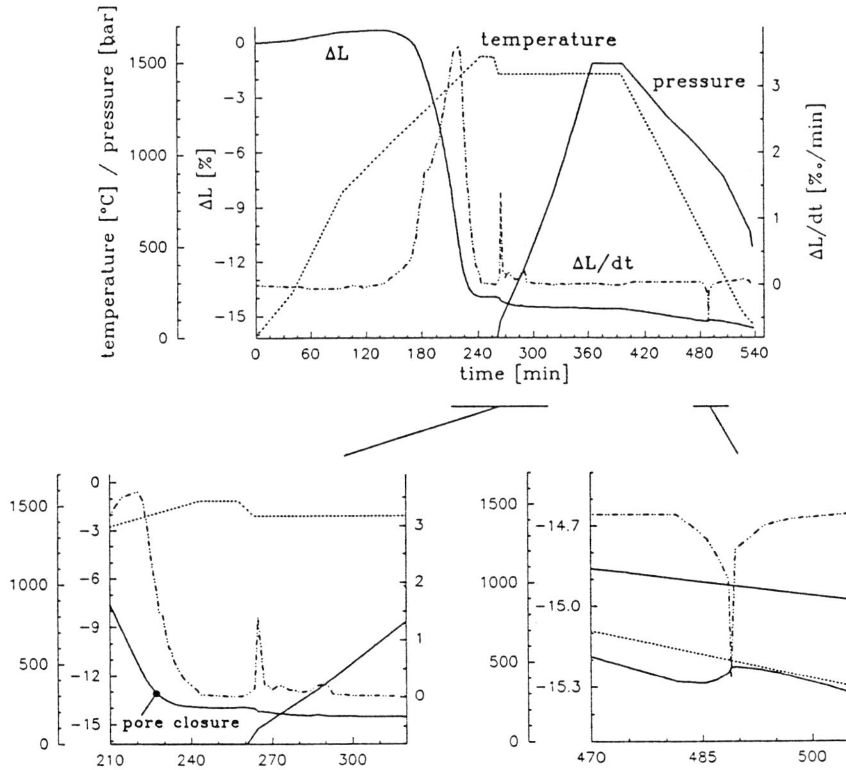

Figure 7.48 Results obtained on Al_2O_3-ZrO_2-compact in a HIP-dilatometer experiment (Kühne); (a) The whole sinter HIP cycle; (b) pore closure and densification by pressurisation; (c) tetragonal-monoclinic phase transformation of the ZrO_2 particles during cooling

95% TD) was reached clearly before entering the isothermal phase. From this shrinkage behaviour essential conclusions for optimising the temperature and pressure regime can be drawn.

7.6.2 Mass and Physico-Chemical Changes

Besides measurement of dimensional changes, a large number of methods for measuring the temperature dependence of materials behaviour and properties is known. Each method involves the analysis of a certain response of a sample at a given time-temperature cycle in a well defined atmosphere. By combining the information gained from different methods a good characterisation of a material is possible.

With *thermogravimetry* (TG) the mass change of a sample is determined in a thermobalance. Melting point, temperatures of dissociation or formation of reaction products can be detected. Such findings can be supported by *differential thermal analysis* (DTA). Measuring temperature differences between the sample and an inert reference specimen allows qualitative determination of the thermal effect of chemical reactions and physical transformations (melting, crystallisation, etc.). *Differential scanning calorimetry* (DSC) is a similar method for quantitative measurements. Here differences in heat transfer between the sample and reference material are used to determine quantitatively endothermic or exothermic reactions.

Evolved gas detection (EGD) is used to analyse whether or not a volatile product is formed during thermal treatment and with evolved gas analysis (EGA) the nature and amount of such volatile products can be determined. Methods for this purpose are *gas chromatography* or *mass spectroscopy* (MS), giving information on evolving volatile products as well as on reactions with sintering atmospheres.

7.7 COMPUTER SIMULATION AND MODELLING

The analytical descriptions of the sintering process generally involve simplifications of the particle shape, particle contacts and neck geometries but they disregard factors important in sintering practice. The significance of their results has to be considered, therefore, as very limited, although several principal statements seem to remain valid, nevertheless. Several attempts at computer simulation have been employed, whose main results are as follows.

The circular neck contour mostly presumed in sintering models (Fig. 7.2) does often not exist in reality. The neck surface undercuts the original particle surface and differs from the tangent-circle approximation. The formation of a dihedral angle at the neck surface-grain boundary intersection has a decisive influence on the neck shape. The neck is filled by material from the near-surface region which occurs by more than one mechanism.

The simple power law functions used to identify sintering mechanisms are not confirmed by computer simulations due to the complicated neck geometry mentioned above. The exponents may vary during neck growth and exponents calculated from experimental slopes may be misleading. For the validity of a sintering equation, the agreement with the experiment is not a sufficient criterion. Sintering rate equations derived for ideal geometries do not apply to pressed powders (as shown also experimentally).

Recently, the sintering and hipping of complex parts with inhomogeneous green density distribution was modelled using the finite element method. The behaviour of the sintering material is described as linearly viscous with density-dependent bulk and shear viscosities and sintering stress. Classical sintering models are revised and they are formulated for multiaxial stress states. If grain-boundary or bulk diffusion is the controlling transport mechanism, the macroscopic constitutive equations are indeed linear viscous. Detailed models were developed, for example for the intermediate stage of sintering, in which the surface of the pore space is (nearly) in equilibrium with respect to the surface and interface energies.

FURTHER READING

W. Schatt, *Sintervorgänge (Grundlagen)*, Düsseldorf, VDI-Verlag, 1992.

G.C. Kuczynski (ed.), *Sintering Processes*, New York, London, Plenum Publ. Corp., 1980.

D.P. Uskokovic, H.E. Palmour III and R.M. Spriggs (eds), *Science of Sintering*, New York, London, Plenum Press, 1989.

A.C.D. Chaklader and J.A. Lund (eds), *Sintering '91*, Solid State Phenomena, Vols 25–26, Trans Tech. Publications, 1992.

H.E. Exner, Principles of Single Phase Sintering, *Rev. Powder Metall. Phys. Ceram.*, **1**, 7–251, 1979.

J.M. Capus and R.M. German, *Sintering*, Vol. 3 of *Advances in Powder Metallurgy and Particulate Materials*, Princeton NJ, MPIF, 1992.

R.M. German, *Powder Metallurgy Science*, Princeton, NJ, MPIF, 1984.

F. Thümmler and W. Thomma: The Sintering Process, *Metall. Rev.* Nr.115, 69–108, 1969.

M.F. Ashby, *Acta Metall.* **22**, 275–289, 1974.

S.B. Swinkels and M.F. Ashby, A Second Report on Sintering Diagrams, *Acta Metall.* **29**, 259–281, 1981.

A. Prümmer, *Explosivverdichtung pulvriger Substanzen*, Berlin, Heidelberg, Springer-Verlag, 1987.

L.I. Trusov, V.I. Novikow and K.N. Lapavok, Powder Metallurgy and Nanocristals, *Science of Sintering*, **23**, No.2, 63–78, 1991.

C. Grescovich and J.H. Rosolovski, Sintering of Covalent Solids, *J. Amer. Ceram. Soc.*, **59**, 336–343, 1976.

R.M. German, *Liquid Phase Sintering*, New York, Plenum Press, 1985.

R.M. German, S. Farooq, and C.M. Kipphut, Kinetics of Liquid Phase Sintering, Mater. Sci. Eng. **A 105/106** 215–224, 1988.

W.A. Kaysser, *Sintern mit Zusätzen*, Berlin, Stuttgart, Gebr. Bornträger, 1992.

W.A. Kaysser and G. Petzow, Present State of Liquid Phase Sintering, *Powder Metall.*, **28**, 145–150, 1985.

G.S. Wright and B. Ogel, 'Supersolidus Sintering of High Speed Steels', *Powder Metallurgy*, **36**, (3), 213–219, 1993.

Symposium of Reaction Synthesis of Materials (Part 1), *Metallurgical Trans.*, **23A**, (1), 7–97, 1992.

H.E. Palmour III, Rate Controlled Sintering for Ceramics and Selected Powder Metals, *Science of Sintering*, New York, London, Plenum Press, 1989, 337–356.

E.Y. Gutmanas, Cold Sintering under High Pressure, Mechanisms and Application, Powder Metall. Int., **15**, No.3, 129–132, 1983.

A. Kühne, R. Oberacker and F. Thümmler, HIP-Dilatometer for Process Development in Powder Materials, *Powder Metall. Int.*, **23**, No.2, 113–119, 1991.

W. Wendlandt, Thermal Methods of Analysis, New York, Wiley, 1986.

H. Riedel and D.Z. Sun, in: Numerical Methods in Industrial Forming Processes, *Numiform '92* (Eds., J.L. Chenot, R.D. Wood and O.C. Zienkiewicz) A.A. Balkema, Rotterdam, 1992, 883–886.

J. Svoboda, H. Riedel and H. Zipse, 'Equilibrium Pore Surfaces, Sintering Stresses and Constitutive Equations for the Intermediate and Late Stages of Sintering, Part I: Computation of Equilibrium Surfaces, Part II: Diffusional Densification and Creep, *Acta metall. mater.*, 1993.

8: Sintering Atmospheres and Equipment

8.1 FUNCTIONS AND REACTIONS OF ATMOSPHERES

Sintering atmospheres, which also include vacuums of different grades, have to fulfil several functions depending on the material to be sintered. These are:

a) To avoid or to control chemical reactions — oxidation, reduction, carburisation, decarburisation, nitridation, and decomposition, including elimination of impurities.
b) To avoid undesired evaporation of the main component or of alloying elements.
c) To remove volatile admixtures and their decomposition products such as lubricants or plasticisers from the injection moulding process.
d) To provide heat transfer by convection in the furnace.
e) To provide the external hydrostatic pressure required in hot isostatic pressing.

The sintering atmosphere must be compatible with the constructional material of the furnace interior and with the heating elements in so far as they are in contact with the gas. The convective heat transfer function is not possible during vacuum sintering. The burning out of binders or lubricants is sometimes performed in a separate presintering step. In any event larger deposits of these products, including their cracking products in the interior of the furnace need to be avoided.

The prevention of oxidation is the most significant function of sintering atmospheres in powder metallurgy. Thus, the majority of metallic and some ceramic products are sintered in protective gases which exclude the ambient atmosphere. Only a few noble metals, the conventional ceramics and some advanced oxide ceramics are sintered in ambient air.

A rough estimate of the reactions in sintering atmospheres is given in Table 8.1, based on the functions listed in a) above. This estimate is valid for a variety of metals with intermediate oxygen and carbon affinity processed by powder metallurgy, especially Fe, Ni and Mo; the last two columns especially for iron and steels. Reduction and decarburisation does not take place with highly reactive metals (Al, Be and group IVa and Va metals in the Periodic Table), whose oxide and carbide stability is too high to be reduced by conventional sintering gases.

Table 8.1 Metal-Gas Interaction at Sintering Temperatures

Oxidation	Reduction	Neutral	Carburisation	Decarburisation
O_2 [1] (air)	Vacuum[2]	N_2 [3]	CO	O_2 [1] (air)
H_2O [1]	H_2	Noble gases (Ar, He)[4]	Hydrocarbons (CH_4, C_3H_8)	H_2O [1] CO_2
CO_2	CO			H_2

1) Common impurity of sintering gases; ingress of air in case of leakage
2) Oxidation may take place dependent on leakage rate
3) Nitridation of metals and alloys with high N_2 affinity
4) Oxidation may take place dependent on O_2 and H_2O impurity levels

The reaction equations related to Table 8.1 can be summarised as follows:

Oxidation (→) *and reduction* (←)

$$Me + \frac{x}{2}O_2 \leftrightarrows MeO_x \qquad (\leftarrow \text{dissociation}) \tag{8.1a}$$

$$Me + xH_2O \leftrightarrows MeO_x + xH_2 \tag{8.1b}$$

$$Me + xCO_2 \leftrightarrows MeO_x + xCO \tag{8.1c}$$

Carburisation (→) *and Decarburisation* (←)

$$Me + 2xCO \leftrightarrows \left\{ \begin{matrix} MeC_x \\ Me_{(xG,\ S.S.)} \end{matrix} \right\}^{\dagger} xCO_2 \qquad \begin{matrix} \text{(Carbide formation)} \\ \text{(Carbon in solid solution)} \end{matrix} \tag{8.2a}$$

$$Me + xCH_4 \leftrightarrows \left\{ \begin{matrix} MeC_x \\ Me_{(xC,\ S.S.)} \end{matrix} \right\} + 2\,xH_2 \tag{8.2b}$$

Decarburisation (→)*

$$\left\{ \begin{matrix} Me\ C_x \\ Me_{(xC,\ S.S.)} \end{matrix} \right\} \begin{matrix} + \frac{x}{2}O_2 \rightarrow Me + xCO\dagger) \quad (8.2c) \\ + xO_2 \rightarrow Me + xCO_2\dagger) \quad (8.2d) \end{matrix}$$

$$\left\{ \begin{matrix} Me\ C_x \\ Me_{(xC,\ S.S.)} \end{matrix} \right\} \begin{matrix} + xH_2O \rightarrow Me + xCO\dagger) + xH_2 \quad (8.2e) \\ + 2\,xH_2O \rightarrow Me + xCO_2 + 2\,xH_2 \quad (8.2f) \end{matrix}$$

The redox-equilibria for several metals of different affinity are shown in Fig. 8.1, demonstrating the necessity of extremely dry H_2 (and, generally, extremely high purity, i.e. freedom from O_2 and H_2O) of any reducing gas, for the possibility of oxide reduction of highly reactive metals. This shows on the other hand, that even traces of O_2 or H_2O oxidise these metals in a cumulative way which can be disastrous especially during prolonged sintering.

* The opposite reaction is not considered in Table 8.1.
† Formation of CO and CO_2 depending on temperature according to Boudouard-equilibrium (see Table 8.3).

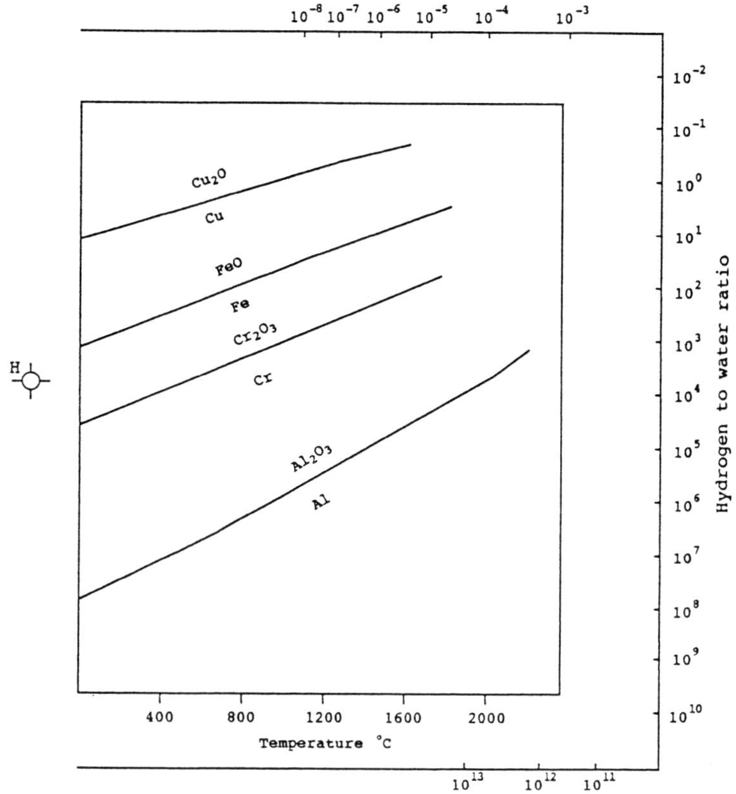

Figure 8.1 Oxidation-reduction equilibria for various metal–metal oxide systems, in terms of hydrogen to water vapour ratio. (Richardson and Jeffes)

Remark: The definite figure for p_{H_2}/p_{H_2O} ratio at a certain temperature is obtained by drawing a straight line from the point H -⊕, crossing the metal/metal oxide line just at the temperature considered

The purity with respect to residual O_2 and H_2O, which act in a similar way (see Table 8.1), is of special significance for all sintering gases and most of the materials to be sintered. The H_2O vapour content of gases is defined normally by the dew point, as given in Table 8.2, which is the temperature of the start of water vapour condensation during temperature decrease; the dryer the gas, the lower the dew point.

Table 8.2 Dew Point of Moist Gases

Dew point (°C)	–60	–50	–40	–30	–20	–10	0	10	20	30
Water vapour (vol–%)	0.0011	0.0039	0.0127	0.0376	0.102	0.257	0.602	1.212	2.308	3.74

The dew point is not a definite measure, however, for the oxidising potential when elementary oxygen is present in addition to the water vapour, which is usually the case in non-reducing gases such as nitrogen and noble gases. Higher concentrations of water vapour also inhibit the sintering of metals such as iron and copper, because the redox-equilibrium (see equation (8.1b)) is not completely on the metal side.

A principal difficulty, especially during steel sintering, is the variation of the carbon potential along the furnace length. It varies considerably along heating, sintering and cooling zones, according to the level of reaction taking place (see equation (8.2)). Careful control is necessary when local carburisation or decarburisation is found as being critical.

From the gases listed in Table 8.1, H_2, CO, CO_2, N_2 and H_2O are of special significance. Economic sintering atmospheres for manufacturing iron and steel as well as non-ferrous metal parts, which are obtained by partial combustion (see section 8.2), consist of these gases.

8.2 SINTERING ATMOSPHERES IN PRACTICE

The first category of atmospheres to be considered are *pure gases*, namely H_2, N_2, and Ar, delivered in steel vessels (200 bar) or tanks in the liquified state for mass production. They are available up to very high purity (≈99,999%).

Pure hydrogen, which is the most strongly reducing atmosphere, is used for sintering iron based metallic soft and permanent magnetic materials (see section 11.7.2). High quality soft magnetic sheets require final annealing in very pure H_2 in order to remove interstitial atoms (C,N) and residual stresses from the lattice. W and Mo are also mostly sintered in dry, pure H_2. Although H_2 is an ideal sintering gas (except for its flammability!) for many applications it is too expensive for the mass production of sintered steel and non-ferrous metal parts. Stainless steel parts, however, are sometimes sintered in pure H_2 with –30 to –40°C dew point.

Pure nitrogen of low dew point is used for sintering aluminium alloys (see section 11.4.1) — being completely inert to these alloys where no reducing potential is required. N_2 based atmospheres containing 3–5 vol-% H_2 ('synthetic' atmospheres) are widely used for sintering conventional iron and steel parts, including stainless steel. N_2 is obtained from the fractional distillation of liquid air, and N_2–H_2 mixtures in any proportion are obtained by simple gas mixing. They have replaced to a great extent dissociated ammonia (see below) for economic reasons. N_2 mixtures with a few vol-% H_2 are non-flammable and provide a sufficiently high reduction potential for many applications in powder metallurgy practice. The carburizing/decarburizing property of such synthetic atmospheres can be adjusted by admixing of hydrocarbons, such as CH_4 or C_3H_8, or by dissociated methanol, according to the equation

$$CH_3OH \rightarrow CO + 2\,H_2 \qquad (8.3)$$

Even the CO-containing exo- and endogases (see below) have been partly replaced by N_2–H_2 mixtures. Many factories producing iron and steel parts have installed tanks with liquid N_2.

Noble gases, such as Ar and He are seldom used in industrial powder metallurgy, but are sometimes used for laboratory work.

Dissociated ammonia

$$2\,NH_3 \rightarrow N_2 + 3\,H_2 \tag{8.4}$$

is obtained by a catalytic reaction of gaseous NH_3 at about 950°C. The high purity, anhydrous NH_3 is delivered in the liquid state. The gas mixture, consisting of 25 vol-% N_2 and 75 vol-% H_2, is very dry with dew points –40°C or even lower, and with residual NH_3 in the range of 100 vpm. The traces of H_2O and NH_3 can be further reduced by absorption. The gas is flammable due to its high concentration of H_2. It is widely used for sintering iron and steel parts with low or moderate carbon content as well as for bronze and brass parts. Dissociated ammonia is also used — to some extent — for sintering aluminium alloys. Pure N_2 is often preferred, however, since H_2 absorption may increase porosity and decrease strength.

Partially burned gases are another very important group of sintering atmospheres for industrial powder metallurgy. They are produced from hydrocarbons or gases containing them — preferably natural gas, propane or other hydrocarbons — by partial combustion with dry air. The principal steps of combustion of the simplest hydrocarbon (methane) are:

$CH_4 + 1/2\,O_2$	$\rightarrow$	$CO + 2\,H_2$	(endothermic)	(8.5a)
$CH_4 + O_2$	$\rightarrow$	$CO^* + H_2 + H_2O$	(exothermic)	(8.5b)
$CH_4 + 3/2\,O_2$	$\rightarrow$	$CO^* + 2\,H_2O$	(exothermic)	(8.5c)
$CH_4 + 2\,O_2$	$\rightarrow$	$CO_2 + 2\,H_2O$	(*complete combustion*)	(8.5d)

*Side reaction with CO_2 formation occur during these steps.

These gases consist, consequently, of different amounts of CO, CO_2, H_2 and N_2. It is usual to distinguish them by their integral heat of reaction into endothermic gas (*endogas*) and exothermic gas (*exogas*). While endogas is produced mainly according to the summarised equation (8.5a) (for details see equation (8.6), exogas is formed according to equations (8.5b and c), while complete combustion is, of course, avoided. The degree of combustion is adjusted by the ratio of combustible gas to air, which is much lower for endogas than exogas production.

For endogas production a Ni catalyst has to be used at a sufficiently high temperature (≈1100°C) to avoid the formation of larger CO_2 content and of unreacted carbon (soot). The reaction equations are (for simplicity again for CH_4):

CH_4	$\xrightarrow{\text{Ni-catalyst}}$	$C + 2H_2$		(8.6a)
$C + 1/2\,O_2$	$\rightarrow$	CO	(high temperature)	(8.6b)
$C + O_2$	$\rightarrow$	CO_2	(lower temperature)	(8.6c)

Table 8.3 Boudouard-equilibrium (equation 8.7) at 1 bar

	450°C	600°C	800°C	1000°C	1100°C
CO_2 (vol-%)	98	77	6	0.7	0.2
CO (vol-%)	2	23	94	99.3	99.8

The CO and CO_2 concentration are determined by the Boudouard-equilibrium

$$C + CO_2 \rightleftarrows 2\,CO \tag{8.7}$$

whose dependence on temperature is given in Table 8.3.

Typical compositions of endogas, synthesized from natural gas, are (vol-%)

H_2: 36–40; CO: 18–20; CO_2: <0.2;
dew point: +5 to −10°C, balance N_2

and of exogas

H_2: ≤17; CO: ≤12; CO_2: ≤10;
dew point: +10 to +30°C (undried), –60°C (dried), balance N_2.

H_2, CO and CO_2 contents are very dependent on the degree of partial combustion, whilst the dew point is adjustable by different methods of H_2O removal: the lowest values are obtained by absorption drying. CO_2 can be removed by alkaline or pressurised water wash, which was at one time used to obtain so-called 'monogas', which has a CO_2 content as low as endogas.

Because of the decarburising property of CO_2, endogas is mainly used for the sintering of steels. Exogas is used more for bronze and brass.

Hardenable steels especially have to have defined carbon content which is included in the powder mixture (see section 11.1.2) and must be kept nearly constant during sintering. Special diagrams have been developed for this aim, showing the relations between the equilibrium C content of the austenite at a given temperature, CO_2 content, dew point of the atmosphere and sintering temperature. Examples are shown in Fig. 8.2 and Fig. 8.3 for endogas, demonstrating that carbon-containing steels have to be sintered in atmospheres with low CO_2 as well as low water vapour content (low dew point).

Endogas and exogas can be produced in separate equipment, or in an integrated device within the sintering furnace, (see section 8.3). This is often preferred in practice. The ranking order of economy for the different sintering atmospheres is given in Table 8.4.

Sintering in high vacuum is performed when any reaction with gases or gaseous impurities should be excluded and when there is no danger of evaporation loss. It is a batch-type process in most cases, although continuous vacuum furnaces have been developed. Vacuum sintering is applied, e.g. for stainless steel, magnets, hard metals and for highly reactive and refractory metals. The quality of the vacuum is defined by the residual pressure as well as by the

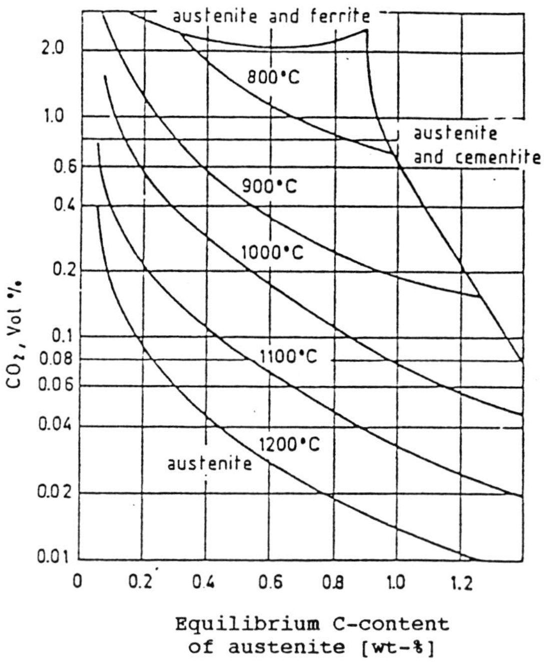

Figure 8.2 Carbon potential of an endogas as a function of the CO_2 content

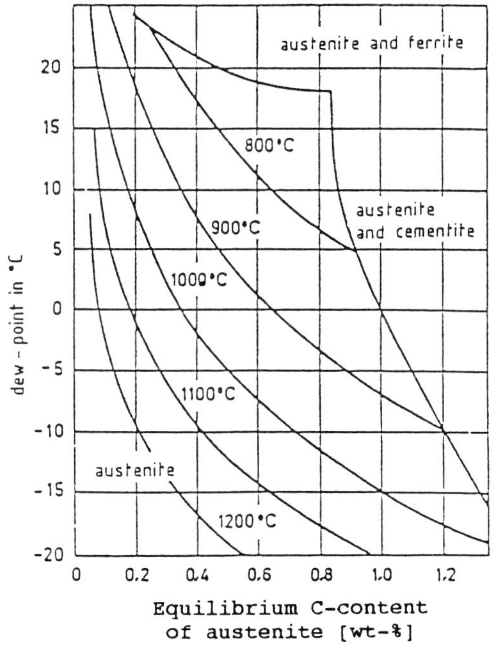

Figure 8.3 Carbon potential of an endogas as a function of the H_2O content

Table 8.4 Relative economy of Sintering Atmospheres (Warga)*

Inexpensive	Exogas produced *in situ*
	Endogas produced *in situ*
	Nitrogen
	Exogas produced in generator
	Endogas produced in generator
	Dissociated methanol containing gases
	Dissociated ammonia
	Hydrogen
↓	Argon
Expensive	Helium

*This ranking is somewhat dependent on the dew point required, which is disregarded here.

leakage rate. Evaporation problems can be suppressed to some extent by back filling with a suitable gas to a partial pressure which equilibrates or exceeds that of the evaporating species. This has been applied in the production of stainless steel in order to avoid chromium losses. Vacuum sintering is an expensive procedure, but is nevertheless in practice often an attractive technical process. Its use has increased, mainly because of the availability of highly sophisticated equipment which enables products of the highest quality to be manufactured.

8.3 SINTERING EQUIPMENT*

Sintering equipment in industrial practice consists mainly of conveyor belt, walking beam and vacuum furnaces. Walking beam furnaces have developed from push-through (manual or automatic) furnaces, which are still in use.

The **conveyor belts** consist of a stainless steel network, carrying the parts to be sintered either directly (larger parts) or stacked in plain sheet iron boxes, (of small parts). The temperature in this type of furnace is limited to 1120–1150°C by the heat resistance of the belt, which is partly in contact with hot gases under oxidative conditions. For economic reasons the belts should obviously have a lifetime as long as possible. They are used mainly for sintering unalloyed or alloyed iron and steel as well as for copper alloy parts. For many of the alloyed steel compositions used temperatures of 1120–1150°C are sufficient for sintering.

Figure 8.4a is a schematic view of a furnace consisting of several zones. In the dewaxing zone (2) the protecting gas is oxygen (air)-enriched in order to burn easily the lubricants used during pressing. In this type of furnace the endogas (air+fuel gas) is produced inside the sintering zone (3) using a catalyst. The gas streams to both sides of the furnace (which is open at both ends) and burns after leaving during contact with ambient air. Zone (4) is the rapid cooling zone, located directly after the sintering zone. Here alloyed steel parts can be cooled with a recirculated, cooled sintering atmosphere in order to be hardened. The parts then enter the convective cooling zone (5).

* For hot pressing and hot isostatic pressing equipment see section 6.1.2.

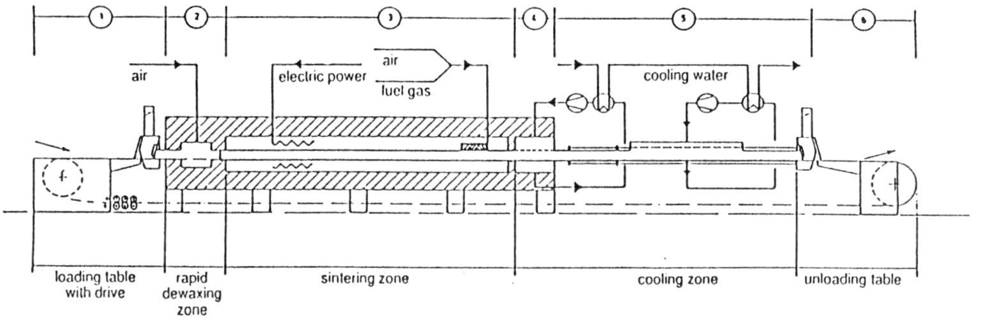

Figure 8.4a Schematic diagram of a conveyer belt sintering furnace consisting of different zones (Mahler, Esslingen)

As well as the production of endogas inside the furnace, separate gas production equipment is also used in industrial practice.

Figure 8.4b is a view of the front end of a conveyor belt sintering furnace, showing the belt and the loaded parts in more detail.

Figure 8.4b Conveyor belt in operation (Mannesmann Demag Hüttentechnik)

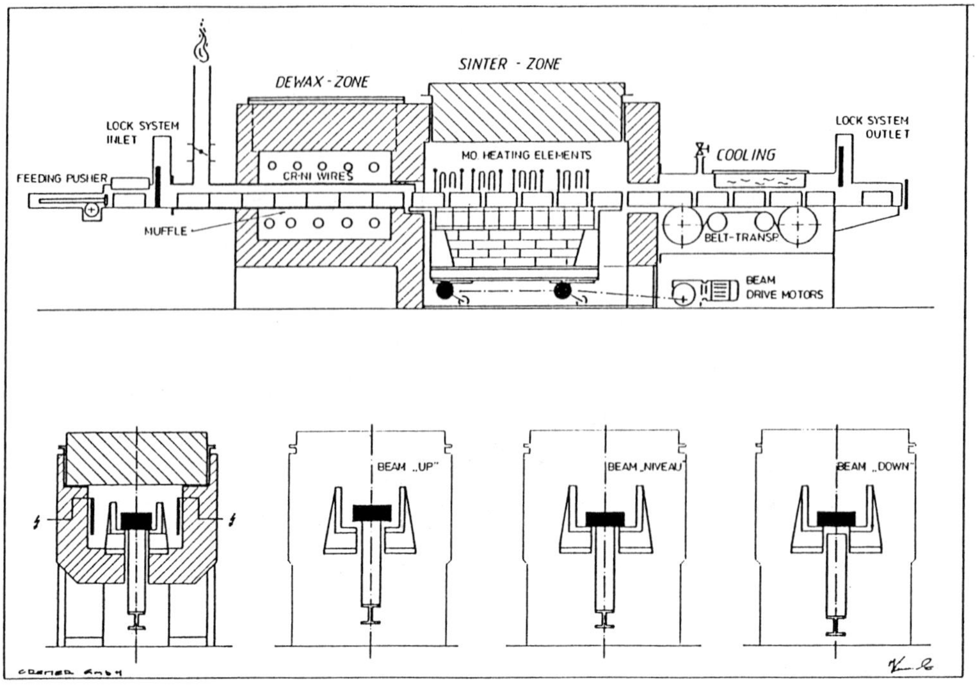

Figure 8.5 Walking-beam furnace system (Cremer, Düren)

In the **walking beam furnace** the parts to be sintered are usually stacked in sheet iron boxes, which are moved discontinously by the beam as shown in Fig. 8.5 (bottom). They are heated by molybdenum elements, which allow for higher sintering temperatures than in belt furnaces (1350°, up to 1600°C for special products.) In each part of the furnace a reducing atmosphere has to be maintained otherwise the heating elements are destroyed. The parts to be sintered enter the furnace by a lock chamber and leave the system at the other end in the same way. Rapid cooling is not possible in this type of furnace since the parts leave the hot zone stacked in boxes. Figure 8.5 (top) is a drawing of a modern sintering furnace of this type which is used, e.g. for sintering of those alloyed steels which require temperatures up to 1300–1350°C.

Vacuum furnaces are integral parts of the sinter–HIP cycle (see section 6.1.2), especially for hard metals, and are used as sintering equipment for special products. They are generally batch type furnaces with heating elements of Mo (≦ 1500°C), Ta (≦ 1700°C) or W (≦ 2400°C). For the production of hard metals, graphite heating elements are used and special dewaxing valves are provided. The final vacuum is about 10^{-2}–10^{-3} mbar, at a leakage rate mostly below $2 \cdot 10^{-3}$ mbar · 1/s. The homogeneity of the temperature field is mostly ± 3–8K, but can be as narrow as ± 2 K over the entire cross section of the sintering chamber. This is especially required for the sintering of high-speed steels. Figure 8.6 shows a vacuum furnace which is distinguished by the vacuum vessel

Figure 8.6 Charging of the sintering muffle of an IPSEN vacuum furnace. Some heating elements are visible in the interior

as well as by the large cross sections of the tube linked with the vacuum pump. The vessel walls are water cooled and are shielded against radiation from the heating elements by graphite felt. Several individual current circuits are operated for optimisation of the temperature field. Single as well as multichamber furnaces are in operation.

FURTHER READING

D. Warga, Controlled Furnace Atmospheres for Sintering, *Powder Metall. Int.*, **23**, (5), 317–321, 1991.

P.R. Wilyman and M. Vandermeiren, Selections of Atmospheres for Sintering, in *Powder Metallurgy, An Overview*, (eds. I. Jenkins and J.V. Wood), London, Institute of Metals, 1991.

9: Porosity and Pore-related Properties

Most of the products manufactured from powders are characterised by pores, since full densification is achieved only in special cases. While in sintered structural parts the residual porosity has no special function, it is in materials with controlled porosity such as filters or bearings (see section 11.8) a functional constituent of the microstructure which is responsible for the properties desired. The means to measure porosity, pore size distribution, pore geometries as well as the relations between porosity parameters and properties are essentials for powder metallurgists. It is not within the scope of this chapter to describe the characterisation of the mechanical, physical or metallographic properties, including quantitative image analysis, which are covered in the standard textbooks.

9.1 DEFINITION AND MEASUREMENT OF POROSITY AND PORE STRUCTURES

The total porosity is the void part of the volume related to the entire volume of any porous material. It can be subdivided in interconnected pores (reaching the specimen surface at least on one side) and closed pores. Figure 7.21 demonstrates the dependence of the amount of closed porosity on total porosity: At total porosity below about 10% the closed pores begin to dominate; below 5%, almost all open pores have been eliminated. Even at higher total porosity, a small proportion of the pores is often not interconnected. Although variations are observed, this general picture is true for many sintered metallic and nonmetallic bodies.

The *total porosity* can be measured for simple geometries by determining mass and dimensions. Generally it is obtained by measuring the apparent (green or sintered) density (ρ_s) by Archimedes principle, i.e. by weighing the specimen in air and subsequently in a liquid (water). This can be applied for any geometry of the part. It is simply,

$$\rho_S = \frac{m}{V} = \frac{m}{m_a - m_w} \cdot \rho_w \tag{9.1}$$

where m = mass and V = volume of the dry specimen, resp., in air, ρ_w = density of the water at its temperature, m_a and m_w are the mass by weighing in air and in water, respectively, with completely sealed pores at the surface. Any intrusion of water into the porosity causes errors in the measurement. Sealing can be done by cellulose lacquer (Zapon varnish) or paraffin, but is not necessary at only closed porosity. The total porosity (P_t) in % is:

$$P_t\,(\%) = \frac{\rho_{th} - \rho_s}{\rho_{th}} \times 100 \tag{9.2}$$

ρ_{th} being the density of the pore-free material. The *percentage of theoretical density* (*TD* [%]) is

$$TD\,[\%] = \frac{\rho_s}{\rho_{th}} \times 100 = 100 - P_t\,[\%] \tag{9.3}$$

The percentage of *open (interconnected) porosity* (P_o) is defined

$$P_o\,[\%] = \frac{V_{P_o}}{V} \times 100 \tag{9.4}$$

For its measurement the mass of the specimen fully infiltrated by the liquid (m_i) has to be considered:

$$P_0\,[\%] = \frac{(m_i - m)}{(m_a - m_w)}\,\rho_w \times 100 \tag{9.5}$$

Small percentages of porosity, are often measured by optical microscopy on ground and polished cross sections. This is a standard technique e.g. for hard metals. The numbers of pores are counted or measured in % of the surface. For the new term, *plain porosity*, see section 9.2.

The size distribution of interconnected pores is measured preferably by *mercury porosimetry*. Mercury is pressed into the porous solid and the intruded volume is measured as a function of applied pressure. It is applicable in the diameter range of about 100 to 0.003 µm at 4000 bar maximum pressure. The method is used in powder metallurgy and ceramics as well as for many other porous materials. The theoretical background is the equation of Washburn,

$$r(p) = \frac{2\,\gamma \cdot \cos\theta}{p} \tag{9.6}$$

giving the relation between the radius of pores r (assumed to be circular) being just penetrated at a given pressure p. γ = surface tension of mercury (γ = 480 MN m^{-1}) and θ the contact angle between mercury and the material to be analysed. Mercury does not wet most materials (θ between 90 and 180°), which is a condition for application of this method. Besides the size distribution also the total volume of accessible pores is obtained.

The method does not provide, however, absolute data of pore sizes, but rather the size distribution of constrictions or of interconnecting channels, resp., between the pores, as shown in the model of Fig. 9.1. When the mercury pressure reaches a certain value, than the channel K_1 with radius r_{K_1} is penetrated. The following pore P_1 with a much larger radius is filled without any pressure

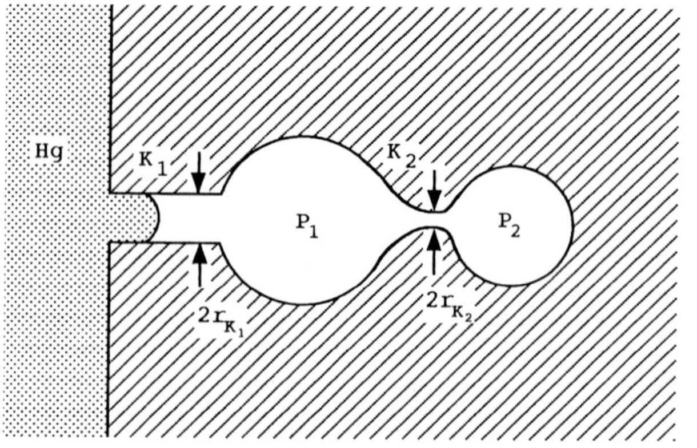

Figure 9.1 Pore model to demonstrate the mercury intrusion process

increase. The same happens with the pore P_2 behind the channel K_2, demonstrating that pores which are accessible only via constricted areas cannot be sized correctly by this method. Therefore, the size distribution curves obtained may be found in a considerably smaller range than those measured by metallography.

Mercury porosity measurements are possible as well on sintered as on cold compacted powders of sufficient green strength. In green compacts it is a technique to separate interparticle void volume from closed pores within particles.

Mercury intrusion and extrusion curves, the latter obtained during pressure release, may differ considerably. Fig. 9.2 show both curves for a special reaction

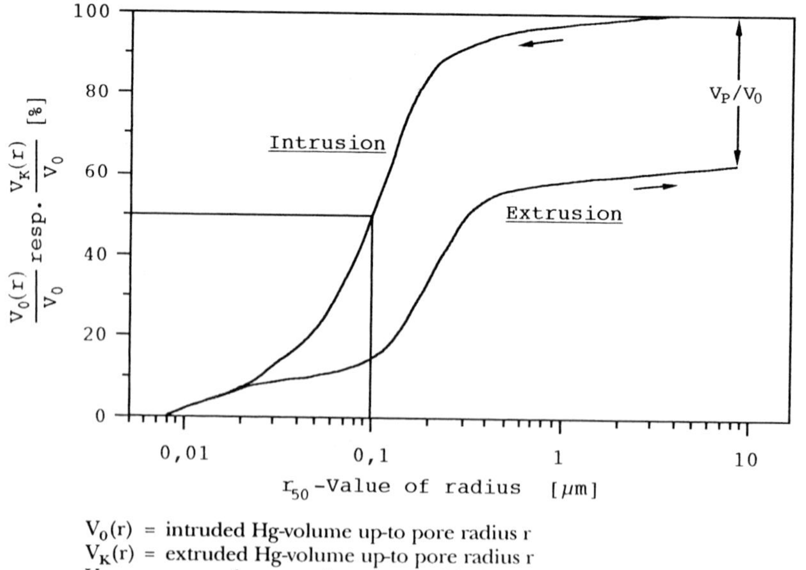

$V_0(r)$ = intruded Hg-volume up-to pore radius r
$V_K(r)$ = extruded Hg-volume up-to pore radius r
V_0 = pore volume totally penetrated by Hg at maximum pressure
V_P = pore volume not drained during extrusion

Figure 9.2 Hysteresis effects in mercury porosimetry intrusion and extrusion curves (Cohrt *et al.*)

bonded Si_3N_4, showing that the extrusion (draining) during slow drop of mercury pressure is not complete. Only pores and channels, whose diameter increases or remains constant toward the surface are drained during the pressure drop. The hysteresis enables a distinction to be made between pore volume behind constrictions and the volume of other pores, channels and constricted areas.

Metals forming alloys with mercury cannot be investigated. The compressibility of the mercury itself is often disregarded, also the correction for deviations from cylindrical pore cross sections.

Widely used are the methods of *quantitative image analysis,* (see section 3.4.3.2 and 9.3) which is the most important tool for the quantitative assessment of pore structures, namely, size distribution, *shape and orientation.* The pore shape especially is very variable depending on the stage of sintering. Linear and planar analytical methods are used, both requiring specimens prepared without any defects and with sharp pore edges. The methods yield results of high reliability in so far as the investigated surface is representative of the entire sample. The results obtained from one representative part of the surface are valid for the bulk material only with non-texturised and homogenously distributed pores.

9.2 MECHANICAL PROPERTIES

The static and dynamic strength data of sintered ductile materials are measured in conventional tensile, impact and fatigue tests, while for semi-brittle products like hard metals and the brittle ceramics the 3 pt (or 4 pt) bend test (transverse rupture test) is more convenient. Many empirical results are available on materials with different porosities, in sintered iron and steel mostly in the density range between 6.5 and 7.5 g cm^{-3}. The results may be summarised as follows: In practice the *tensile strength* can be often interpolated linearly in this density range, while the *elongation at fracture and impact strength* exhibit a stronger dependence on porosity, as seen from Fig. 9.3. This holds for a number of sintered steels, whilst others show somewhat different behaviour in strength with increasing slope at high densities. From Fig. 9.3a can be seen that elongation and impact strength are greatly improved in the high density range, as achieved by sinterforging. The results suggest that elongation and dynamic properties are more dependent upon notch effects than the strength values. The ductility at porosities higher than 15% is very small, the scatter of data being mostly higher than that for strength. The linear dependence of strength on porosity is not theoretically based, however. More basic investigations lead to a strength dependence such as

$$R_m = R_{m,0}\cdot[\exp - K P] \qquad (9.7a)$$

and

$$R_m = R_{m,0}\cdot(1 - P)^n \qquad (9.7b)$$

with R_m = tensile strength of the porous material, $R_{m,0}$ = tensile strength of the pore-free material, K = constant, P = fractional porosity, n = density exponent. K has been found to lie between 2 and 10, often 4, whilst n ranges between 3 and

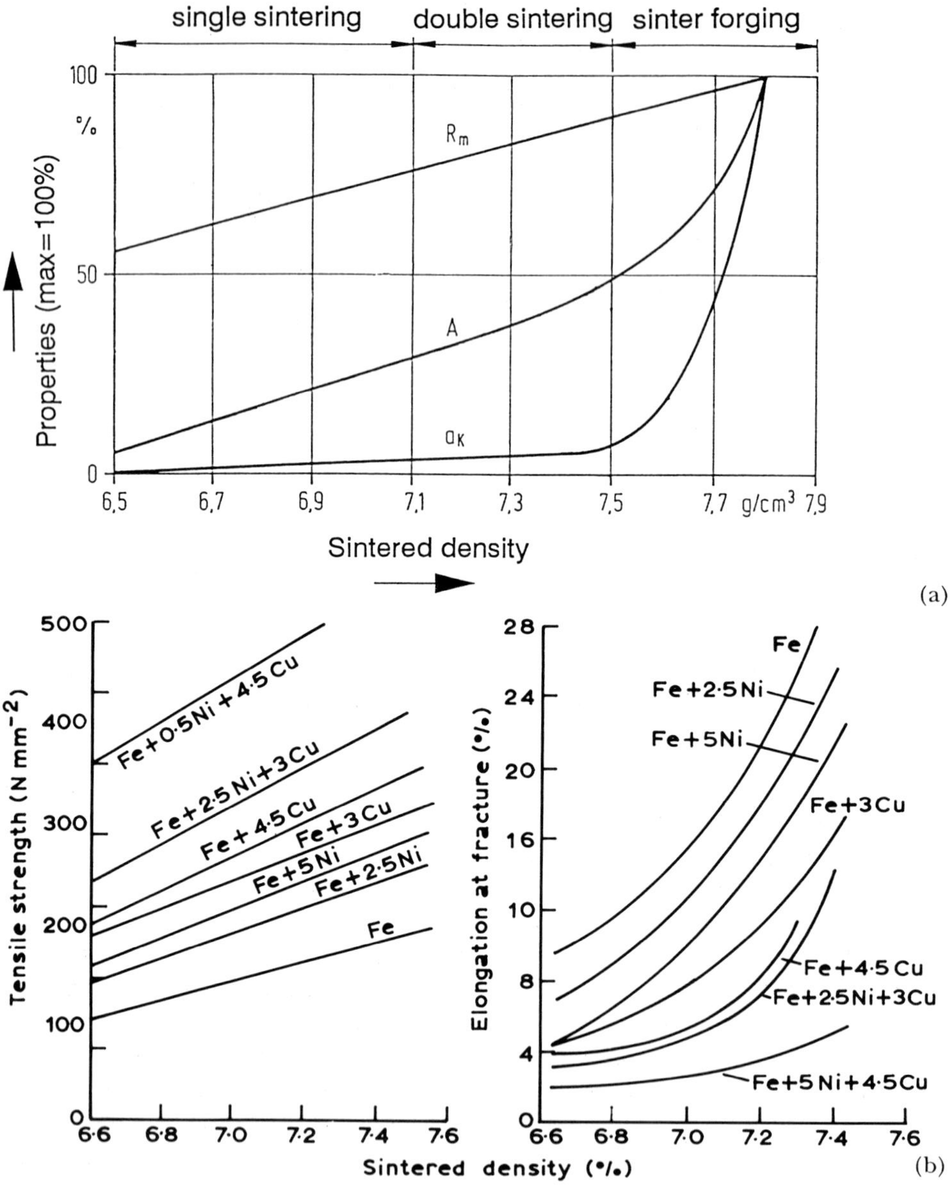

Figure 9.3 Porosity dependence on mechanical properties of sintered steels a) schematic for tensile strength (R_m), elongation at fracture (A), impact strength (a_K) (Bockstiegel, Huppmann) b) measured data for tensile strengtrh (left) and elongation at fracture (right) (Zapf and Dalal)

6, being influenced by notch effects, which are dependent mainly from pore size and shape. Therefore, mathematical expressions like this are only rough approximations, since they do not consider the pore geometries (see section 9.3). Sintered products with a constant density exhibit a decreasing strength

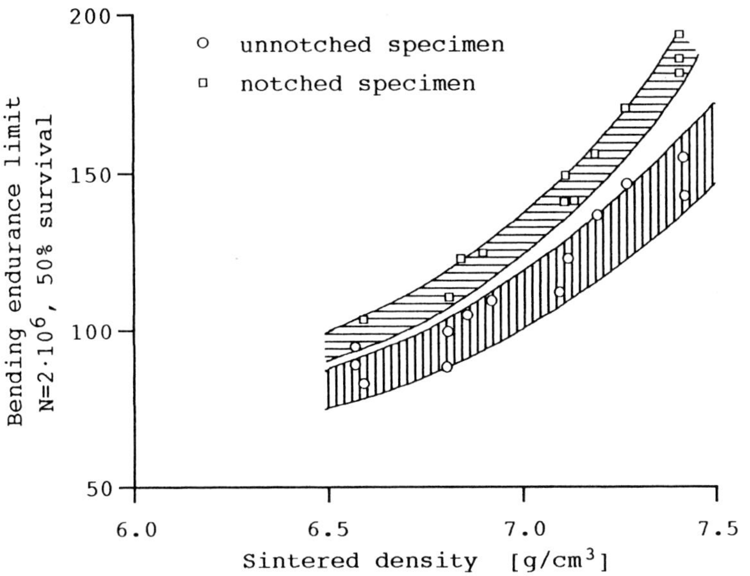

Figure 9.4 Porosity dependence of bending endurance limits of unalloyed and some low-alloyed sintered steels. (The results of many steels investigated are within the scatter bands shown (DIN 30912, Pt 6), but higher endurance limits have been measured in newly developed alloys)

when the pore shape is more irregular, while spherical pores have been found as to be relatively harmless.

It has been shown recently of sintered iron, that a newly defined 'plain porosity' provides improved correlation to mechanical strength data that the total porosity. Plain porosities are determined from fractured surfaces and represent the part of the surface where sintered contacts have not formed. Image analysis from microscopic screens or from photographs are used for quantitative evaluation.

Fatigue properties are likewise strongly dependent on porosity. Fig. 9.4 summarizes *endurance limits* on notched and unnotched sintered steel samples. The pores act as initiation sites for fatigue cracks, but the influence of notches is not as strong as with conventional steels since the porosity diminishes the external notch effects. This has been found at constant amplitude and at random loadings as well and indicates a lower notch sensitivity of porous compared to wrought materials. With notch factors > 2 and densities ≥ 7.1 g cm^{-3}, similar endurance limits are obtained for wrought steels, nodular cast iron and various sintered steels, with similar scatter of the data.

The fatigue properties and the related crack propagation rates, however, are far from being solely a function of porosity and pore chracteristics, but also of alloying constituents, homogeneity, post-sintering treatment and, last but not least, of design criteria. A stress distribution analysis based on the service loading conditions is necessary for careful fatigue life assessment.

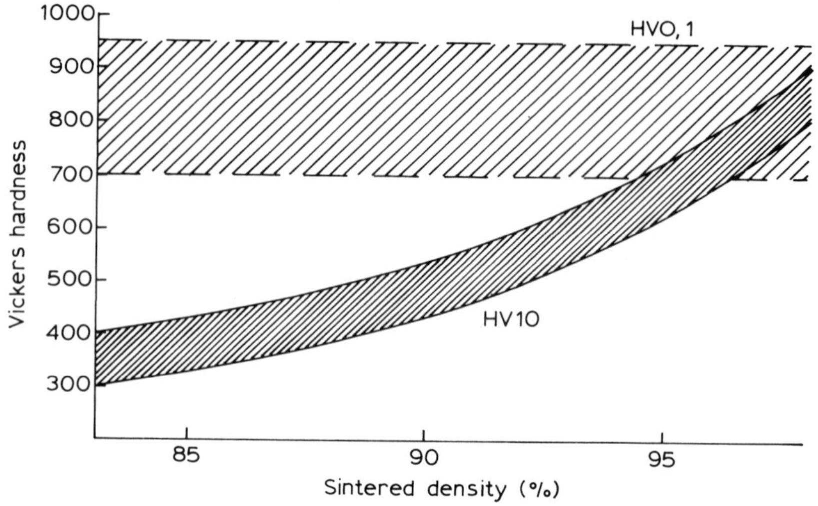

Figure 9.5 Porosity dependence of Vickers hardness HV 10 and HV 0.1 of hardened sintered steels (DIN 30912, Pt 3)

The *hardness* of sintered iron and non-iron metal parts is mostly measured with the Brinell or Vickers test; in cemented carbides Vickers or Rockwell tests are preferred. For hardness measurement on porous materials the strong dependence of the results on test load has to be considered. Conventional *macrohardness* indentations include the solid crystallites as well as the pores, leading to a decreasing influence of pores when the test load decreases. In *microhardness* measurement with very low test loads single grains or single phases in a multiphase-material can be tested and no pores are involved. Fig. 9.5 shows hardness data of both types in hardened sintered steels.

9.3 PHYSICAL PROPERTIES

The effect of porosity on physical properties is different depending on the property considered, and can be stated in the following summarised from: The *linear thermal expansion coefficient* does not depend on porosity*, since the thermal expansion is independent of the shape and on the reduction of cross section by void volume. Properties being controlled solely by the mass of a given composition, like *heat capacity or magnetic saturation* (but not other magnetic properties!) are also not related to pore parameters, disregarding the property of the gas inside the pores.

The group of so-called *field properties*, describing the materials response in thermal, electric or magnetic fields (*thermal and electrical conductivity, magnetic*

* Nevertheless, some experimental results appear to contradict this statement.

permeability) depend in a complicated manner on porosity and pore geometries as well. One of the simple equations often used for thermal conductivity is

$$\lambda_P = \lambda_O(1 - P) \tag{9.8}$$

where the porosity P is a volume fraction, λ_P and λ_O are the conductivities of the porous and the pore-free material, resp. This and other equations derived may be used for a rough estimate. More concise mathematical treatment is based on a model derived for bi-phase materials, considering the pores as the second phase and modelling the pore shape by ellipsoids. Different types of pores have different shape factors ranging from 0.0 (lamella), 0.33 (spheres) to 0.5 (cylinders). A decision as to which type of pore has to be considered for calculation should be made by quantitative image analysis. For texturised porosity an orientation factor $\cos^2\alpha$ has to be considered. For thermal conductivity the following equation has been derived in the case of isolated (closed) pores:

$$\lambda_p = \lambda_O (1 - P)^X; \; X = \frac{1 - \cos^2\alpha}{1 - F} + \frac{\cos^2\alpha}{2F} \tag{9.9}$$

For interconnected pores the following equation is valid:

$$(1 - P)\,(\lambda_O - \lambda_p)\,\left[\frac{2\,(1 - \cos^2\alpha)}{\lambda_O + \lambda_p} + \frac{\cos^2\alpha}{\lambda_p}\right] = P\,(2 - \cos^2\alpha) \tag{9.10}$$

is valid. F and $\cos^2\alpha$ are shape and orientation factors for the pores, resp., the former occurs in the equation only for isolated pores. α is the angle of pore orientation related to the direction of the applied field. The equations 9.9 and 9.10 yield similar results over a certain range of porosity and are applicable also in principle for other field properties.

The dependence of *elastic modulus* on porosity can be treated in a similar way. Mechanical properties (see section 9.2) which are dependent on notch effects or include plastic deformation do not follow these lines.

9.4 CHARACTERISATION OF PERMEABLE SINTERED MATERIALS

Special testing methods have been developed for materials, whose interconnected porosity has a decisive function, namely in filters, bearings, porous electrodes etc., involving the pore size characteristics and the permeability of gases or liquids.

When a steadily increasing gas pressure is applied to one surface of a porous plate immersed in a test liquid (e.g. isopropanol) gas bubbles occur at the opposite surface. The pressure difference (Δp) to be applied for generation of the first bubbles depends on pore size defined as the largest equivalent capillary circular diameter (d) in the test piece:

$$d = \frac{4\gamma}{\Delta p} \tag{9.11}$$

$$\Delta P = p_g - p_l \tag{9.12}$$

γ = surface tension of the liquid p_g = applied gas pressure, p_l = gas pressure in the test liquid at the moment of bubble formation, which is given by:

$$p_l = g \cdot \rho \cdot h \tag{9.13}$$

g = gravity constant, ρ = density of the test liquid, h = height of the liquid above the test specimen. This *bubble test*, shown in Fig. 9.6 is representative only when no large single pores or defects are present. It provides a relative figure of pore size, being not identical with the size of solid particles which may pass the porous structure during filtration. Due to the complicated pore structure, and deviations from circular cross section, even smaller particles (factor 0.2–0.4) will be retained. This figure can be determined by the *glass-pearl test*, (pearls suspended in water), in which the diameter of the largest particle passing the porous test piece is measured.

The standardised test for *gas permeability* (α) is shown in Fig. 9.7. It can be determined from the gas flow ($\dot{V}$) at the pressure applied in front of the specimen and depends on filter thickness (S), dynamic gas viscosity (η) and the surface exposed (A). A small pressure difference (Δ_p) measured is equivalent to a high gas permeability:

$$a = \frac{\dot{V} \cdot S \cdot \eta}{A \cdot \Delta_p} \tag{9.14}$$

Equation 9.14 can be used as a test method for filters within the laminar flow region only, where α takes a constant value. The test can be performed with hollow cylinders as well by using modified test equipment and calculation procedure.

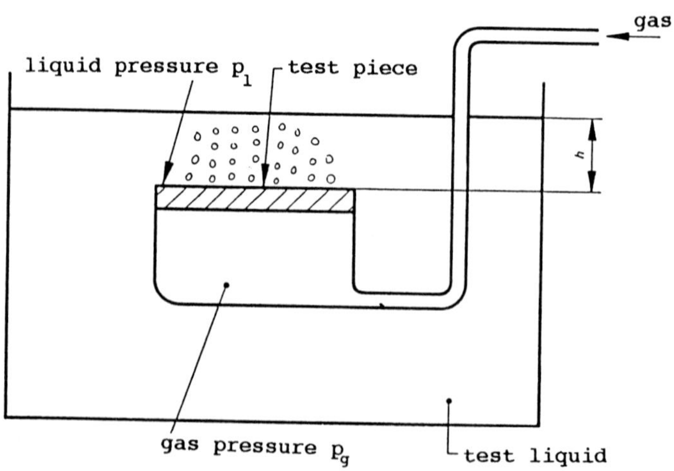

Figure 9.6 Experimental device for the gas bubble test (DIN ISO 4003)

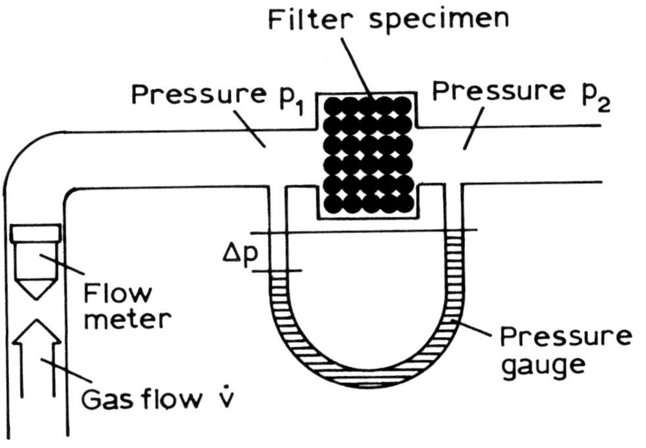

Figure 9.7 Experimental device for the permeability test of filters (ISO 4022, DIN 30911 pt 6)

FURTHER READING

Metals Handbook 9th Edition, Vol. 7, Powder Metallurgy, 194–208, Metals Park, OH, American Society for Metals, 1984.

DIN-Taschenbuch 247, *Pulvermetallurgie*, Berlin, Köln, Beuth-Verlag, 1991.

H. Cohrt, F. Porz and F. Thümmler, Determination of Pore Structure in Reaction Bonded Silicon Nitride, *Powder Metall. Int.*, **13**, 121–125, 1981.

B. Schulz, Thermal Conductivity of Porous and Highly Porous Material, *High Temperatures, High Pressures*, **13**, 649–660, 1981.

M. Šlezár, E. Dudrová and E. Rudnayová, Plain Porosity as a Microstructural Characteristic of Sintered Materials, *Powder Metall. Int.*, **24**, 232–237, 1992.

C.M. Sonsino, Fatigue Design for Powder Metallurgy, *Powder Metallurgy*, **33**, 235–245, 1990.

10: Production Routes in Practice

The entire field of powder metallurgy including special ceramics requires many types of processing technologies, which are collected in Table 10.1. These variants provide very different potential in densification, microstructural and properties development and range from the simple cold pressing and sintering to the expensive and complicated HIP- or Sinter-HIP-cycles. The table covers the methods used for mass production, except powder rolling (see section 6.1.1.5) and except very special procedures such as shock compactions, application of magnetic fields, ultra rapid sintering, reactions sintering, etc. The sintering temperatures of several materials in practice are summarised in Table 10.2.

Several *post-sintering* or *secondary operations* including *finishing* are used in powder metallurgy mass production especially with sintered machine parts, which are summarised as follows.

- *Sizing* is a re-pressing operation at room temperature, which is generally used for precision parts, but is limited to strength levels of 500–550 MPa. Separate presses and tools are mostly used. The process involves limited plastic deformation and densification in the near-surface regions and improves dimensional tolerances towards those of the cold compacted part, depending on the strength level. Also the surface roughness is decreased. The actual level of dimensional accuracy during sintering requires sizing for many of the precision parts. When surface profiling is included the process is called *coining.*
- *Impregnation (Infiltration)* with an organic such as oil or polymers, mostly *in vacuo,* is the final treatment at many types of bearings, in order to provide self-lubricating properties. Alternatively they may contain solid lubricants added by mixing. Impregnation provides some increase in mechanical properties and improves corrosion resistance by sealing the pores. The machinability of sintered parts is also improved.
- *Machining,* (turning, milling, grinding, bore-drilling etc.) enlarges the possibilities of intricate shaping, when the final shape cannot be provided by conventional compaction. The combination of the powder metallurgy route and automated machining has been introduced to a considerable extent and permits the production of more complicated parts with close dimensional tolerances. The machinability can be improved by incorporating additives like lead, graphite, sulphur or certain sulphides like MnS in the powder mixture or by impregnation with polymers. *Deburring* is necessary for many of

Table 10.1 Production variants in powder metallurgy, including special ceramics and semi-finished products, starting from powder ready for use

1 Single sinter process	2 Double sinter process	3 Indirect shaping	4 Pressureless (loose powder) sintering	5 Hot pressing (pressure sintering)	6 Hot isostatic pressing (HIP)[3] a With can	6 b Can-free[4]	6 c Can-free sinter-HIP[4]	7 Powder forging	8 Infiltration
Compaction (die or isostatic, injection moulding, slip casting, extrusion)	1st Compaction	Compaction	Shaping (powder pouring, vibration)	Compaction (die or isostatic) or pressureless shaping	Compaction (die, isostatic or pressureless shaping		Compaction	Compaction (die)	Compaction (die or isostatic)
Presintering[1] (lubricant removal)	1st Sintering	Presintering[1)] (increase of strength)			Canning (gas-tight in thin-walled sheet)		Unpressurized sintering in HIP-equipment to pore closure (TD ≫93%)		Sintering (porosity has to remain inter-connected)
		Machining (final shaping)							
Sintering (with or without liquid phase)	2nd Compaction 2nd Sintering	Sintering (with or without liquid phase)	Sintering	Hot pressing (graphite die)	HIP	HIP of sintered product (≥ 93%TD) with closed porosity to ≫98%TD	Application of pressure at temperature, HIP	Sintering or heating-up only (eventual inductive)	
Sizing[2)]	Sizing2)							Hotforging (die, ≫ 95%TD)	Infiltration (metal or polymer)
Finishing[5)]	Finishing[5)]	Finishing[5)]	Finishing[5)]	Finishing[5)]	Finishing[5)]	Finishing[5)]	Finishing[5)]	Finishing[5)]	Finishing[5)]
Production examples: engine parts, magnets, contacts, filters, cemented carbides, special ceramics	High-strength/ high-density enginer parts	Cemeted carbides, as parts or semifinished products, special ceramics	Filters, diaphragms	Hardmetal drawing tools, special ceramics and PM products	Special ceramics and PM products, as parts or semifinished	Cemented carbides, special materials, defect healing	Cemented carbides, special ceramics	Highest strength alloyed engine parts	Heavy duty electrical contacts, bearings, high strength engine parts

Table 10.2 Sintering temperatures in powder metallurgy products incl. special ceramics

PM Product or Special Ceramic	Sintering temperature [°C]
Al-alloys	590–620
Bronze	740–780
Brass	890–910
Iron, C-steels, low-alloyed steels (Cu, Ni)[6)]	1120–1150
Low-alloyed steels (Cu, Ni, Mo; Distaloy)[6)]	1120–1200
High-alloyed ferritic and austenitic steels (Cr, Cr–Ni)	1200–1280
Hard magnets (Alnico)	1200–1350
Hard metals (Cemented Carbides)[1)]	1350–1450
Molybdenum and Mo-alloys	1600–1700
Tungsten[2)]	2000–2300
Heavy Metal (W-alloy)	~1400
Ferrites (soft and hard)	1100–1300
Silicon nitride (with different additives)[3)]	1750–2000
Silicon carbide (with different additives)[4)5)]	1750–2100
Alumina[5)]	1400–1800
Zirconia (with different additives)[5)]	1400–1750

1) TiC-based hard metals (Cermets) up to 1600°C.
2) about 3000°C, when direct sintering is used.
3) highest temperature under pressurised N_2-atmosphere or in powder bed.
4) low temperatures for liquid phase sintering.
5) low temperatures for highly active powders developed recently.
6) for 'high temperature' sintering see section 11.1.2.3

NOTES TO TABLE 10.1 (previous page)

1) Partly in same equipment as final sintering. Large amounts of plasticiser (≈ 40 Vol.%) have to be removed from injection moulded or extruded parts, lasting many hours or even days.
2) Only at strength level below 550 MPa, but for most of the precision parts, generally with special presses and tools.
3) Maximum gas pressure 1000 to 2000 bar. A special variant is pressure-assisted sintering, to avoid dissociation (e.g. Si_3N_4 in 20–50 bar N_2). This pressure is less effective to enhance densification.
4) Sinter-HIP is economically advantageous for cemented carbides, compared with can-free post-HIPing after sintering. Also properties improvement, compared with 6b, has been observed in special ceramics.
5) For finishing procedures see text.

the machine parts. It is used to smooth edges caused by compacting or machining step.

- *Corrosion protection* can be provided temporarily or permanently. Dewatering fluids, which provide as temporary protection, are used for transport and storage of the parts. The oxidation by overheated and pressurized steam, (at 500–520°C) resulting in an 5–10 µm dark-blue-grey Fe_3O_4 layer, gives a more efficient, but not really permanent protection. A less effective improvement in corrosion resistance can be achieved by simple heating in air at 200–250°C (*blueing*), caused by the formation of a very thin Fe_3O_4-film. Steam treatment is an inexpensive and commonly used method of improving the surface properties of sintered steel. Permanent protection can be achieved by electro- or electroless plating of metals like, Zn, Cd, Cu, Ni, Cr and others. The pores have to be sealed first, however, (e.g. by polymer impregnation) except in high-density parts. Organic coatings are also very protective against corrosion.

- *Heat treatment* and *surface hardening*. Most of the well known steel hardening methods are applicable to sintered steels, and are often applied in alloyed steels to optimise strength and ductility. The most frequently used process, however, is surface hardening by carbo-nitriding, using solid, liquid or (preferably) gaseous media. The surface hardness and penetration to the interior is very dependent on the interconnected porosity, and the hardness gradients are less steep than in wrought steels. Ion-nitriding processes are also applied. When molten salt baths are used, penetration of the pores by the liquid has to be avoided, either by foregoing impregnation (see above) or by using high density parts ($D \geq 7.1$ g cm^{-3}). Otherwise a severe risk of corrosion exists.

Surface hardening is also obtained by steam treatment, mentioned above. The oxide layer formed is less susceptible to adhesive wear attack than the unprotected surface. The penetration depth of oxide formation depends on the sintered density and increases strongly below 7.0 g cm^{-3}.

Other post sintering operations, namely hot forging, hot isostatic pressing, coating of hard metals and others are mentioned in the relevant sections.

FURTHER READING

J.M. Capus and R.M. German (eds.): Secondary Operations, Quality and Standards, Vol. 4 of *Advances in Powder Metallurgy and Particulate Materials*, Princeton NJ, MPIF, 1992.

Powder Metallurgy Design Manual, Princeton NJ, MPIF, 1989.

Design Solutions with Powder Metallurgy, Princeton NJ, MPIF, 1993.

11: The Products of Powder Metallurgy and their Applications

This chapter is a short description of the most important powder metallurgy products, completed by a paragraph on advanced ceramics which are manufactured in many cases using the same or similar techniques to powder metallurgy. Advanced ceramics enlarge the application of sintered materials considerably and powder metallurgists should be well aware of this important field.

The powder metallurgy route generally has to compete with other technologies, like machining of wrought semifinished products, special fine casting, forging technologies, electron-beam and arc melting (esp. for high melting-point metals), etc. The larger the production run of a single part, the more economic is the powder metallurgy route, because of the steep decrease of unit cost with increasing number of pieces produced. (Fig. 11.1.) This is due to the high costs of compaction tools, especially for complicated geometries. In many cases the cost-decrease curve for competitive technologies is less steep. The curve for the powder metallurgy route may change, of course, when expensive sintering technologies or finishing procedures (see Chapter 10) are required. There is a great need for continuing economies as well as quality assurance of powder metallurgy production; these are keys for future growth or even for long-term survival of many powder metallurgy products.

11.1 IRON AND STEEL

Sintered iron and steels are the most important mass-produced powder metallurgy parts and are dominant in tonnage. The iron powder shipment in North America amounted in 1992 to 246 365 tonnes, in Western Europe 89 500 and in Japan (roughly) 100 000 tonnes. Economic reasons are decisive in most cases for the choice of the powder metallurgy route (Fig. 11.1). Close dimensional control is one of the most important issues for sintered parts (see Chapter 10). The properties of sintered iron and steels depend heavily upon their composition, manufacturing route, porosity and finishing. The porosity dependence of the mechanical behaviour is strongest for dynamic and long-term properties and is more distinct for the elongation at fracture than for tensile strength (see section 9.2).

Fig. 11.2 shows the classification of sintered parts according to their relative densities. It ranges from low-density parts (see section 11.8) to products nearly

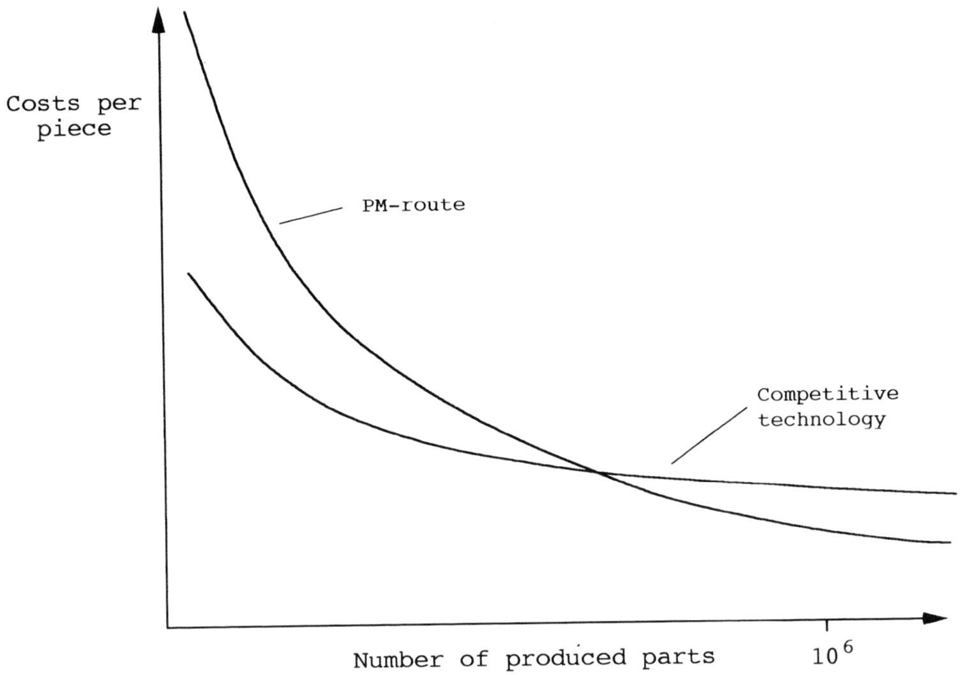

Figure 11.1 Costs per piece against number of produced parts for powder metallurgy route and for competitive technologies (schematically)

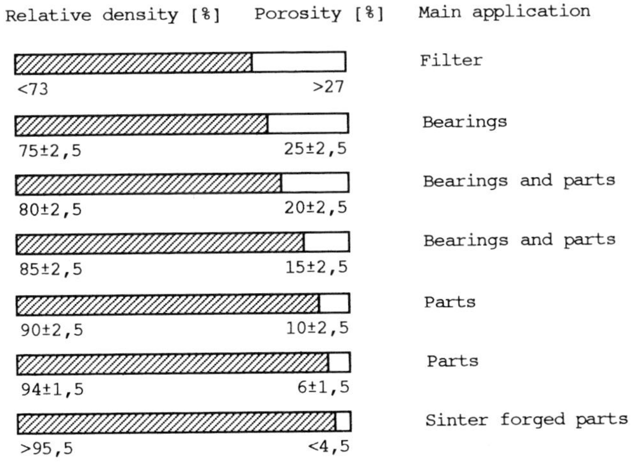

Figure 11.2 Classification of sintered parts according to their relative density or porosity, respectively (DIN 30910)

free of pores, including sinterforged and infiltrated parts as well as parts containing solid lubricants. Fig. 11.2 includes also non-iron parts.

11.1.1 Plain Iron

The term sintered iron is used for the products containing essentially no carbon and no alloying elements; compositions with at least one of these elements are named sintered steel. 'Unalloyed' sintered steels often contain, as well as some carbon, 1.2–1.5% Cu for shrinkage compensation. These steels represent an important part of iron powder metallurgy (estimated as 65–75%) since they are very economic. The strength level of low carbon unalloyed steels doesn't exceed 200–250 MPa, depending on sintered density, with ~6% elongation at fracture. This is adequate for many applications. When the carbon content is increased to 0.6%, the strength ranges up to 430 MPa with 1.5% elongation at a density of 6.9 g cm^{-3}. These sintered steels are used as parts, including bearings in mechanical engineering and the car industry (Fig. 11.3).

11.1.2 Alloyed Steels

Alloyed steels have been developed in a large variety of compositions and provide strongly enhanced mechanical and, as high-alloyed steels, improved chemical properties.

11.1.2.1 Alloying Techniques

The principal alloying techniques are treated in Chapter 4. A few comments are added with respect to alloyed steels. *Powder mixtures* are the easiest way to provide the desired concentration of alloying elements such as for e.g. Cu, Ni and Mo. The sintering process often leaves behind heterogeneities (see 7.3.2.2). Carbon admixtures homogenise easily due to the high carbon diffusivity. Metals forming very stable oxides (Cr. Al) are not used as elemental powders since their oxide films strongly inhibit sintering. They can be applied, however, as master alloy powders, like Fe–Al for Alnico-magnets. *Diffusion bonded (diffusion stabilized) powder mixtures* are widely used in different compositions with a few percent of Cu, Ni and Mo and were first manufactured as 'Distaloy' powders by Höganäs AB, Sweden. The problem of copper segregation combined with some dimensional scatter has been improved by using this type of powder. Diffusion stabilised powders homogenise only partly during sintering but show very interesting mechanical properties (Table 11.1) irrespective to the heterogeneous microstructure (Fig. 11.5). *Homogeneously alloyed (pre-alloyed) powders* are generally used for sinter-forging and highly alloyed products such as high-speed or stainless steels, since complete homogeneity is essential in these cases. A substantial improvement of the compressibility has been achieved by low concentrations of interstitial elements and residual oxygen, being obtained during atomisation by protecting the melt from oxygen pick-up.

11.1.2.2 Strengthening Mechanisms

Besides alloying and density increase, microstructural optimisation (grain refinement, pore rounding, pore filling) can be used to improve the strength of sintered steel parts. Microstructural optimisation includes heat treatment, similar to that for conventional steels and a combination of these principles is often practised. Nevertheless, alloying is the most important measure for strengthening and the following mechanisms can be distinguished:

a) Solid–solution strengthening (s.s.s.), e.g. in Fe–Ni, Fe–Cu, Fe–Ni–Mo, Fe–Si, Fe–P. These elements stabilise partly the α–, partly the γ-range, thus, α–, α + γ or γ-alloys are obtainable. High densification rate occurs when sintering takes place in the α-range (Fe–0.6 P, Fe–3 Si). The technique provides tensile strength values up to 400–500 MPa with usually high elongation at fracture.
b) Precipitation and dispersion strengthening, e.g. in Fe–Cu, Fe–Cu–Ni, Fe–Cr–Cu, Fe–Al_2O_3, microalloyed steels. This principle is seldom applied as a single mechanism. Combined with s.s.s. strength levels up to 800 MPa have been obtained.
c) Transformation strengthening, e.g. Fe–Cu–C, Fe–Cr–C, Fe–Mo–C, Fe-Cu–Ni–Mo–C, Fe–Mo–Cr–Mn–C etc. These carbon-containing alloys can be hardened like conventional steels. Optimised heat treatment and combination with s.s.s. leads to tensile strength values of about 1200 MPa, in sinter-forged steels up to 2000 MPa.

11.1.2.3 Low Alloy Steels

Not all of the alloying elements used in conventional steel technology are used in powder metallurgy since the high surface reactivity makes the application of high oxygen affinity powders (such as Cr, Al, Mn) difficult. The problems are diminished, however, by using master alloys. On the other hand, elements causing difficulties during hot rolling of conventional steels (like phosphorus) are applicable in powder metallurgy. A selection of important low alloyed steels is summarized in Table 11.1. Some of these alloys are sintered at 1120–1150°C (see section 8.3 and Table 10.2), others at e.g. 1250°C. This 'high-temperature' sintering is becoming increasingly common, mainly in USA, for high-performance applications in car industry. It results in improved strength (mainly fatigue) properties, probably as a consequence of more complete reduction of residual oxides. However, an increase of cost has to be taken into account.

The influence of the single elements can be characterised as follows: *copper* forms transient liquid phase at conventional sintering temperatures and provides s.s.s. It is used for shrinkage compensation (see 7.3.3). At high copper content precipitates appear. *Nickel* forms s.s. with high ductility and toughness but is less strengthening than copper. The strengthening effects of *Nickel and Copper* are essentially additive. *Phosphorus* is a very strong s.s. hardener, is added normally as Fe_3P and forms Fe–P and Fe–P–C low melting point eutectics. Stabilising of the α-range allows α-sintering at 1120°C; due to the high Fe-self-diffusivity in this phase good pore rounding and (a disadvantage) high shrinkage occurs. Carbon provides the hardenability but at the same time in unalloyed sintered steels low ductility. It is more important in alloyed steels, as with Mo–C, Ni–Mo–C, Cu–Ni–Mo C and Mn–Cr–Mo–C.

Table 11.1 Composition, tensile strength (R_m) and elongation at fracture (A) of selected sintered steels at different densities (ρ)

Composition	Heat treatment	ρ ≈ 7,0 g cm^{-3} R_m (MPa)	A (%)	Higher density ρ (g cm^{-3})	R_m (MPa)	A (%)
Fe	Sintered	180	15			
Fe 1.5 Cu	Sintered	270	10			
Fe 5 Ni	Sintered	330	9	7.4	410	18
Fe 3 Cu 2.5 Ni	Sintered	440	5	7.4	550	14
Fe 0.45 P	Sintered	370	12			
Fe 3 Cu 2.5 Ni 0.6 C	Sintered	650	2.5			
Fe 1.75 Ni 1.5 Cu 0.5 Mo 0.5 C[1)]	Sintered	550	2	7.4	650	3
Fe 1.75 Ni 1.5 Cu 0.5 Mo 0.5 C[1)]	Sintered + heat treated	1000	1	7.4	1100	1
Fe 4Ni 0.5 Cu 0.5 Mo 0.5 C[4)]	Double pressed and sintered	>800	<1	7.3	>1000	<1
Fe 1.5 Mo 0.8 C[3)]	Double pressed and sintered, heat treated			7.5	1400	<0.5
Fe 2Ni 1Mo 0.6C[4)]	Double pressed and sintered, heat treated			7.4	1800	
Fe 0.4 Mn 0.4 Cr 0.4 Mo 0.6 C[2)]	Sintered + forged; forged + heat treated	550	3	>7.7 >7.7	960 1800	12 3.5

1) from diffusion bonded powders
2) from powder mixtures using master alloys
3) from pre-alloyed powders
4) from partially pre-alloyed powders

Sintered iron and steel parts are used mainly for the car and truck industry, providing 50–70% of all produced parts, depending on country. The figure for Japan is over 80%. A selection of parts is shown in Fig. 11.3. Powder metallurgy parts are used in many types of electric and other engines, business equipment and domestic appliances. A variety of mostly small and complicated parts is now manufactured by metal injection moulding (see section 6.1.1.3). For heavy-duty applications, sinterforged parts are often employed. They consist of Mn- Cr- Ni- and Mo-alloyed (sum: 0.8–3%) steel with 0.2–0.7% carbon, exhibiting often a higher fatigue resistance than conventionally sintered steels. Some of them are illustrated in Fig. 11.4. The connecting rod shown in this figure is now in mass production for an 8-cylinder car engine. A not fully homogenised Distaloy is shown in Fig. 11.5, a nearly homogeneous alloyed steel in Fig. 11.6 as metallographic section. The combination of the powder route with fully automatized machining as well as with welding has considerably increased the scope of intricate shaped parts manufactured by powder metallurgy.

11.1.2.4 High Alloy Steels

Ferritic, martensitic and austenitic *stainless steels* are manufactured by powder metallurgy, the compositions being similar to those produced by melting, casting and forming. A composition 18% Cr, 12% Ni, 2% Mo, balance Fe is often used. The pre-alloyed powders are produced by water or gas atomisation. Due to the high oxygen affinity of chromium the sintering atmosphere (vacuum or high-purity hydrogen) has to have a very low oxygen partial pressure. Stainless steel tubes are produced in large quantities by cold isostatic pressing of gas atomised powders and subsequent hot extrusion. Metal injection moulding

Figure 11.3 Selection of sintered parts for use in passenger cars and trucks (Krebsöge)

Figure 11.4 Sinterforged parts, including gear parts and connecting rods for car engines, for heavy-duty applications (Krebsöge)

(MIM, see section 6.1.1.3) has been applied to the mass production of very small stainless steel gear parts, weighing below 1 g, which are used in new electric toothbrushes with a twisting brush. Also filters with controlled porosity (see section 11.8) of stainless steel are produced.

High-speed-steels (HSS) contain 10–20% Co, Cr, W, Mo and V, to a great extent as carbides (except Co), depending on carbon content. They are partly manufactured by the powder metallurgy route, starting from atomised powders. Finished as well as semifinished products are produced by die or cold isostatic pressing and sintering *in vacuo.* A close temperature control (certain compositions require ± 2 K) is necessary, since the amount of liquid phase formed is strongly dependent on temperature (see 7.3.3, supersolidus sintering). Insufficient densification or grain coarsening, occur at too low or too high sintering temperatures respectively. Also the HIP route is applied (see Table 10.1, column 6a).

Their main advantage is the very homogeneous distribution of highly dispersed carbides, which cannot be obtained by melting and casting, especially

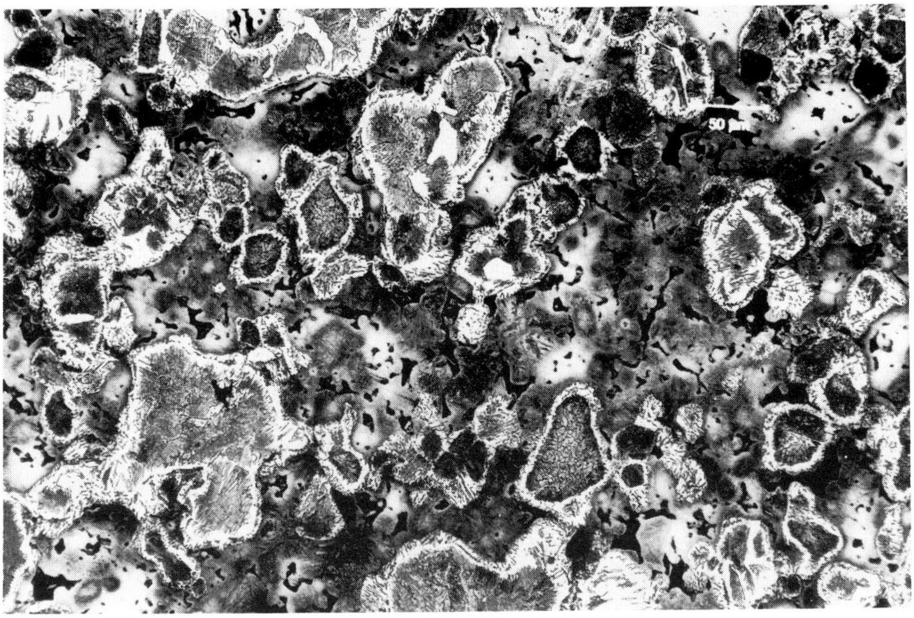

Figure 11.5 Distaloy AE with 0.5% C, sintered 30 min at 1120°C, Endogas (Höganäs AB)

Figure 11.6 Nearly homogeneous alloyed Mo–Cr–Mn sintered steel

with high carbide content (Fig. 11.7). This leads to more isotropic properties, better grindability and for several applications to increased tool life-time, compared with conventionally manufactured products. The powder metallurgy route also enables an increase in the carbide content (e.g. VC) compared to cast and wrought HSS qualities.

A sub-group of sintered steels, currently under development, are based on plain iron or iron alloys can contain 10–20 vol-% of hard phases, such as NbC, TiC or Al_2O_3. They are introduced by mechanical alloying, together with sintering aids (phosphorus, boron). The sintered products exhibit good mechanical

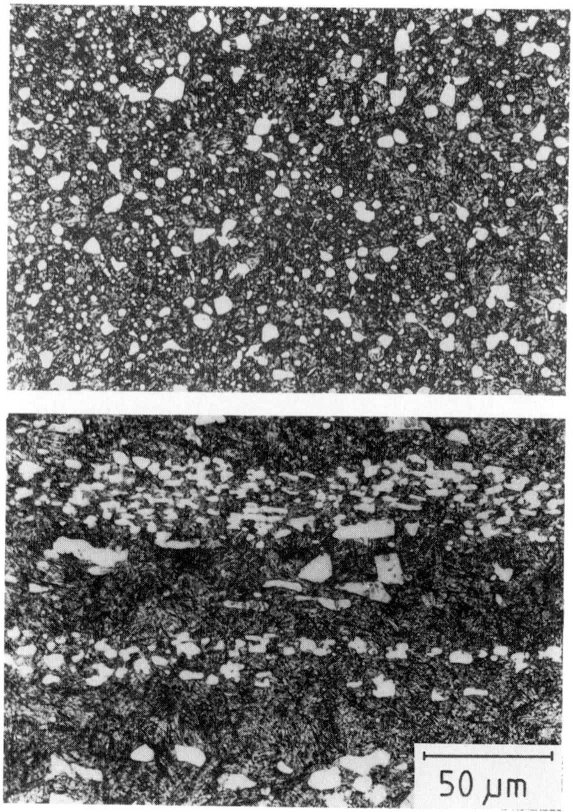

Figure 11.7 Carbide distribution in T15 high speed steel (Krebsöge); *top*: processed by the powder metallurgy route; *bottom*: conventionally cast and forged.

and wear properties and may perhaps compete with HSS for special application.

11.2 COPPER AND COPPER-ALLOYS

Copper-based structural parts rank behind iron and steel parts with about 10% of their tonnage. Pure sintered copper is manufactured to a limited extent as high conductivity material for special applications. Sintered bronze parts with controlled porosity belong to the oldest powder metallurgy products and are used as filters, bearings etc. (which is outlined in section 11.8). The most common composition of sintered bronze is Cu 10%–Sn, often with some tenth %P. Also Pb bronzes are used as bearings, supported with steel shells for strength improvement. Sintered brass with Zn content between 10 and 40% is manufactured mainly as engine parts by using water atomized, alloyed powders. Some vaporisation of Zn occurs during sintering, which can be decreased by maintaining a certain Zn vapour pressure within the furnace. Other, less important sintered Cu alloys are German Silver (18% Zn, 18% Ni, balance Cu) for decorative parts and carbon-containing alloys (Cu/C; Cu/Pb/Sn/C) as friction materials.

11.3 NICKEL- AND COBALT-ALLOYS (SUPERALLOYS)

Heat- and oxidation-resistant nickel- and (to a lesser extent) cobalt-alloys, often designated as 'superalloys', have been developed with many compositions and variation of properties. Three principles of strengthening can be combined in as nickel-based alloys, namely by solid-solution hardening (Co, Cr, Mo, Nb), γ' precipitations (Al, Ti, forming Ni_3Al and complex phases), and oxide dispersions (e.g. Y_2O_3). The latter leads to the so-called ODS alloys. While the majority of superalloys have been manufactured in the past by melting and casting, the powder metallurgy route provides two major advantages: (1) the refinement of the microstructure with equalizing the γ' phase precipitation at high γ' content, (2) the introduction of a dispersed phase as the third hardening principle as mentioned above. While the γ'-phase provides a high strength level in the low and medium temperature range, (<600–800°C) the dispersion hardening is effective especially at higher temperatures. The ODS alloys represent the greatest potential for further development (see section 11.9).

A precondition to achieve acceptable long-term strength properties is the use of low-oxygen powders (50–100 ppm see section 2.1.1.2). Higher oxygen content may oxidise the reactive elements in the γ' phase making them partly ineffective. Clustered oxide inclusions may also promote crack propagation under mechanical stresses. Powder handling and processing, especially for gas-turbine disk production occurs preferably in clean rooms under a protective gas. It includes encapsulated HIP-consolidation, usually followed by die forging, machining and heat treatment. High creep strength requires a relatively coarse matrix grain size (for minimizing grain boundary sliding), while for high fracture strength after long time exposure grains elongated in the direction of the tensile stress are necessary. Thus, microstructural optimization and minimising oxygen content are most essential for powder metallurgy superalloy production.

11.4 LIGHT METALS

11.4.1 Aluminium Alloys

Due to the stable and dense Al_2O_3 layer on each powder particle it is impossible to sinter compacts of pure aluminium powder to high densities. The sintering temperature is limited to about 630°C, near to the Al-melting point. Nearly no atom mobility takes place in the Al_2O_3-layer at this temperature. Alloyed powders, mainly atomized, are therefore used, forming a liquid phase which penetrates the oxide films at Al–Al_2O_3 phase boundaries, leading to adequate sintering behaviour. The use of rapidly solidified powders (see section 2.1.1.2) is of special significance for many of the alloys produced, especially with higher strength. Some compositions are similar to those in conventional Aluminium technology, namely Al–Cu–Mg, Al–Zn–Cu–Mg and others. Powders of them are easily compressible, and are sintered under pure nitrogen or vacuum, giving tensile strength levels between 120 and 200 MPa, and after heat treatment 230–320 MPa. They are used as engine parts. High-strength alloys, namely Al–Li, Al–Mg–Co, Al–Fe–Co, Al–Fe–Ce with tensile strengths of 450–600 MPa at RT have been developed mainly for aircraft and space applications. Al–Li (2.5–4%) alloys, mostly with minor amounts of Zr, Cu or Mg are of low density (2.6 g cm^{-3}) and high stiffness. Ternary and quarternary Al–Fe-alloys are used at elevated temperature applications, whilst others are especially corrosion resistant.

Aluminium alloys compete to some extent with sintered steel parts (see section 11.1) and high-strength polymers. Advantages are their good strength:density ratio, high ductility, corrosion resistance in ambient air and the possibility of easy surface treatment. The development of the high-strength aluminium alloys, as mentioned above, and the expected larger use of aluminium in the car industry will certainly push forward the aluminium powder metallurgy in future. (For dispersion strengthened aluminium see 11.9 for intermetallic compounds see 11.11.)

11.4.2 Titanium and Titanium-Alloys

The manufacture of titanium and its alloys is preferably by melting and investment casting. Complicated parts, however, whose final shaping requires a lot of machining are manufactured by the powder metallurgy route. The powders used are mostly vacuum rotational atomised (see 2.1.1.2) with a large amount of spherical particles, which are preferably HIP-consolidated in capsules (e.g. 920°C, 2 kbar, 1–3 hrs) to almost 100% of TD. This technology can be used for titanium and its alloys, such as Ti Al6 V4. There is no large difference in tensile strength and ductility (1000–1100 MPa; 20%) compared with conventional wrought alloys. The processing of rapidly solidified powders, however, yields improved properties.

Titanium alloys provide very high strength:density ratio, which is of highest significance for aerospace industries. Very precise specifications have to be fulfilled (as usual) for applications in this field, especially with respect to long-term strength and endurance limits. Any metallic or non-metallic impurities which may act as nuclei for crack propagation under service conditions have to be avoided during powder manufacturing, handling and processing as well.

11.4.3 Beryllium and Magnesium

Fine-grained beryllium with a certain ductility is available only by the powder metallurgy process. The preferred consolidation method of the mostly flake-type powders (see sections 2.1.2.1 and 2.1.3.3) is by hot pressing in vacuum, followed by extrusion, rolling or machining. Its very low density, combined with high strength and high modulus lead to applications in space technology and ordnance as gear and control parts. The biggest beryllium parts ever produced (diameter ~2m) were developed for the early Mercury space capsules as heat shields. Although considerable amount of work has been performed in developing high-strength and corrosion-resistant magnesium alloys by powder metallurgy, their practical significance is very limited. They will not be discussed here.

11.5 HIGH MELTING POINT METALS

When high melting point is defined to occur at temperatures above 2000°C, Hf, Nb, Ta, Mo, W and Re as well as some noble metals have to be considered. Powder metallurgy of noble metals is not very significant, the recent development of ODS platinum, however, is mentioned in section 11.9.3. Most important are W and Mo, followed by Ta and Nb. Despite its melting point of 1857°C, the powder metallurgy of pure chromium will be mentioned in this section.

11.5.1 Tungsten and Molybdenum

Molybdenum especially for large sheets is manufactured partly by conventional metallurgy. Tungsten – due to its high melting point of 3420°C – is exclusively produced by powder metallurgy. Both metals follow a similar route of processing, invented by Coolidge as early as 1910–1913. The temperatures necessary for sintering and subsequent hot working are higher for tungsten than for molybdenum. For manufacturing of rods (as intermediate products for wire-fabrication) the metal powder (see section 2.1.2.1) is die or isostatically pressed, presintered and finally sintered at 75–85% of the melting temperature (in K) in furnaces with cylindrical tungsten-heaters under pure hydrogen. This indirect sintering has replaced sintering under direct low voltage current at about 3000°C, since strong density gradients at the rod ends are unavoidable in this process.

The sintered rods (with e.g. 20 mm diameter and a few percent of residual porosity) are completely brittle at room temperature in the case of tungsten. Pure molybdenum has a brittle – ductile transformation near room temperature. They are swaged with shaped dies of hardened steel, beginning at 1700–1500°C, for tungsten up to about 3mm ∅, followed by drawing in hard-metal dies, and at smaller diameters in diamond dies. Dozens of single, consecutive working steps at continuously decreasing temperatures, down to 400°C or even

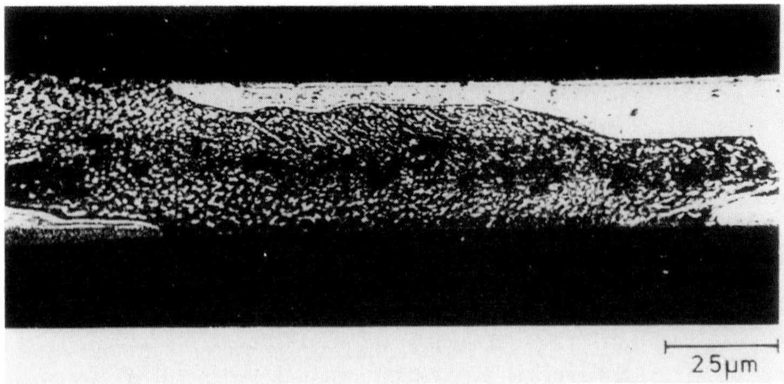

Figure 11.8 Microstructure of non-sag tungsten wire, obtained by sintering of doped tungsten powder, as recrystallised under service conditions (Plansee, Reutte)

room temperature are necessary to obtain fine wires. Wires for lamp bulbs are generally doped (see 2.1.2.1) in order to increase the recrystallisation temperature and to achieve a microstructure after recrystallisation with grain boundaries to a large extent parallel to the wire axis (Fig. 11.8). This increases the life time of the coiled lamp filament by diminishing the possibility of grain boundary sliding under its own weight at the extremely high service temperatures (non-sag [NS]-wire). Lamp-filaments recrystallise completely at the service temperature.

Tungsten and molybdenum, including W–Re alloys, are also used as anodes in X-ray analysis and other electrical and electronic equipment, as they have to sustain extreme thermal loadings. They are also used with different rhenium content as thermocouple wires for temperatures up to 2500°C. W–Re (up to 26% Re) and Mo–Re alloys exhibit improved workability and good high-temperature strength.

A liquid-phase sintered tungsten alloy is the 'heavy metal', containing 1.5–5.5% Ni, 1.5–4% Cu or alternatively 3% Fe (Fig. 11.9). It reaches nearly full density (17.0–18.5 g cm^{-3}), after sintering at 1200–1400°C, has a considerable ductility and can be machined by conventional tools. The tensile-strength-level is 750–950 N mm^{-2} with 2–10% elongation at fracture. Its uses include radioactive shielding, including isotope containers and inertial applications such as vibration dampeners, counter weights, small volume rotating masses penetrators for ordnance application, etc.

11.5.2 Tantalum and Niobium

At present tantalum metal is far more important than niobium. Both metals are processed by vacuum electron-beam or arc melting or by the powder metallurgy route. The powders are die pressed and presintered, followed by indirect or direct sintering at low voltage and current intensities up to 50000 A at the end of the process. Effective purification from residual H, halogens, C, N and O

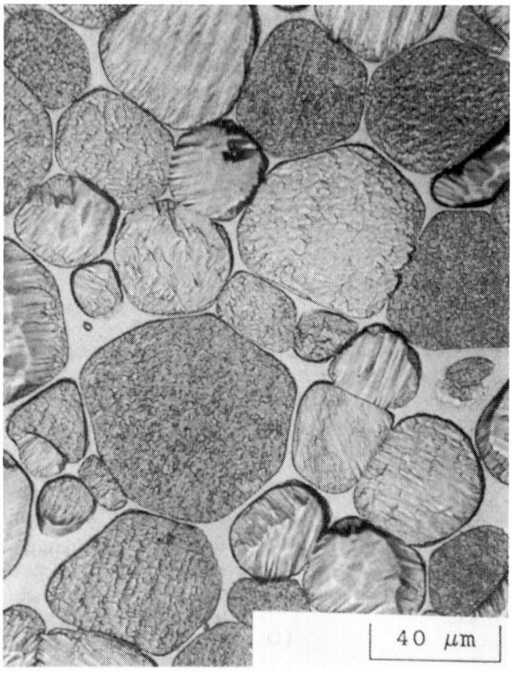

Figure 11.9 Microstructure of tungsten-based 'heavy metal' (German *et al.*)

takes place during sintering, stepwise in this sequence from 800°C up to 2000°C in tantalum. Sintering to high density is possible only after the purification has been finished and needs temperatures of 2500–2700°C for Ta and only 1600°C for Nb. Both metals are processed mainly to sheets and wires. Tantalum is used mainly as a corrosion-resistant material in the chemical industries, as high temperature heating elements and in electrolytic capacitors as electrodes, which are covered by anodic oxidation with Ta_2O_5 as the dielectric. Niobium is the base for metallic superconductors in the form of Nb_3Sn. High-purity Ta and Nb which contain less than 100ppm of interstitial impurities (C, O and N) exhibit good ductility at room temperature.

11.5.3 Chromium

Chromium has only recently been manufactured as a structural material in chemical engineering. Its outstanding corrosion resistance against acids, alkalis, reactive gases and slags created a challenge for powder metallurgy to produce high-purity chromium as finished or semifinished products. Starting from electrolytic powder (average size 40–50 μm) normal cold pressing, sintering under H_2 and subsequent forming after canning in vacuum-tight steel containers yield fully dense products. Material of this purity is brittle at room temperature but exhibits 5–20% elongation at fracture at 200°C with a tensile strength of 200–300 MPa.

11.6 HARD METALS (CEMENTED CARBIDES)

Hard metals represent, together with sintered steels, the most important powder metallurgy products. They consist at of least one hard compound and a binder metal or alloy. The basic metallic hard compounds, are carbides of the IVa, Va and VIa-group metals of the periodic table of elements, preferably WC and TiC (see section 2.2). The binder metal is mostly cobalt ranging between 3 and 25 wt%. Hard metals were invented in 1923 by Schröter (Osram), who first manufactured WC–6% Co. To date, WC-Co represents the largest group in the entire hard metal field. The function of the binder metal is twofold: (1) to provide a liquid phase during sintering resulting in a high densification rate and a high final density (99–99.7%); (2) to develop a bi- or multiphase microstructure with high bend strength and fracture toughness. The efficiency of the binder phase reaches an optimum at complete wetting (angle ~0°) of the hard compound and with temperature dependent solubility of the hard phase in the binder. This leads to considerable rearrangement due to capillary forces and solution reprecipitation processes during sintering (see section 7.3.3). Due to its very metallic nature WC has an excellent thermal conductivity and possesses a limited plasticity under service conditions. Strength and toughness properties are strongly improved with increasing metal content, while the hardness decreases and crack propagation is effectively inhibited by the ductile binder phase. Thus, WC–Co is a unique hard metal system.

11.6.1 Transition Metal Hard Phases

Before treating hard metals in more detail it is worthwhile to provide some basic information on metallic hard phases themselves. Many of them especially carbides and nitrides are interstitial compounds, (Hägg 1930) i.e. the nonmetallic atoms are placed in the octahedral sites of a hypothetical cubic face centred metal basic lattice in a 1:1 ratio. This occurs when $r_x/r_{Me} < 0.59$ (r_x = atomic radius of the nonmetallic atom) and results in the formation of the simple cubic MX–(Na Cl) structure, e.g. TiC, TiN, ZrC, ZrN, VC, VN, NbC, NbN, TaC, TaN, UC, UN (Fig. 11.10a). At higher r_x/r_{Me} ratios, more complex structures with other compositions (e.g. Fe_3C, Cr_3C_2) are formed. Also other degrees of filling-up of the octahedral sites than by the 1:1 ratio are possible, resulting in compounds such as M_4X, M_2X, or MX_2, but the most important is the MX-type. By inclusion of the nonmetallic atom in the basic (hypothetical) metal lattice the distance of the metal atoms increases and does not allow direct contact with each other, and this results in a decrease of the metallic bond and a building up

Table 11.2 Some properties of transition metal carbides, nitrides and borides

Carbides				Nitrides				Borides			
Composition	Crystal structure	Melting point T_s [K]	Micro-hardness *MV*	Composition	Crystal structure	Melting point T_s [K]	Micro-hardness *MV*	Composition	Crystal structure	Melting point T_s [K]	Micro-hardness *MV*
TiC	Cubic	3420	2988	TiN	Cubic	3478	2000	TiB_2	Hexagonal	3253	3370
ZrC	Cubic	3803	2925	ZrN	Cubic	3253	1500	ZrB_2	Hexagonal	3313	2252
HfC	Cubic	4163	2913	HfN	Cubic	3255	1640	HfB_2	Hexagonal	3523	2900
VC	Cubic	3083	2094	VN	Cubic	2633	1520	VB_2	Hexagonal	2373	2800
NbC	Cubic	3753	1961	NbN	Hexagonal	2573	1396	NbB_2	Hexagonal	3273	2600
TaC	Cubic	4153	1599	TaN	Hexagonal	3360	1100	TaB_2	Hexagonal	3373	2500
Cr_3C_2	Rhombic	2163	1350	CrN	Cubic	1773	1093	CrB	Rhombic	2323	1250
								CrB_2	Hexagonal	2473	2100
								β-MoB	Rhombic	2623	2500
								MoB_2	Hexagonal	2373	1200
Mo_2C	Pseudohex. (Rhombic)	2683	1499					Mo_2B	Tetragonal	2413	2500
WC	Hexagonal	2993	1780					α-WB	Tetragonal	2673	3700
W_2C	Pseudohex. (Rhombic)	3003	3000 (?)					W_2B	Tetragonal	3043	2420

Table 11.3 Tungsten carbide-based hard metal groups according to application (partly after ISO 513, draft)

Symbol	Application categories	Range of composition (wt-%)	Hardness (HV)	Bend strength (MPa)
K	Metals with short chips (preferably cast iron) some non-ferrous alloys, plastics, wood	WC–Co (Co 4–12%) 0–3% TiC 0–4% (Ta, Nb) C	1300–1800	1200–2200
P	Metals with long chips (steel, steel castings, non-ferrous metals)	WC–TiC–Co WC–TiC–(Ta, Nb) C–Co (Co 5–14%) (TiC–(Ta, Nb) C up to > 50%)	1300–1700	800–1900
M	Multi-purpose applications	WC–TiC–Co WC–TiC–(Ta, Nb) C–Co (Co 6–15%) (TiC–(Ta, Nb) C 6–12%)	1300–1700	1350–2100
G	Non-machining applications (rock drilling, stone cutting, metal forming, wear parts, abrasive grits)	WC–Co (Co 6–30%) (TiC–(Ta, Nb) C 0–2%	800–1600	2000–3000

of 'non-metallic', mostly covalent bond. The electrons are no longer belonging only to a metal–metal bond. The part of covalent, directional bond is responsible for the brittleness and hardness as well as for the high melting points of these compounds.

This group of materials combines high melting point, high hardness, high wear resistance and high elastic modulus. This is of practical significance mainly for carbides and borides. Most of the carbides, however, are not very oxidation resistant. Several silicides, which do not have a high hardness, are highly oxidation resistant because of the formation of a protective silica surface layer in an oxidising environment. The same lattice structure with similar parameters leads to the possibility of extended solid solutions in many quasi binary or ternary systems (Fig. 11.10b). Some properties of this group of materials are shown in Table 11.2. TaC and HfC can be seen as the materials with the highest melting point.

11.6.2 Conventional Hard Metals

The manufacture of hard metals occurs according columns 1, 3, 6b or 6c in table 10.1. Rod-like pieces such as massive drilling tools etc. are shaped by cold powder extrusion. Mixing and milling (leading to extensive coating of each hard particle by the binder metal) is generally performed with organic liquids in attritors (see section 2.1.1.1), the milling product is dried and granulated in spray dryers, sintering after compaction occurs usually in vacuum furnaces. At temperatures above 1300°C, a certain pressure of Ar is often provided in order to reduce evaporation of the binder metal. Indirect shaping is advisable for smaller lots of pieces, when the expenses for the tool for direct shaping are too high. Special products such as working rolls for precision milling and other high-performance tools require lower residual porosities (0.01–0.03 vol%) than obtained by conventional sintering. This is achieved by HIP or Sinter-HIP, enhancing also the strength, especially at a lower binder content.

The conventional hard metals are classified by ISO standards as shown in Table 11.3, distinguished according to their fields of application, which require different compositions. Cast iron is cut with tungsten carbide, steels and some non-iron metals and alloys with complex-carbide hard metals, since tungsten hardmetals do not resist the high temperatures (~1000°C) during steel cutting. The most important alloying carbide for WC is TiC (often together with minor amounts of TaC/NbC). It enhances the high temperature properties, namely hardness, oxidation – and wear resistance and diminishes the welding tendency with the *in-situ* formed chips due to the decreased diffusivity and wetting behaviour with iron. TiC, however, decreases the bend strength and toughness of the WC-hardmetal. The microstructure of a WC and WC-TiC hard metal is shown in Fig. 11.10c+d. While WC-Co is bi-phase, WC-TiC-Co has a three-phase microstructure.

11.6.3 Coated Hard Metals

Hard metals with special *wear resistant coatings* such as TiC, TiN, Ti(CN), Al_2O_3 and Al-oxinitride, have a large and still increasing significance (thickness 5–10 μm). The combination of a relatively tough matrix with a very hard surface layer is advantageous especially for cutting purposes. In most cases multilayer coatings are used (Fig. 11.11), the inner layer having the best bonding to the substrate. The multilayer system displays better toughness than single layers of comparable thickness. Chemical vapour desposition (CVD) is mostly used which requires, however, a relatively high substrate temperature (800–1000°C), which is disadvantageous for edge stability. Physical vapour deposition (PVD), e.g. sputtering, allows lower substrate temperatures (~500°C), but works more slowly. Plasma assisted CVD works also at 500°C and combines the advantages of CVD and PVD. The majority of the tool tips used are coated, although they can be fixed to the shaft only mechanically and cannot be re-ground after usage. Working rate and life time of coated tools are increased by factors between 2 and 6, depending on the application.

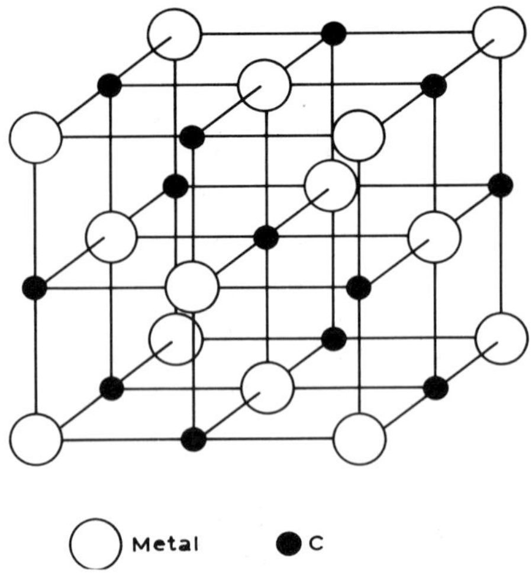
Metal
C

(a)

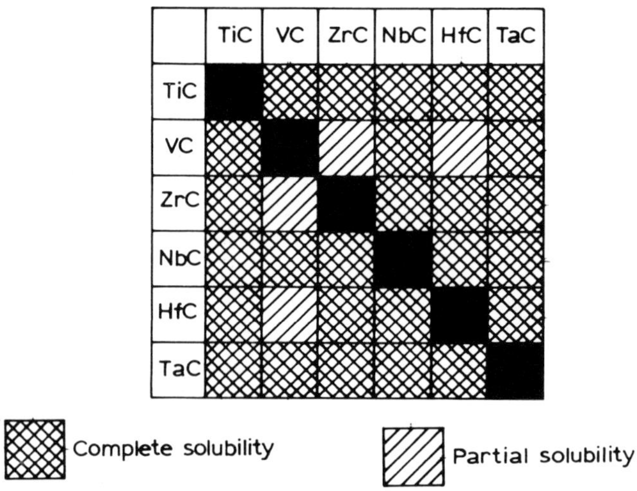
TiC
VC
ZrC
NbC
HfC
TaC
Complete solubility
Partial solubility

(b)

10 μm

(c)

(d)

(e)

Figure 11.10 Lattice structure and solid solubility of hard compounds and microstructure of hard metals: (a) Rock salt structure of metallic hard compounds; (b) Solid solubility of hard compounds; (c) WC–20%Co; (d) WC–TiC–NbC–9%Co; (e) Conventional and fine grained hard metal (fractured surface) (Krupp-Widia)

Recently developed coatings of diamond are deposited in a CVD process under high-temperature, high-pressure conditions. They may further promote the use of coated tools. Coatings of cubic boron nitride (CBN) are also under development, especially for cutting steel. This is not possible with diamond due to the carbon affinity of steels under cutting conditions.

11.6.4 Recent Developments

A very important recent development yields improved microstructures, namely *micrograined* tungsten carbide *hard metals* with carbide grain sizes <1.0 µm. (Fig.

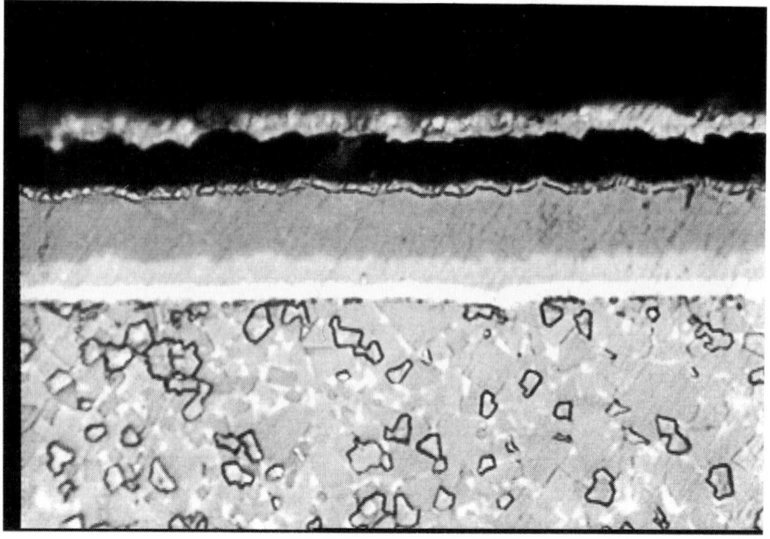

Figure 11.11 Cross section of a CVD-multilayer-coated hard metal × 2000 (Hertel, Ebermannstadt)

11.10e) They are manufactured using ultrafine WC powders with particle sizes <0.1 μm and show improved cutting performance e.g. on hardened steel, high temperature alloys and under interrupted cutting conditions. Combined with high-purity raw materials they provide a substantial decrease in failure both in the microstructure and at the surface and, consequently, performance enhancement. Nowadays even nanosized powders (less than 50 nm) can be successfully used for developing micrograined hard metals, by introducing substantial grain growth inhibition. Such hard metals may in future replace conventional qualities to a considerable extent, since they are harder and more wear resitant.

Tungsten-free hard metals are based on MoC_{1-x} or on TiC both of which have much lower densities than WC-products. *TiC and Ti(CN) with Ni or Ni-Mo-binders* (up to 22 wt%) are of special interest due to their hot hardness and chemical stability, resulting in low diffusion controlled wear. The first introduction of TiC-based hard metals occurred as early as 1931, but their strength level was very low at this time. They are simply named 'Cermets'* and have been continuously improved and introduced with a considerable market share at first in Japan, but now in many other countries. With optimised microstructures and composition bend strengths of about 2500 MPa have been achieved. Special wear- and corrosion-resistant hard metals are based on *Cr_3C_2 with Ni-binder.*

The gap in properties between cemented carbides and high-alloy tool steels has been bridged by *materials with about 50 wt % TiC or TiN* embedded in a heat-treatable steel matrix. Such materials have been developed for special drilling and milling purposes.

* This term is justified since TiC is less metallic in nature than WC.

Table 11.4 Properties of selected hard metals (Krupp-Widia, Essen)

Composition (wt%)				Micro-grained			Cermet	
WC	94.0	85.3	75.0	92.3	78.5	60.0	(TiMo) carbonitride:	85.7
other carbides (TiC, TaC, NbC)	–	2.7	–	1.7	10.0	31.0	other carbides:	0.8
Co	6.0	12.0	25.0	6.0	11.5	9.0	Co/Ni:	13.5
Density (g cm^{-3})	14.9	14.2	12.9	14.9	13.0	10.6	7.0	
Hardness HV 30	1580	1290	780	1780	1380	1560	1600	
Bend strength (MPa)	2000	2450	2900	3000	2250	1700	2300	
Elastic modulus (GPa)	630	580	470	640	560	520	450	
Fracture toughness (MPa m$^{1/2}$)	9.6	12.7	14.5	10.8	10.9	8.1	7.9	
Thermal conductivity (W m^{-1} K^{-1})	80	65	50	46	60	25	11.0	
Thermal expansion coefficient, 293–1073K (10^{-6} K^{-1})	5.5	5.9	7.5	6.2	6.4	7.2	9.4	

(a)

(b)

Figure 11.12 Variety of hardmetal parts: (a) Cutting and milling tool tips (indexable inserts, coated and uncoated); (b) Parts for non-cutting applications (Krupp-Widia, Essen)

Attempts have been made to *replace the Co-binder* by Fe/Ni or Fe/Ni/Co-alloys in WC hard metals, resulting in improved hardness and lower abrasive wear than WC-Co. Martensitic, ferritic or austenitic binder phase with tailor-made properties are available with these binders.

TiB_2–Fe hard metals have been developed to sintered densities above 99% TD, using fine-grained raw materials and vacuum sintering. The microstructure contains some Fe_2B and Ti_2O_3 which can be avoided by additives of Ti or TiH_2.

Figure 11.13 Hardmetal rollers for ribbed wire production (Plansee, Reutte)

The interesting strength and toughness properties of these materials suggest some future applications.

The properties of several important hard metals are collected in Table 11.4, demonstrating especially the influence of binder content, micrograined structure, alloying carbide content and the properties of a TiMo-carbonitride based 'cermet'.

11.6.5 Applications

The application of hard metals (Table 11.3) in technology is extremely widespread: Cutting, milling and drilling tools for metals and non metallic materials (plastics, wood etc.), saws, chisels, drawing dies for wires, rods, deep drawing and cold working, wire flattening rolls, many types of wear parts, etc. More than half of the production is for cutting purposes. Large applications are established in the mining industry (salt, coal, oil), for working of rocks (tunnel construction) and concrete. A variety of parts is shown in Fig. 11.12, and Fig. 11.13. Hard metals compete with high speed steels (see section 11.1.2.4) on the one hand and ceramics (see 11.11), diamond and cubic boron nitride tools on the other hand.

11.7 MAGNETIC MATERIALS

Magnetic materials may be subdivided into two groups, namely: Soft magnetic materials having low coercivity (*Hc*) and low remanence (*Br*) combined with high permeability (μ) and magnetic saturation, while hard (permament) magnets have to be resistant against demagnetisation by exhibiting high coercivity, high remanence and the maximum product of both $(BH)_{max}$. This is shown schematically in the hysteresis loop Fig. 11.14. Both types of magnets are manufactured by powder metallurgy but conventional melting and casting technologies are also established, especially for soft magnetic alloys and the classical Alnico permanent magnets. Soft and hard ferrites are generally sintered and belong to the field of ceramics (see section 11.11.2).

11.7.1 Soft Magnetic Materials

The most important metallic soft magnetic materials which are used as parts are pure Fe, Fe–Si, Fe–P, Fe–Si–P, Fe–Ni, Fe–Co and Fe–Co–V alloys. They are pressed as elemental powder mixtures or prealloyed powders (Fe–Si) and sintered in hydrogen of high purity to homogeneous solid solutions. Homogeneity and a stress-free condition are essential for good magnetic properties. One advantage of powder metallurgy processing is the complete absence of segregations. Very low impurity content, especially of N and C (below 0.001–0.005% each) is necessary to achieve highest permeability and low coercitivity, since both elements form interstitials leading to lattice distorsions which inhibit easy movement of Bloch walls in magnetic fields. Fe–Ni alloys of highest permeability have a Ni content of 77–79% and often a few percent of Mo (Permalloy). Iron with 3–4% Si has a relatively high electrical resistivity and consequently low eddy-current loss.

A considerable proportion of soft magnets are pressed carbonyl powders, embedded in a polymer matrix. These materials are used in unsintered form,

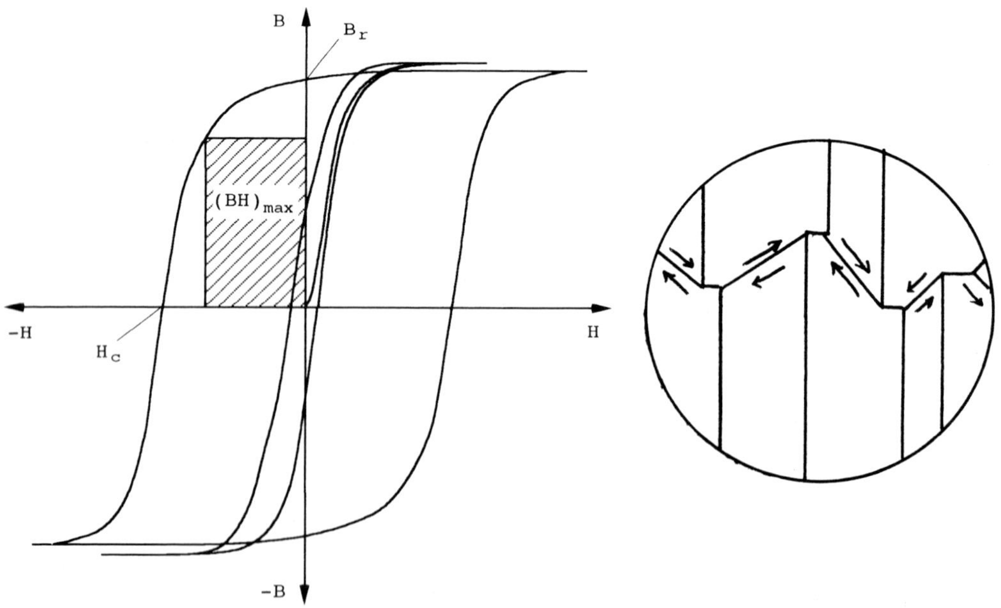

Figure 11.14 Behaviour of soft and permanent magnets in magnetic field (cf. text): (a) Hysteresis loop; (b) Magnetic domains and Bloch walls

Table 11.5 Properties of some soft magnetic alloys and powder-polymer composites

Composition	Static permeability μ μ_4	μ_{max}	Coercivity (A/cm)	Saturation induction (T)	Curie-Temp. (°C)
High purity Fe*	2.000	40.000	0.06	2.15	770
Fe 3% Si	1.500	10.000	0.20	2.0	750
Fe 50% Ni	5.000	35.000	0.08	1.55	440
Fe 80% Ni	25.000	60.000	0.035	0.80	400
Fe Co	1.000	8.000	1.4	2.35	950
Fe Co V	1.500	10.000	0.65	2.35	950
					Losses *(mW/cm³)*
Fe in Polymer-matrix†	175	–	3.9	1.95	55
FeSi in Polymer-matrix†	20	–	1.5	1.0	15

* peak values; † Prometheus Sintermetalle, Graben

although a thermal treatment for stress release is advisable in order to obtain optimised magnetic properties. Sintered soft magnets are widely used in electro-engineering, communication technology and business machines (computers). Permalloy is used as laminated components in transformers. The powder composites are applied as magnetic cores in electrical converters. They represent the majority in the fields of alternating current engineering technology including high frequency communication and compete with ferrite cores. The properties given for the sintered materials in Table 11.5 are dependent on purity and processing parameters. The properties, especially saturation induction of unsintered powder cores, depend strongly upon the percentage of polymer and the basic metal powder.

11.7.2 Hard (Permanent) Magnetic Materials

Three physical principles are used to achieve permanent magnets with high $(BH)_{max}$ values:

1. Multiphase microstructure with fine and homogeneously distributed ferromagnetic precipitates, in e.g. Alnico and Co–Pt alloys.
2. Single domain particles in which Bloch-walls are absent (in the ideal state). Theoretically best are elongated single domains (ESD) with particle lengths of about 0.1 μm. Single domains are achieved to a certain extent in sintered permanent magnets, as fine precipitates (Alnico) or in very fine grained microstructures (Co–rare-earth [RE] phases, ferrites, etc.).

3. Non-cubic phases with very high magnetic anisotropy (Co–RE phases [Co_5Sm], Fe–Nd–B magnets).

There are several reasons for the importance of powder metallurgy in this field. The microstructures achieved by sintering are more suitable than by melting and casting and provide higher mechanical strength than equivalent cast materials. Very little or no machining is necessary and, last but not least, the superior economies, especially in mass production of smaller pieces due to low wastage and low recycling percentage of the powder metallurgy process.

Alnico magnets are manufactured partly by melting and casting, partly by powder metallurgy. They are pressed from powder mixtures of Fe, Co, Ni and an Al–Fe master alloy and sintered in H_2 at about 1300°C. Anisotropic magnetic behaviour, i.e. improved properties in one direction, can be obtained by cooling in a magnetic field causing texturised precipitates. The best qualities with highest $(BH)_{max}$ (~70kJm^{-3}) and H_C require careful heat treatment.

ESD-powders are manufactured to a small extent from Fe and Fe–Co alloys by electrolytic deposition on a mercury cathode, followed by pressing, vacuum distillation and heat treatment in order to optimize the shape of the dendritic powders. The final powder is pressed in a magnetic field with a lead powder additive and used as unsintered permanent magnets.

Cobalt–(RE) magnets are important in the form of intermetallic hexagonal phases (RE)Co_5 and $(RE)_2Co_{17}$. Their extremely high magnetic anisotropy leads to a strong resistance to demagnetisation and very high $(BH)_{max}$ values. The phases especially Sm Co_5 and Sm_2Co_{17}, synthesised in a separate step from the elements, are ground to powders (1–5 μm) near single domain size, compacted and sintered in pure protective gas. Particle alignment in a magnetic field during die pressing provides anisotropic magnetic properties. High density whilst maintaining the smallest possible grain size has to be achieved, otherwise too many Bloch-walls appear. Complex phases like $Sm(CoCu)_5$ and $Sm_2(Co,Cu,Fe,Zr)_{17}$ are still better than the phases mentioned above. They may be heat treated to get fine precipitates for pinning Bloch-walls, which cannot be completely avoided in sintered magnets. Due to the high cost of pure rare earth oxides, attempts have been made to introduce the cheaper RE-Mischmetall, which contains a mixture of unseparated rare earths. The properties of such magnets are somewhat reduced however.

Nitrogen containing compounds, such as $Sm_2Fe_{17}N_x$ also exhibit very high $(BH)_{max}$, but decompose above 600°C.

The most recent development are *FeNdB magnets*, invented as late as 1984 in Japan. The tetragonal phase $Fe_{14}Nd_2B$ provides 30–50% higher $(BH)_{max}$ values than (RE) Co magnets. Their processing is similar to the latter, the nominal composition being about 30–34% Nd (with small amount of Dy), 1–1.3% B, balance Fe.

The capacity of advanced permanent magnets in terms of $(BH)_{max}$ values is summarized in Fig. 11.15. This can be compared with classical Alnico magnets whose $(BH)_{max}$ values, are 50–70KJ m^{-2} and coercitivities 50–60 KA m^{-1} in their best qualities. The hard ferrites, also included in this figure have partly replaced the Alnico qualities since they are rather inexpensive in raw materials and production. Fig. 11.15 includes also non-sintered plastic-bonded magnets which have recently reached about 10% of the total quantity of permanent magnets. It shows also the enormous progress resulting from the development of RE–Co and FeNdB magnets, which permits very small volumes of magnets. SmCo magnets require only 15–20%, and NdFeB magnet 10% of the volume of hard ferrites at the same magnetic energy required. They make a substantial contribution to miniaturisation of many types of electro-magnetic devices and machines, being very important, for example, in communications technology.*

11.8 HIGH POROSITY MATERIALS

A variety of parts with controlled interconnected porosity is manufactured for several applications. They are:

Bearings in which the porosity serves as a lubricant reservoir. Bearings of bronze, (preferably 90/10), plain iron and Fe–Cu belong to the oldest mass products of powder metallurgy (since 1925) and contain between 15 and 35% of interconnected porosity. After pressing, sintering and eventual sizing, they are infiltrated with oil or special solid polymers. Also graphite or MoS_2 admixed to the powder before processing serve as lubricants especially for low sliding speeds.

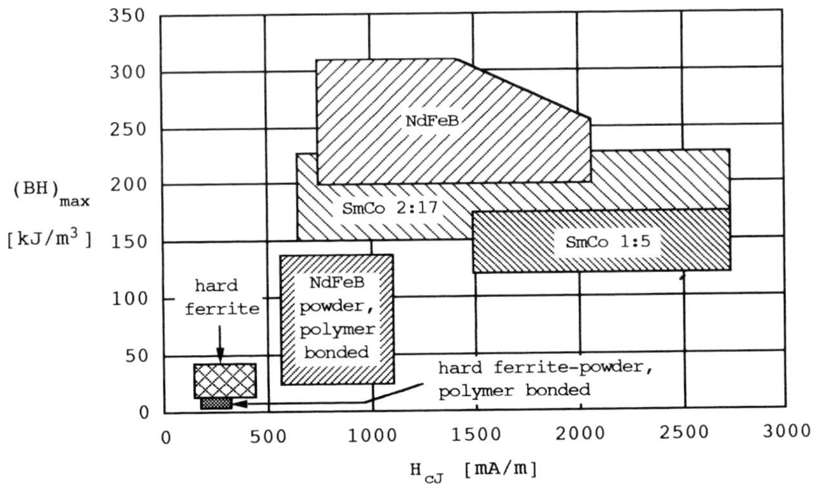

Figure 11.15 Magnetic properties of sintered and polymer-bonded permanent magnets (Krupp-Widia, Essen)

* Not mentioned in this paragraph is the thin-film powder technology for magnetic recording media.

Figure 11.16 Selection of porous metal bearings (GKN Powder Metal Ltd)

Sintered bearings retain their lubricant often during their entire life time, i.e. they are 'self-lubricating'. The rate of oil supply depends on temperature and increases with increasing speed of rotation. They are widely used in household equipment, electric engines, mechanical engineering and in the car industry under moderate loading conditions. Fig. 11.16 shows a selection of porous metal bearings.

For heavy duty applications, non-porous bearings with steel backs are widely used. Cu–Pb or Cu–Pb–Sn powder mixtures are sintered at atmospheric pressure on a steel strip and rolled to near theoretical density and the required thickness.

Products of the lowest density classes (see Fig. 11.2) are manufactured as *filters for gases and liquids* and other applications, mentioned below. The required controlled permeation rates and separating effects from solid particles are provided by using powders with a narrow range of particle size. Bronze filters are manufactured from spherical particles by loose sintering in graphite dies, where sintering takes place with very little shrinkage by neck formation only (Fig. 11.17 and section 7.1.3). Filters of austenitic stainless steel, titanium and nickel-based alloys are obtained from non-spherical, mostly atomized powders by cold pressing and sintering. Sheets and thin-walled, hollow cylinders require special technologies. The pore size distribution is a function of the powder particle size and the processing parameters. Monosized spherical particles provide good sintered strength and the easiest control of pore sizes, avoiding closed pores. Irregular shaped particles, however, provide a greater amount of porosity. A collection of high porosity parts is shown in Fig. 11.18. They compete with filters of ceramics and glasses and are applied widely in gas, gasoline, oil, aqueous liquid and molten plastic filtration, as flame arrestors, distributors, air bearings, conveyor systems, separators, sound absorbers, etc. The porosity ranges between 25 and 60%, with filters from sintered fibres up to 90%, and pore sizes between 1 and 200 μm. Special applications include the separation of gaseous $U^{235}F_6$ from $U^{238}F_6$ for the production of highly enriched U^{235} and porous metal stainless steel de-icing panels on the surface of aircraft wings. However, these have now been replaced by perforated sheets. Characterisation methods for high-porosity materials are described in section 9.4.

Electrodes for batteries and fuel cells with high internal surface area (porosity 70 to 80%) are the carriers for the electrochemically active material, e.g. in Ni-Cd batteries. Nickel powders are sintered on nickel plates using loose powder, slurry or powder rolling technology (see 6.1.1.5). The resulting porous structure is impregnated by a Ni salt, which is transformed to $Ni(OH)_2$ forming the positive electrode. The negative electrode is $Cd(OH)_2$. The porous layer should carry as high a quantity as possible of the active hydroxide. Another application for highly porous metals is as *catalyst carriers* in several chemical industries, representing an obviously expanding field.

Metallic foams are products with pore volumes of about 90% or even more. Aluminium foams are made from cold compacted conventional powders with

Figure 11.17 Microstructure of a bronze filter, showing sintered necks between particles. Nearly all pores are interconnected

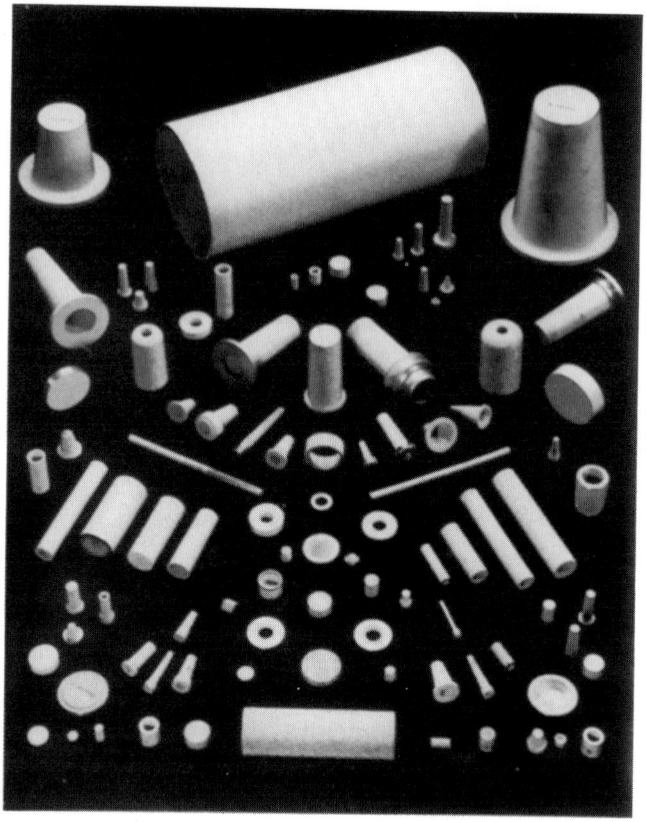

(a)

homogeneously distributed gas-evoluting media (e.g. metal hydrides). When heat-treated near its melting point, the material foams up, forming essentially closed porosity, even at this high pore volume. Conventional shaping (rolling, extrusion, forging etc) at sufficient low temperature before heat treatment enables a variety of geometries to be produced. The porosity of the final product is controlled by the amount of gas-producing media and the heating rate. Foamed metals may be used as light-weight structures ($D = 0.3 - 1$ g cm^{-3}), for fire protection, energy absorption, as barriers etc. *Metallic felts*, e.g. from nickel are available by nickel-coating of carbon fibres followed by decarburisation by sintering in wet hydrogen. The tubular fibres obtained may provide felts up to 95% of porosity.

11.9 COMPOSITES

11.9.1 Principal Aspects

It is not easy to present a clear cut definition of the term 'Composite' although it is used very often. Composites are multiphase materials, whose single phases

(b)

Figure 11.18 (above and opposite) Selection of bronze and stainless steel parts for gas and liquid filtration, separation and distribution purposes: (a) GKN Powder Metal Ltd; (b) Krebsöge

often, but not always, belong to different groups of materials, namely metals, ceramics or polymers. Most materials specialists, however, agree upon several exceptions to this widest possible definition. *Not* included in this term are:

- Materials, whose multiphase character has been achieved solely by heat treatment or simple cooling of homogeneous alloys, such as steels, cast iron, hardened aluminium alloys, etc.
- Materials containing a second phase only as undesired impurity, which may be or may not be avoided by process optimisation. Many of the technical materials belong to this category.
- Doped materials with minor amounts of second phases (say 1/100–1/10%), are seldom called composites.

The bi-phase or multiphase structure of composites exists during the entire processing route, in contrast to the classical hardened alloys. Consequently, infiltrated alloys with two metallic phases (see 11.9.5) as well as plated materials (layered composites) belong to this field, the latter being outside the field of powder metallurgy.

The wide variety of shapes of at least one phase of the composite allows it to subdivide into particulate, whisker or short fibres and continuous fibre-containing composites, respectively. Whiskers and fibres are generally used for mechanical reinforcement, as in many cases is particulate matter such as that found in dispersion strengthened alloys. Compositions, concentration and shape of the phases involved as well as the microstructures obtainable provide a larie amount of possibilities in composite development. Powder metallurgy plays a decisive but not an exclusive role. Particle-, whisker- and short fibre-reinforced materials are mostly manufactured by powder metallurgy, although alternative processes such as melt stirring are also used. Continuous fibre reinforced composites are processed by other means, usually by manufacturing preforms in which a matrix is built-up by spray deposition or melt infiltration. This section therefore excludes continuous fibre reinforced materials.

Composites manufactured from powders may cover the whole concentration range from one phase to the other. In ceramic–metal composites, the matrix may change from metallic to ceramic in nature, depending on concentration and processing route. Thus, a subdivision between metal-matrix composites (MMC) and ceramic matrix composites (CMC) is usual, the latter being mentioned shortly, in 11.10.3. The hard metals, as a 'closed field' of composites with special significance have been treated in a special section (see 11.6). MMC's used as friction and antifriction materials are mentioned in 11.9.6. Considering mechanical properties, which are decisive in all types of reinforced materials, the structures of phase and grain boundaries (wetting behaviour, boundary layers) are of special significance.

11.9.2 Cermets

Cermets consist of at least one ceramic and one metallic component and are mostly MMCs. In a more general sense, metallic hard compounds belong to the ceramics and, thus, the hard metals to the cermet group. In a more limited definition cermets contain non-metallic hard compounds only. They are manufactured mainly by mixing of the components, compaction and consolidation.

The phase dominating in vol% acts as matrix, the other as a dispersed phase, as far as the conventional technology of mixed powders is concerned. By special processing, namely CVD of metals on larger ceramic particles followed by consolidation, a minor metal constituent can be provided as a matrix. At similar concentration both phases provide three-dimensionally interconnected networks and no matrix can be defined. These principal microstructures can be seen in Fig. 11.19. Generally, the homogeneity of the phase distribution strongly depends on the care devoted to processing.

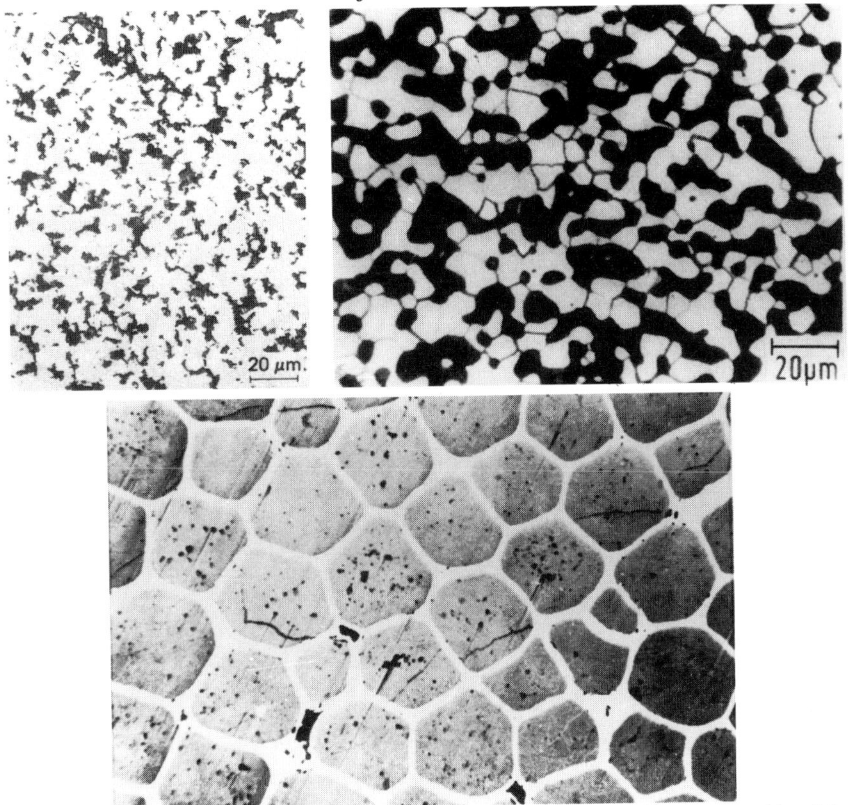

Figure 11.19 Microstructure of cermets: (a) with matrix and dispergent (Ag- 12wt% CdO); (b) with two interconnected phases; no matrix can be defined, (Mo–40wt% ZrO_2) (Plansee); (c) with idealised microstructure (UO_2–11wt% Mo), the minor component being the matrix

Cermets, as the oldest type of MMCs, were developed long ago, the system Fe Al_2O_3 even before World War II, in order 'to combine the good properties of metals with those of ceramics'. The condition and properties of the metal–ceramic interface were completely disregarded by early researches. Oxides and metals display bad wetting due to the electron-clouds around the large oxygen ions, which repell negative metal surfaces and often result in strength lower than that of the components.* Thus, conventional oxide cermets are applicable only at moderate strength requirements. Cermets based on ZrO_2–Mo have found application as protective tubes for thermocouples and as extrusion dies for Cu and Al alloys. They are easily machined. U_3O_8–Al and U_3Al–Al with Al claddings are world-wide used as dispersion-type fuel elements in high-flux, low-temperature research reactors. Silver-based cermets with 10–15% CdO are important electrical switching contacts for moderate and heavy loadings. Because

* 1) This is effective at least with larger particles (> 1–5 μm) in oxide cermets and has been overcome in dispersion-hardened systems (see 11.9.3), especially by mechanical alloying (see 2.1.1.1). Dispersion hardening does not play any role in coarse cermet microstructures.

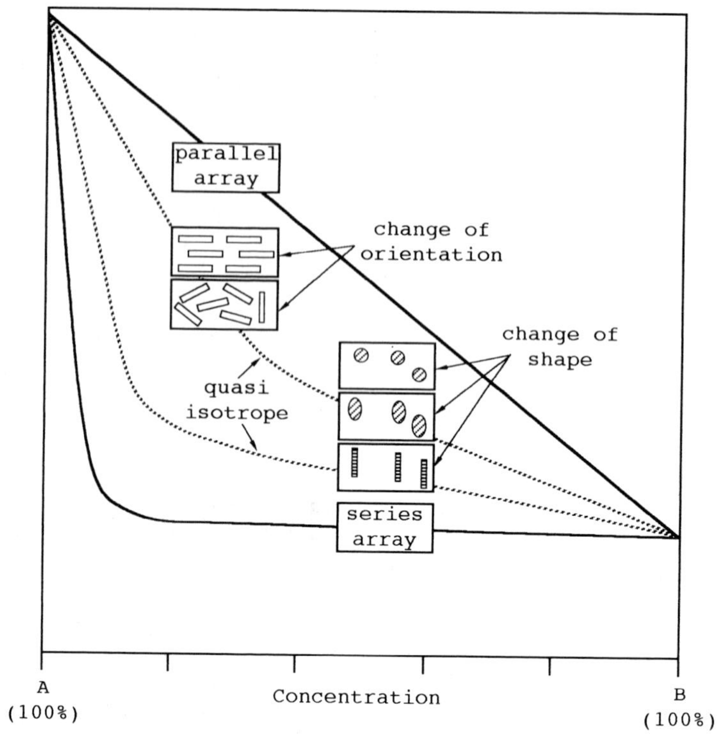

Figure 11.20 Dependence of field properties of dispersion type composites on composition (vol.%) at different microstructures (Ondracek *et al.*)

of toxicity problems, CdO has been replaced now by other oxides, mainly SnO_2. They are manufactured by powder mixing and conventional pressing and sintering or by internal oxidation of Ag–Cd or Ag–Sn, respectively, alloy sheets.

Most of the properties are strongly dependent on the microstructure, especially when there are great differences of properties of the single phases. Several properties can be calculated for dispersion type composites considering volume concentration, shape-factor, type of distribution (textured or random) of the dispersed phase, and properties of the phases. This gives a broad variability of properties, even at one concentration, as shown in Fig. 11.20. Tailor-made materials are available in principle, as far as processing technologies allow, to meet closely the microstructural requirements. Strength properties cannot be calculated in this way because they depend strongly on the nature of metal-ceramic interfaces.

11.9.3 Dispersion Strengthened Metals and Alloys

The principle of dispersion strengthening (dispersion hardening) is to inhibit the dislocation motion in a ductile matrix by finely dispersed particles. While in

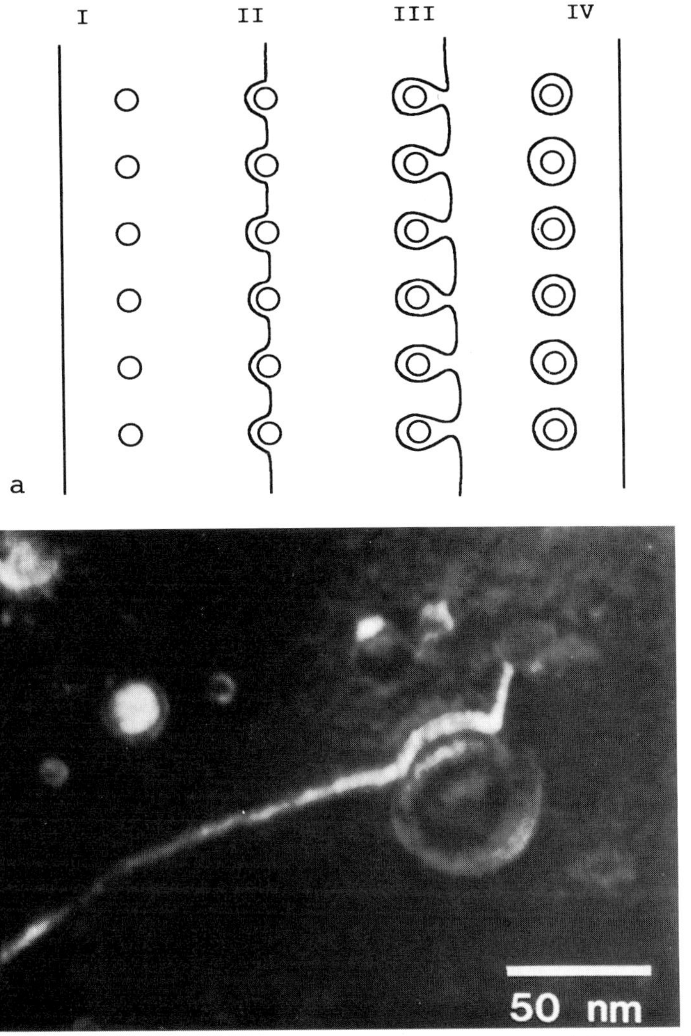

Figure 11.21 Retardation of dislocation motion by dispersed particles (a) The Orowan mechanism (schematically); (b) Dislocation line attached to a single dispersed particle (Arzt)

precipitation hardening the precipitates are coherent or semicoherent with the matrix, dispersion hardening is based generally on incoherent particles. The hardening mechanism is according to Orowan, i.e. the interaction of particles with dislocations is not by particle cutting but by deforming the dislocation lines leading to the formation of closed rings around the particles. The strengthening effect is given by the particle distance a (representing the mean free path of the dislocations) and the resulting increase in yield strength ΔR_p is

$$\Delta R_p = K \frac{g\,b}{a} \tag{11.1}$$

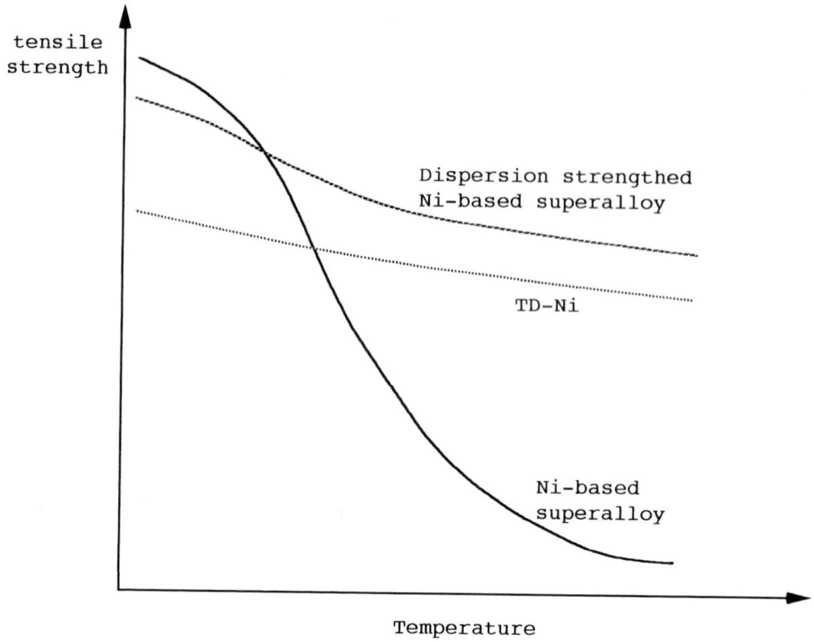

Figure 11.22 Strength dependence on temperature for alloys with different hardening principles (schematically)

with: K = constant, g = shear modulus of the matrix, b = Burgers uvector. The efficiency of a certain concentration of the dispersed phase is optimized when its particle size is small combined with an even distribution. Fig. 11.21 explains the principle and demonstrates the interacton of a dislocation line with a single particle. Additional strengthening effects occur by refinement of the grain and subgrain size of the matrix by retardation of grain growth, by increase of the matrix dislocation density and by matrix stresses due to thermal expansion mismatch.

Since the particles are stressed by the dislocations in motion they have to have an adequate strength, otherwise they are mechanically destroyed during deformation. High melting point and microstructural as well as chemical stability under production and service temperature conditions, i.e. no reactivity or solid solubility are required, otherwise their efficiency is reduced by partial disappearance of the dispergent or Ostwald ripening, respectively. Oxides of high thermodynamic stability, like ThO_2 and Y_2O_3 meet these requirements even in high melting metals. Dispersion strengthening may be applied for most of the common metals and alloys. It is effective, in contrast to classical precipitation and solid solution hardening, up to high temperatures (Fig. 11.22).

The first dispersion hardened material was *SAP* (= *sintered aluminium powder*), aluminium with 8–15% Al_2O_3, as invented in 1950. It could not be established in the market. A recent development is Al with a few % Al_4C_3, and some Al_2O_3, the former being built-up during reaction milling of the aluminium–carbon black powder mixture. After consolidation a high-strength aluminium alloy is achieved, also suitable for service-temperatures up to, say, 450°C, which may be defined as 'high-temperature' for Al-based materials. Also SiC-particle reinforced aluminium has received much attention within the group of high-strength aluminium alloys. Nickel with 2% ThO_2 was a forerunner of *oxide-dispersion strengthened (ODS) alloys.* It was produced at first by Du Pont using coprecipitation of Ni and Th salts with further processing and sintering in a reducing atmosphere. Nickel-alloys with a few % Y_2O_3 are now favoured in this group. They combine the three strengthening principles mentioned above and are oxidation- and creep-resistant. Their significance has increased steadily during the last decade, although their manufacturing and machining is expensive and classical welding is not possible due to destruction of the microstructure. Their main application is in gas turbines for military aircraft. Ferritic ODS–iron–chromium alloys with minor additives of Al and Ti and 0.5 wt% Y_2O_3 may represent a more economic alternative to the austenitic nickel-based alloys.

Of considerable importance are *ODS-platinum metals,* especially Pt and Pt–Rh-alloys with less than 0.5wt% ZrO_2. Their long-term rupture strength is dramatically improved compared with the oxide-free platinum metals. They are used in the modern glass industry as crucible material for glass melts up-to 1400°, mainly for the production of optical glasses and glass fibres. Very recent development include ODS–Nb and Ta alloys with attractive high-temperature properties. For the entire group of dispersion hardened (particle strengthened) materials the mechanical alloying process (see section 2.1.1.1) is highly significant, unless coprecipitation methods can be used.

11.9.4 Fibre- and Whisker-Reinforced Metals

Efficient reinforcement of metals, alloys, ceramics, polymers and carbon has been developed by uni- or multidirectional implantation of whiskers, (mostly SiC), short or continuous fibres, (C, SiC or Al_2O_3), being used mostly as multifilament fibres. The reinforcing effect occurs due to the extraordinary high strength of whiskers and fibres with diametres below a few μm (see section 2.5). The whisker- and short fibre-reinforcement of metals can be achieved as effectively by powder metallurgy as by melting and casting technology. The health hazards associated with handling SiC whiskers, their high price and the only small difference between the property enhancement between whiskers and particles have shifted the emphasis recently more to particle or to chopped fibre rather than whisker reinforcement of metals, especially aluminium. For reinforcement of advanced ceramics, see 11.10.3 and Table 11.7.

11.9.5 Infiltrated Alloys

The infiltration process (see section 7.3.3.3 and Table 10.1 column 8), which may be regarded as a special kind of liquid phase sintering is the base for a small group of powder metallurgy products, mainly W–Cu, W–Ag, Mo–Cu and Mo–Ag. These composites consist of two metallic, mutually insoluble components. An effectively presintered compact of the high-melting metal is infiltrated by the (good wetting!) lower melting metal. When dip-infiltration is used, the capillary forces have to provide the complete penetration of the pore network. The more common procedure is to put the exactly calculated amount of infiltrating metal on the top of the compact, which is sintered, preferably *in vacuo.* These alloys are used as heavy-duty electrical contacts. Tungsten or Molybdenum provide the matrix resistant against heat, wear and arc erosion, copper or silver the phase of good conductivity. Copper-infiltrated Fe and Fe–C compacts are of interest as high strength parts with good machinability, and Fe–Pb and Fe–polymer composites as low-friction materials, e.g. for bearings.

11.9.6 Others

Diamond tools, which have been manufactured for several decades, are composites of 2–35 vol% natural or synthetic diamond with grain sizes between 10 and 500 μm embedded in a metal matrix such as Cu, bronze or Co-alloys. Also coated diamond is used in order to improve bonding to the metallic matrix. They are manufactured mostly by hot pressing to nearly full density and are used for grinding and cutting of hard alloys and ceramics and to a large extent for concrete and rock-working. The maximum service temperature is around 800°C; at higher temperatures graphitisation occurs. For grinding of hardened steels cubic BN composites may be preferred due to their superior stability against iron at elevated temperature.

Composites, mainly based on Cu, bronze or Fe with various amounts of non-metallic components, are used as *tribological (friction and antifriction) materials.* Most of the compositions used are manufactured by conventional powder metallurgy. *Oxide and Graphite-containing composites* serve in a wide variety of engineering as high-performance friction materials for brakes and clutches in heavy vehicles, excavators, military equipment, aeroplanes and other heavy engines. Their metal content ranges up to 85 wt%, graphite up to 20 wt% and oxides such as Al_2O_3 and SiO_2 below 10 wt%. While metal-free, fibre-based materials in former years mainly asbestos, may resist temperatures up to 400–500°C, the MMCs are usable up to 1000°C. Even higher service temperatures (1200°C) are achieved by high-strength carbon-fibre-reinforced carbon (CFC), a rather expensive material, now utilised in aircraft breakes. The MCC friction materials are produced with a wide range of friction characteristics, which are obtained by variations in the non-metallic components as well as in the metal matrix.

Similar composites, but without mineral oxide additives, are used as antifriction materials containing solid or liquid lubricants. Graphite, polytetrafluorethylene, MoS_2 or Pb are utilized as solid lubricants in bearings (see 11.8) under dry service conditions up to 50 vol.%. They are applied in heavy engines under conditions which cannot be fulfilled by oil-lubricated bearings, e.g. high temperatures up to 500°C and in a corrosive environment. Composites based on graphite with metal (mainly copper) powder content up to 30 wt.% are widely used as carbon brushes in electric motors. Silver based composites with some graphite or with 20–40% nickel are used as contact materials.

11.10 ADVANCED CERAMICS

11.10.1 Principal Aspects

Only a few decades ago ceramics were produced mainly as constructional material (bricks), sanitary ware, house building, insulation, refractory and corrosion-resistant products. Today ceramics with enhanced properties are high-tech materials in many industries. The 'classical' ceramics are mainly silicates with complicated chemical composition and lattice structures, ionic bonding and multiphase, coarse and often porous microstructure. High-tech ceramics are oxides or non-oxide materials of simple composition and lattice structure, with a fine-grained and dense microstructure. The non-oxide ceramics, mainly covalent in bonding, exhibit excellent high-temperature properties. Advanced ceramics are defined normally as excluding classical refractories and insulators. Their main applications are in electronic, telecommunication and electrical engineering industries which represent about 70 or even 80% of the entire field. These, often called 'functional ceramics', are used under conditions where, in contrast to the 'structural ceramics', other non-mechanical properties are decisive for successful application. Structural (engineering) ceramics are used as components in engines, in chemical devices, as wear parts, high temperature parts, etc. and represent about 10–20%, depending on the country. Table 11.6 classifies the advanced ceramics according to their present applications.

Functional ceramics are seen to be concentrated more in the lower part, while engineering ceramics are in the upper part of Table 11.6. They are often subject to high mechanical and high thermal loading. The load mostly leads to long-term multiaxial stresses which change with time. In the case of high temperature application, the mechanical stresses are superimposed by thermal stresses, while the materials properties are changing with temperature and time. Problems arising under such severe conditions are often difficult to solve with brittle materials. This is one of the reasons why the progress in introducing engineering ceramics in technology is slower than formerly expected. Furthermore, many possible applications in engineering are not yet thoroughly

Table 11.6 Classification of advanced ceramics

Main function	Properties required	Applications (examples)
Thermal	High-temperature and thermal shock resistance, thermal conductivity (high or low, respectively)	High-temperature components, burner tubes, nozzles, heat exchangers, heating elements, non-iron metallurgy, insulating parts, thermal barrier coatings
Mechanical	Long-term, high-temperature resistance, fatigue, thermal shock, wear resistance	Wear parts, seals, bearings, cutting tools, engine, motor and gas turbine parts, thermal barrier coatings
Chemical, biological	Corrosion resistance, biocompatibility	Corrosion protection, catalyst carriers, environmental protection, sensors, implants (joints, teeth, etc.), filters, membranes, fuel cell materials
Electrical, magnetic	Electrical conductivity (high or low, respectively), semiconducting, piezo-thermoelectricity, dielectrical properties	Heating elements, insulators, magnets, sensors, IC-packages, substrates, ferroelectries, piezoelectrics, optoelectrics, dielectrics, varistors, solid electrolytes, super conductors
Optical	Low absorption coefficient	Lamps, windows, fibre optics, IR-optics
Nuclear	Irradiation resistance, high absorption coefficient, high-temperature resistance, corrosion resistance	Fuel and breeding elements, absorbers, shields, matrix for waste conditioning

evaluated, because design in brittle materials is much less developed than design with metals and alloys. The strength of ceramics at nearly all of the applications is limited by brittle fracture and not by plastic flow, since dislocations do not move at stresses, at which cracks or crack-like failures grow. This requires quite different design criteria than with metals.

The physical and especially the mechanical properties of ceramics depend not only on chemical composition but strongly on their microstructure. This includes grain size distribution, residual porosity, microcracks, grain boundary phases, especially glassy phase, inclusions, etc. The optimisation of the microstructure including grain boundaries is often a decisive tool to get good properties with low scatter.

11.10.2 Magnetic and Electronic Ceramics (Functional Ceramics)

One of the most important groups is *magnetic ceramics*, namely *hard and soft ferrites*, whose technological developments goes back to the nineteen-forties and fifties. Their general formula is $MeO \cdot Fe_2O_3 = Me^{++}\ Fe_2^{+++}O_4$ ($Me^{++} = Fe^{++}$,

(a)

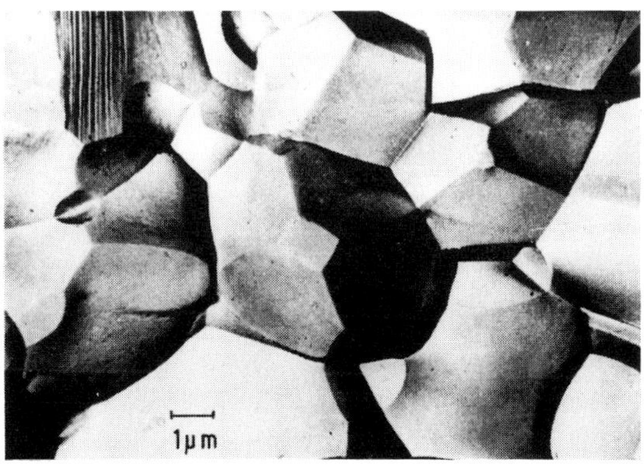

(b)

Figure 11.23 (a) Collection of soft ferrite parts (Härdtl); (b) Microstructure (REM) of a Mn–Zn-ferrite

Co^{++}, Ni^{++}, Mn^{++}, Zn^{++}) for the soft ferrites, having a cubic structure of inverse spinel. They are not ferr*o*-, but ferr*i*magnetic, where only the magnetic moments in Me^{++} ions are parallel, whilst the Fe^{+++} moments are antiparallel, i.e. 'self-compensated'. Hard ferrites are $MeO x{\cdot}Fe_2O_3$ (x = about 6, $Me^{++} = Ba^{++}$, Sr^{++}) with a hexagonal structure, having high magnetic crystal anisotropy. Both

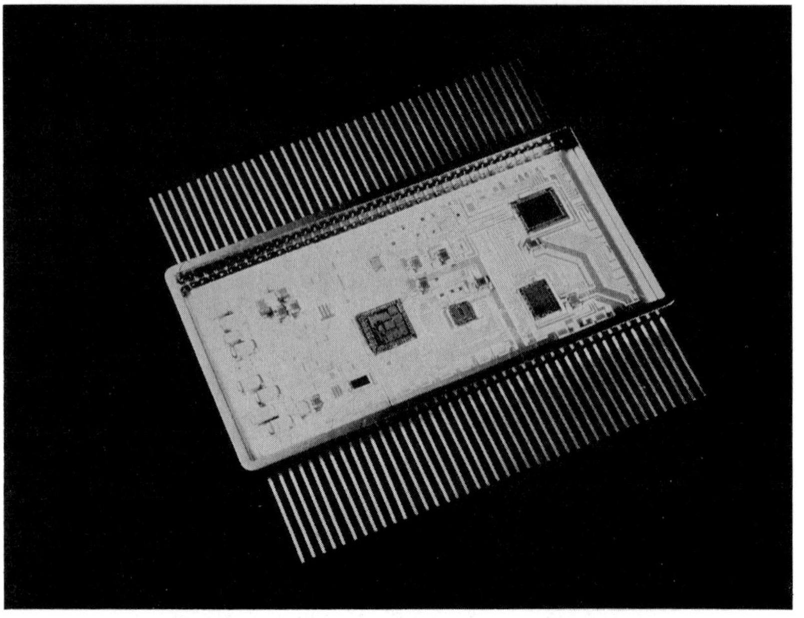

Figure 11.24 Integrated circuit with Al_2O_3 substrate (with thin-film conducting tracks, resistors and capacitors [dark]) (Valvo)

types of ferrites are electrical insulators (10^{-1}–10^{-4} Ωm) and are inexpensive in manufacturing due to cheap raw material prices and relatively simple technology. Soft ferrites have the advantage of very low eddy current losses and are widely used in communication technology, e.g. as cores for radio and television. Hard ferrites are, together with the metallic Alnico-types (see section 11.7.2) the most widely used permanent magnets. Fig. 11.23a is a collection of mass produced soft ferrite parts; Fig. 11.23b shows the microstructure of a Zn–Mn–ferrite.

Thin plates of Al_2O_3 are widely used as *substrates for electronic circuits,* (Fig. 11.24) and are produced by the tape casting technology. In the so-called thick-film technology of the conducting layers a purity of ~96% and surface roughness below 0.3 µm is used. For the thin-film technology 99.6% Al_2O_3 with even lower surface roughness is necessary. Another substrate material is AlN, having higher thermal conductivity than Al_2O_3.

Ceramics with *special electrical and optical properties* are ABO_3 compounds of perowskite ($CaTiO_3$) structure. *A* is a bivalent ion such as Ca^{++}, Sr^{++}, Ba^{++} or Pb^{++}, B is Ti^{4+}, Zr^{4+} or Sn^{4+}. The 4-valent ions may be easily displaced in the lattice, resulting in polarisation effects responsible for ferro-, piezo- and pyroelectricity as well as high dielectric constants. Piezoelectric and dielectric ceramics are widely used, the latter in miniaturised capacitors. Perovskite ceramics may exhibit also electro-optical effects, i.e. their optical refraction constant is influenced by electrical fields. This effect is used for light switching

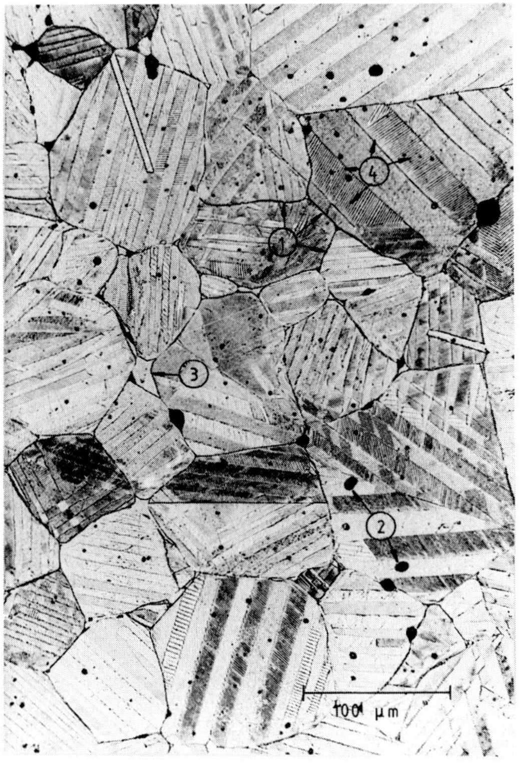

Figure 11.25 Microstructure of $BaTiO_3$, etched, showing grain boundaries (1), pores (2), second phase (3) and ferro-electric domaines (4) (Härdtl)

within microseconds. Sufficient transparency has been obtained with doped Pb $(TiZr)O_3$ ceramics. All these materials are electrical insulators. The microstructure of $BaTiO_3$ is seen in Fig. 11.25, and different pieces of transparent ceramics in Fig. 11.26.

Semiconducting ceramics are based e.g. on ZnO which is an electronic semiconductor. Its properties can be changed by doping (e.g. with Bi, Co, Mn or Sb) yielding high resistivity ($\sim 10^{10}\ \Omega$) in the low voltage range ($\ll 50$ V) and low resistivity at high voltages ($\ll 500$ V). This is a function of inhomogeneous conduction, especially of the behaviour of the grain boundary phase. These varistors (variable resistors) are used for overvoltage protection. The insulator $BaTiO_3$ may be transformed to a semiconductor by special doping, leading to ceramics with strong positive temperature coefficients (PTC-resistors). They are used in self regulating heating elements, temperature sensors etc. A well known application of semiconducting Y_2O_3-doped ZrO_2 is as an oxygen sensor in the *Lambda-sonde* for measuring the oxygen concentration in car exhaust gases. The material is a pure anion-semiconductor, giving voltage differences between both ends of a sample, when they are in contact with different oxygen

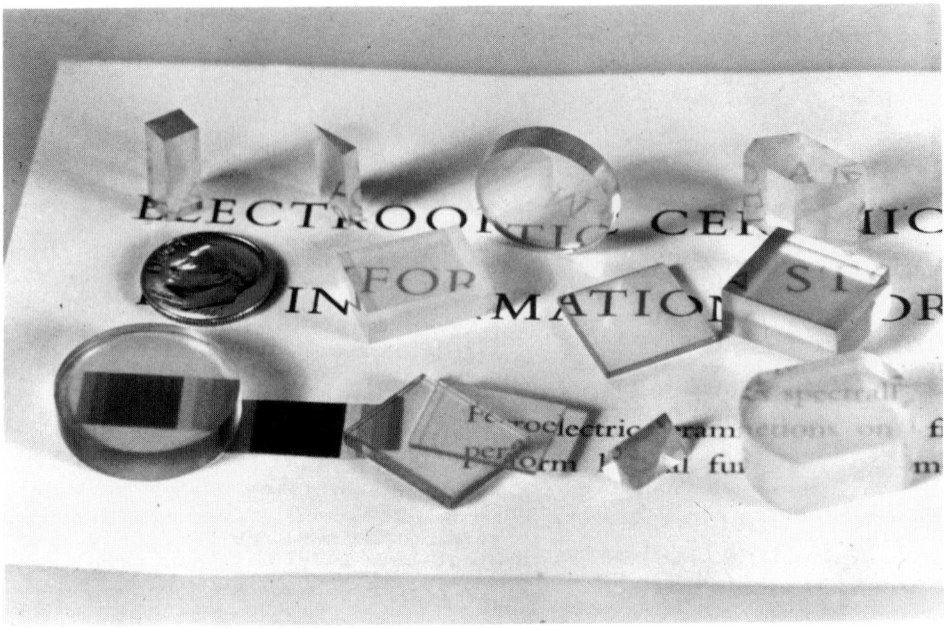

Figure 11.26 Transparent ceramics for eletro-optical application (Härdtl)

concentrations. The electrical signal is used for regulation of the fuel-air mixture in order to optimise the catalytic efficiency for minimising air pollution. Many other types of ceramic sensors have also been developed.

Superconducting ceramics, discovered as late as 1986/87, are complex oxides such as (LaSr) CuO_4 and $YBa_2Cu_3O_7$, the latter having a transition temperature of 92 K. Even > 100 K transition temperatures have been reported. This spectacular invention may be of the highest future significance with respect to using inexpensive liquid nitrogen as the cooling medium instead of expensive liquid helium, used for the metallic superconductors with low transition temperatures. The technical realisation of loss-free energy transportation, however, is still doubtful due to the difficulties in manufacturing wires and coils from brittle materials.

In *nuclear energy* UO_2 and UO_2-PuO_2 ceramics are well established as fissionable and breeding materials, preferably as pellet stacks in metallic tubes. The oxide form is preferred due to the high melting point, the good compatibility with the fuel can and cooling media and the favourable irradiation behaviour.

11.10.3 Structural (Engineering) Ceramics

Most of the engineering ceramic parts used at the present time are monolithic, either oxides or non-oxides. Important oxide ceramics are Al_2O_3 and ZrO_2, and for thermal insulating applications also Al_2TiO_5, while non-oxide ceramic parts

Table 11.7 Selection of monolithic and composite advanced ceramics for structural applications

Material or matrix (vol%)	Theoretical density (g cm^{-3})	Young's modulus (GPa)	Bend strength[g] (MPa)	Fracture toughness (K_{ic}) (MPa m$^{1/2}$)
Monolithic				
Si_3N_4	3.19	160–300	250[h]–1200	3[h]–10 (15)[i]
SiC	3.21	300–400	400–800	3–5
Al_2O_3	3.98	400	400 (1000)[a]	3–4
Al_2TiO_5[k]	3.77	13	40	–
Monolithic or reinforced	5.56			
ZrO_2	6.10 (TZP)	205[b]	600–900 (1500)[b]	5–10 (15)[i]
Reinforced				
Al_2O_3–SiC, (20–30)[c]	3.6	400	850	8.5
C–C, (30)[d,e]	1.9	70	1000	–
SiC–SiC, (45)[d,f]	2.9	250–270	580	12–18
Glass–SiC, (30–65)[d,e]	2.6	120–140	600–900	10–35

[a] Peak values; [b] Partially stabilized; [c] Whisker; [d] Fibre; [e] One-dimensionally reinforced; [f] Two-dimensionally reinforced; [g] Property depending on composition, microstructure and residual porosity; [h] low values for porous RBSN; [i] latest development (1991); [k] thermal insulator material

are mainly based on Si_3N_4 or SiC (Table 11.7). Each formula in this table represents a group of materials, their microstructural and mechanical properties depending on composition, consolidation method, porosity, etc. The properties of most of these products have been improved continuously during the last two decades. A large increase in strength and fracture toughness has been achieved by developing composite materials, namely transformation-toughened ZrO_2 and Al_2O_3, SiC whisker-reinforced ceramics (mainly Al_2O_3) and fibre-reinforced products (e.g. C–C and SiC–SiC).

The strength, measured mostly as bend strength, is not represented by definite, single values, but is subjected to statistical distribution according to Weibull (Fig. 11.27). The Weibull modulus m in equation 11.2, indicating the width of distribution (P = failure probability, σ = fracture strength, $\overline{\sigma}$ = mean strength,

$$P = 1 - \exp\left[-\left(\frac{\sigma}{\overline{\sigma}}\right)^m \frac{V}{V_0}\right] \qquad (11.2)$$

V = stressed volume, V_0 = unit volume) is a very important figure, ranging in monolithic ceramics between 10 and 15. The latest progress in the development of Si_3N_4 yields m between 20 and 40, which may be considered to be a breakthrough for application in the car industry, e.g. in valve systems.

A definite strength without scatter is (theoretically) defined by m approaching ∞. Each single strength value is determined by one defect (microcrack, pore, inclusion etc) becoming critical at a certain stress intensity resulting in catastrophic crack growth. The defects (flaws) are statistically distributed in the volume as well as at the surface, so that the strength distribution is in many cases a function of surface quality. It can be often improved by polishing or glazing but can be deteriorated also by unqualified machining. The strength generally decreases with increasing size of the part since the probability of the presence of large defects increases with increasing size.

Defects in ceramics are less critical when the fracture toughness (K_{IC}) becomes high. K_{IC} is, therefore, another very important figure. Since strength and fracture toughness, respectively, are improved by different microstructural features (strength by fine microstructure and minimized flaw size, fracture toughness by coarse and elongated grains which provide crack-bridging), maximum values of both cannot be obtained in the same material.

The brittleness of ceramics is their worst disadvantage. It is a function of the strongly directed ionic or covalent bonding. Therefore, the most important developmental goal is to optimise the microstructure in order to overcome the inherent brittleness and to achieve an adequate reliability and lifetime during application. One can summarize these aims as follows:

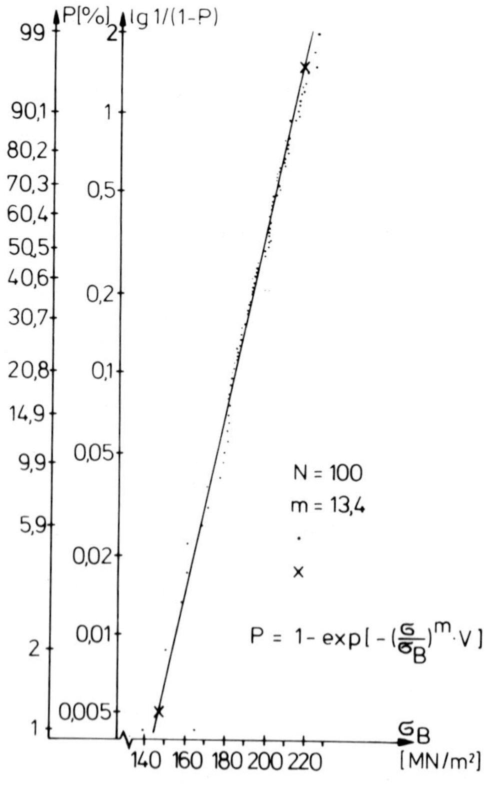

Figure 11.27 Weibull-distribution of bend strength of reaction-bonded silicon nitiride (RBSN) ● measured data, x calculated data

- Improving the strength level in order to increase the safety margin to catastrophic failure. However, this can only be achieved at the expense of fracture toughness, and *vice versa.*
- Increasing the Weibull modulus in order to decrease the failure probability.
- Increasing the fracture toughness (K_{Ic}) in order to inhibit the crack propagation up to catastrophic failure.
- Inhibiting stress induced slow crack growth in order to avoid crack growth up to the critical size (i.e. catastrophic failure) at long-term application.
- (For high temperature application): Minimizing surface or volume damage by oxidation, to guarantee microstructural stability during high-temperature-long-term exposure and to minimize creep deformation. All these phenomena may lead to growth of failures followed by catastrophic cracking after reaching critical size.

The most important R&D principles pursued in order to approach optimised materials are: Transformation toughening, whisker- and fibre-reinforcement and powder processing optimisation.

Transformation toughening is based on the tetragonal (*t*) -monoclinic (*m*) transformation of ZrO_2, accompanied by volume increase*. t-ZrO_2 particles in a stabilized cubic (*c*) ZrO_2 or other ceramic matrix (Al_2O_3) introduce compressive stresses during t–m-transformation which effectively inhibit the propagation of microcracks. At very small particle size (<< 1 μm) the transformation is shifted below room temperature and is induced only by stresses, namely in front

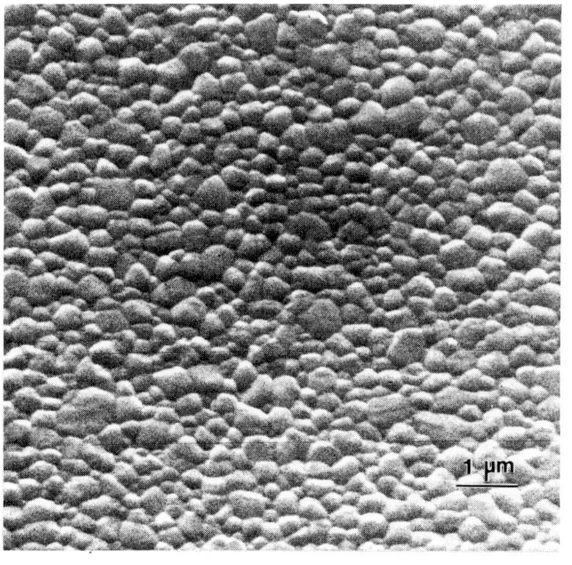

Figure 11.28 Microstructure of an ultrafine-grained *t*–ZrO_2 ceramic (Zh. Ding)

* Modifications of ZrO_2: m < 1423 K; t 1423–2623 K; c > 2623 K

of any propagating crack. Also ultrafine grained monophase t-ZrO_2 is a desired quality. With both materials high toughness, high strength ceramics for near-room-temperature applications have been developed. The microstructure of *t*-ZrO_2 is demonstrated in Fig. 11.28.

Whisker- and especially fibre-reinforced ceramics have the highest toughening potential in principle and are only at the beginning of their development. The strength of whiskers and fibres (see section 2.5) ranges between 2000 and 3000 MPa, depending on their diameter and length. Effective reinforcement requires 20–50 volume % of them which can be introduced (with continuous

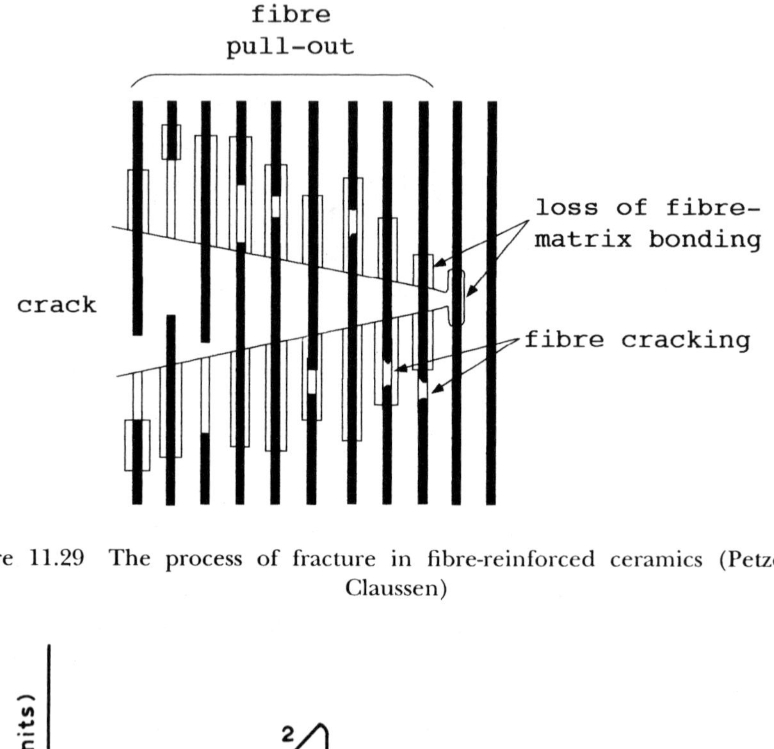

Figure 11.29 The process of fracture in fibre-reinforced ceramics (Petzow and Claussen)

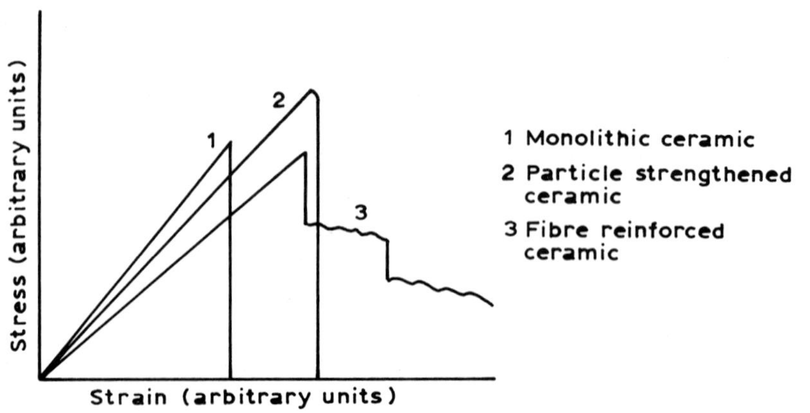

Figure 11.30 Stress-strain diagram for continous fibre-reinforced ceramics compared with other types. (Curve 3 stands for a variety of pseudo-plastic behaviours observed in these materials)

Figure 11.31 Collection of ceramic parts, Al_2O_3, ZrO_2 and SiC (dark). Sealing rings and disks, bearings, plungers, balls, parts for textile industry, etc. (Friedrichsfeld GmbH, Mannheim)

fibres) one- or two-dimensionally and with short fibres and whiskers mostly randomly. During fracture the cracks are bridged for a certain time by the fibres, which must be partly pulled out before they become fractured (Fig. 11.29). This leads to unique stress–strain diagrams (Fig. 11.30), especially with continuous fibre reinforcement. Catastrophic failure only occurs in this case after considerable pseudo-plasticity, depending on pullout which is controlled by the fibre-matrix phase-boundary strength. SiC-whisker-reinforced Al_2O_3 is already in use for cutting tools, representing one of the very few applications of whiskers. Continuous carbon-fibre-reinforced carbon (CFC) and SiC fibre reinforced SiC are used as structural material in engineering and space technology. *In-situ* reinforcement by oriented grain growth, as obtained by heat treatment of special Si_3N_4 qualities, is another way to improve mechanical properties.

Optimisation of powder processing with submicron range powders is of enormous significance for enhancement of monolithic ceramics, at least for strength. It aims to optimise green particle packing to avoid differential sintering (see section 7.2.3.2) and to achieve fine-grained microstructures with minimised defect sizes. Powders with definite particle size distribution seem to be more advantageous than monosized powders, but it is impossible to make a definitive statement with respect to the 'best powders'. The progress achieved recently in properties of monolithic ceramics (Table 11.7) is to a great extent due to the availability of improved powders as well as to advanced processing.

Figure 11.32 Honeycomb ceramics (Cordierite) for catalyst-carriers

The present situation in application engineering ceramics is characterised by increased use as engine parts, wear resistant materials including cutting tools (Al_2O_3- and Si_3N_4-based), corrosion resistant and high temperature parts, such as burner tubes and nozzles, combustion chambers etc. A collection of ceramics applied as wear resistant parts is shown in Fig. 11.31. In the car and chemical industries, the catalyst carriers (honeycombs, Fig. 11.32) are well established, as also are the portliners of Al_2TiO_5 in the exhaust channel of a few car types, combined with a turbocharger. Ceramic turbochargers (Si_3N_4-based) have been in use for a few years in Japan, and are probably not yet optimised. Ceramic valve systems and other parts are in field tests in several countries, yielding considerable advantage due to their light weight. A very difficult field of development (since the late sixties) is the ceramic gas turbine (Si_3N_4- or SiC-based) at gas inlet temperatures up to 1350°C for use in trucks or cars. Real technical success has not yet been achieved, but prototypes are in operation and strong developmental efforts continue in several countries.

Considerable success has been reported with different types of *ceramic coatings*, namely as wear resistant (for cutting tools see 11.6.3), corrosion resistant and thermal barrier layers (gas turbine parts), which are widely used in engineering.

11.10.4 BIOCERAMICS

Long-term development projects in many countries have led to the application of pure, fine-grained, high density Al_2O_3-ceramics as medical implants, especially as parts of hip joints (ball and sometimes pan) and as artificial teeth. Besides the strength, endurance and corrosion-resistant properties the tribological behaviour and bio-inertness of Al_2O_3 are most important. The ball of the hip joint is mostly combined with a pan of polyethylene, depending on the prosthesis design, the shaft being generally metallic (Ti- or Co-alloy). The *in-vivo* mean life time of such implants has been increased substantially during the last decade.

11.11 INTERMETALLICS AND MISCELLANEOUS

Several other groups of materials have been developed or are produced by powder metallurgy. They include:

- *Intermetallics,* such as nickel and titanium aluminides, which are now under intensive development for future application as high-temperature parts e.g. as valves for car engines, and for gas turbines. Their thermal stability, mechanical properties and corrosion resistance range between those of superalloys and ceramics. Their application temperature may be expected to be up to 1200°C. Some of them can be processed to a certain ductility, combined with good high-temperature strength and oxidation resistance. Favoured phases are Ti_3Al and TiAl. $MoSi_2$ is an electrical conductive intermetallic with extremely high oxidation resistance allowing application as heating elements up to 1800°C. It has been considered recently as HIP-consolidated, highly dense (>99%TD) and fine grained material for structural applications. Above 1000°C some plasticity has been observed. $MoSi_2$ has been investigated as reinforced with ultrafine boride dispersoids, which improve low as well as high temperature properties and allows superplastic deformation. Intermetallies will possibly represent in future an important class of materials.
- *Dental amalgams* from Ag–Sn-powders. Mixing of Ag–Sn alloy powder with Hg results in liquid-phase room-temperature reaction sintering (formation of Ag_3Hg_4 and Sn_7Hg) to solid and corrosion-resistant amalgams. They have been used for many decades for dental fillings, but are now replaced to a great extent by ceramics or polymers.
- *Noble metals,* like Au, Pt, Pd, as powders or pastes for metallisation of ceramics for electronic parts, for conducting tracks on glass and ceramics and as sintered parts for special purposes, especially as ODS-alloys (see section 11.9.3). For Ag see sections 11.9.2 and 11.9.6.

FURTHER READING

I. Jenkins and J.V. Wood (eds), *Powder Metallurgy, an Overview*, London, The Institute of Metals, 1991.

I. Jenkins and J.V. Wood (eds), *Powder Metallurgy, Selected Case Studies*, London, The Institute of Metals, 1991.

G. Dowson, *Powder Metallurgy, the Process and its Products*, Adam Hilger JOP Publ. Ltd. 1990.

L.F. Pease III, R.J. Sansoucy (eds.): Advances in Powder Metallurgy, 1991, Vol. 1–6, MPIF, Princeton, N.Y., 1991.

DIN-Taschenbuch 247, *Pulvermetallurgie*, Beuth-Verlag, Berlin, Köln, 1991.

Properties and Selection: Iron, Steel and High Performance Alloys, *Metals Handbook*, Vol 1, 10th Edition, ASM International, 1990.

H.I. Sanderow, *High Temperature Sintering; New Perspectives in Powder Metallurgy*, Vol 9, Princeton, NJ, MPIF, 1990.

R. Oberacker, Konventionelle und neue Legierungssysteme für hochfeste Sinterstähle, *Powder Metallurgy International*, **22** (1990), 54–60.

E. Pink, L. Bartha (eds), The Metallurgy of Doped, Non-sag, Tungsten, Elsevier Applied Science, 1989.

Tungsten 1990 (5th Internat. Tungsten Symposium, Budapest), MPR Publ. Services Ltd., 1991.

W. Schedler (ed.), *Hartmetalle für den Praktiker*, VDI-Verlag, Düsseldorf, 1988.

Advances in Hard Materials Production, MPR Publ. Services Ltd, 1992.

K.J.A. Brookes, World Directory, *Handbook of Hard Metals*, 5th Edition, International Carbide Data, MPR Publ. Services Ltd., 1992.

H.H. Stadelmaier, E. Th. Henig, G. Petzow, A Chronic of the Development of Iron-Based Rare-Earth High-Performance Magnets, *Z.Metallkunde*, **82** (1991), 163–168.

C. Lall, *Soft Magnetism – Fundamentals for PM and MIM*, Monograph in PM Series Nr 2, Princeton NJ, MPIF, 1993.

J. Pfeiffer, R. Ebeling, W. Erwens: Plastic Bounded Permanent Magnets, *Powder Metallurgy International*, **24** (1992), Nr.3, 180–185.

V.T. Morgan, 40 Years of Porous Metals, *Metal Powder Report*, Jan. 1986, 39–43.

R.Eck, H.P. Martinz, Powder Metallurgical Chromium, *Materials Science and Engineering* A 120 (1989), 307–312.

H.D. Kunze, J. Baumeister, J. Banhart, and M. Weber, P/M Technology for the Production of Metal Foams, *Powder Metallurgy International*, **25**, (4), 182–185, 1993.

A. Walder and M. Leroy, Industrial Production and Application of High Porosity Metallic Felts, *Metal Powder Report*, 26–30, Jan. 1988.

Composits, *Engineering Materials Handbook*, Vol I, ASM International, 1987.

P. Kumor, K. Vedula and A. Ritter (eds), *Processing and Properties for Powder Metallurgy Composits*, TMS, 1988.

G. Ondracek, The Quantitative Microstructure — Field Property Correlation of Multiphase and Porous Materials, *Reviews on Powder Metallurgy and Physical Ceramics*, **3**, Nr 3 and 4, 205 (1987).

K.-H. Härdtl and F. Thümmler, Keramische Materialien, in *Neue Werkstoffe* (ed. A. Weber) VDI-Verlag, Düsseldorf, 1989.

F. Thümmler, Engineering Ceramics, *J. Europ. Ceram. Soc.*, **6**, 139–151, 1990.

PM Aero 91, Proceedings, MPR Publ. Services Ltd., 1992.

W.J. Huppmann and K. Dalal, *Metallographic Atlas of Powder Metallurgy*, Verlag Schmidt, Freiburg, 1986.

Standard Test Methods for Metal Powders and Powder Metallurgy Products, Princeton NJ, MPIF, 1993.

J.M. Capus, R.M. German (ed.), *Advances in Powder Metallurgy and Particulate Materials*, Princeton NJ, MPIF, 1992.

Vol 5: PM Steels.
Vol 6: Non-Ferrous Materials
Vol 8: Properties of Emerging PM Materials
Vol 9: Particulate Materials and Processing

International Plansee Seminars, Reutte /Tyrol (latest: May 1993), Proceedings, ed. by Metalwerk Plansee GmbH, Reutte/Tyrol, Austria.

Index